CHEMISTRY RESEARCH AND APPLICATIONS

GALLIUM NITRIDE

STRUCTURE, THERMAL PROPERTIES AND APPLICATIONS

CHEMISTRY RESEARCH AND APPLICATIONS

Additional books in this series can be found on Nova's website
under the Series tab.

Additional e-books in this series can be found on Nova's website
under the e-book tab.

Chemistry Research and Applications

Gallium Nitride

Structure, Thermal Properties and Applications

Kiera Olivia Peak
Editor

nova publishers
New York

LIBRARY OF CONGRESS CATALOGING-IN-PUBLICATION DATA

Gallium nitride : structure, thermal properties, and applications / Kiera Olivia Peak.
pages cm. -- (Chemical engineering methods and technology)
Includes bibliographical references and index.
Includes index.
ISBN 978-1-63321-387-6 (hardcover)
1. Gallium nitride. 2. Electronics--Research. I. Peak, Kiera Olivia.
TK7871.15.G33G36 2014
621.3815'2--dc23

2014023751

Published by Nova Science Publishers, Inc. † New York

CONTENTS

PREFACE

Gallium nitride (GaN), a member of the family of III-V semiconductor compounds, has been investigated intensively and have found extensive applications. Recent development has suggested that GaN may be used as an excellent host material for light-emitting devices (LEDs) that operate in the blue and ultraviolet region, due to its wide and direct band gap. Nano-materials often present novel properties that are different from the corresponding bulk materials due to their unique structures, high specific surface area and quantum confinement effects. This book discusses several topics on the structure, thermal properties and applications of GaN. Some of these topics include the effects of radiation damage in GaN and and related materials; low-dimensional GaN; structural and dielectric properties of the GaN and silica nanoparticles investigated by molecular dynamics; and GaN bulk and nanostructures for spintronic application.

Due to their relatively high average bond strengths, GaN and related materials are inherently radiation hard when compared to the more conventional Si and GaAs. For example, experimental evidence suggests that GaN is several orders of magnitude more resistant to radiation damage than GaAs, i. e., it can withstand radiation doses at least two orders of magnitude higher than those degrading GaAs of similar doping level. In Chapter 1, the authors review the literature on different types of ionizing radiation on GaN, InGaN and AlGaN materials and devices, including high electron mobility transistors and light-emitting diodes. A comparison of theoretical and experimental threshold displacement energies is given, along with a summary of energy levels introduced by different forms of radiation, carrier removal rates and role of existing defects. In terms of heterostructures, experimental data shows the radiation hardness decreases in the order AlN/GaN >AlGaN/GaN > InAlN/GaN, consistent with the average bond strengths in the Al-based materials. Future areas for research are identified, including the strong asymmetry in carrier removal rates in n- and p-type GaN and the interaction of radiation defects in doped nitrides with the high dislocation in this material.

AlGaN/GaN high electron mobility transistors are gaining commercial acceptance for use in high power and high frequency applications, but the degradation mechanisms that drive failure in the field are not completely understood. Since some of these mechanisms are current or field driven, reliability studies must go beyond the typical Arrhenius accelerated life tests. In Chapter 2, the authors summarize recent work on electric field or current driven degradation in devices with different gate metallization, device dimensions, electric field mitigation techniques (such as source field plates), and the effect of device fabrication processes for both dc and RF stress conditions. Some of the mechanisms now established include gate or Ohmic metal reaction, hot carrier trap generation, stress-induced crack generation and oxidation reactions involving the AlGaN layer.

Recent developments in the low-dimensional GaN materials have revealed their novel structural and physical properties which can be exploited for applications in electronics and optoelectronics. In Chapter 3, the authors review the work performed on the low-dimensional GaN materials, including clusters, nanotubes, nanowires, nanoribbons and sheets. Specifically, they will review their uniqueness in structure and electronic structure compared to those of the bulk GaN. A special example is discussed about the structure evolution from atomically thin layers to bulk. The stability of these low-dimensional materials is also discussed in light of the available thermodynamic data. It is followed by a summary of synthesis methods used for these GaN nanostructures together with their morphology and structural characterization. The authors also review the electronic, magnetic and optical properties of various GaN nanostructures ranging from pure to chemically doped or passivated GaN nanostructures. Lastly, their potential applications such as the use of GaN-based devices in light emitting diodes (LED) are discussed. The review ends with a summary of the current achievements, challenges and future directions in the research and development of low-dimensional GaN materials.

Gallium nitride is a kind of well-known semiconductor with a direct and wide band-gap of 3.4 eV. It has wide applications in white light-emitting diodes, blue lasers, and field effect transistors because of excellent optoelectronic properties, high mobility, and super thermal stability. The nitride nanomaterials inherit the excellent bulk properties while possess novel properties of nanomaterials, such as quantum effects, being the promising nanostructures for future applications on renewable energy. In chapter 4, the authors first introduce various GaN nanostructures, including syntheses and growth mechanisms. Then several popular characterization techniques are introduced. The future outlook of the GaN nanostructures is talked about at the end of the chapter.

Electrical contacts are vital for the fabrication of modern electronic and optoelectronic devices. A junction forms when a metal is in intimate contact with a semiconductor. This junction is either an ohmic or a Schottky contact. Typically, the latter, being more technologically important, exhibits distinct rectifying characteristics and has thus attracted much more interest due to a wide range of practical applications. Currently, Schottky contacts on GaN are still an active topic because the properties of contacts are found to be strongly dependent on not only the GaN materials, such as quality, back ground carrier concentration, and surface status as well as their growth techniques and conditions, but also the deposition techniques and parameters of contact materials as well as post-deposition treatments. Chapter 5 provides an overall review of GaN based Schottky contacts using metals or transparent conductive oxide (TCO) as contact materials. Particularly, the influence of GaN epilayer qualities and preparation conditions such as surface treatment, deposition parameters, selection of contact materials etc., on the microstructure, surface status, and electrical characteristics of Schottky contacts is addressed. Finally, comparison with different materials and the optimized fabrication conditions for selected device applications are discussed.

GaN metal-oxide-semiconductor field-effect transistors (MOSFETs) with recessed gate on AlGaN/GaN heterostructure are described in Chapter 6, in which the drain and source Ohmic contacts were fabricated on the AlGaN/GaN heterostructure and the electron channel was formed on the GaN buffer layer by removing the AlGaN barrier layer. General consideration to develop an enhancement-mode (E-mode) GaN MOSFET and the basic evaluation methods are described. GaN MOSFETs with a silane-based SiO_2 insulator deposited by plasma-enhanced chemical vapor deposition (PECVD) were demonstrated on an AlGaN/GaN heterostructure. E-mode operation up to a gate voltage of 15 V was confirmed at an ignorable gate leakage current. To investigate the effects of the fabrication process, including the dry recess process, the dielectric, the annealing, the dielectric thickness and the surface treatment, on the device performance, devives were fabricated and evaluated. The authors found that the electron mobility and the interface state density were improved by a thermal treatment at 1000°C in nitrogen ambient. Trenching effect at the bottom near the sidewall was observed when positive photoresist was utilized and the inductively coupled plasma (ICP) power was low. The recess profile was improved by adopting SiO_2 as the etching mask and increasing the ICP power. It indicates that the nitrogen vacancies are mainly responsible for the negative threshold voltages rather than the silicon contamination. The maximum field-effect mobility of about 150 cm^{-2} V^{-1} s^{-1} and the minimum interface state density of about 1×10^{11} cm^{-2} eV^{-1} were obtained from the optimized process conditions. It demonstrates that the optimization of the ICP recess conditions and the removal of the etching damage to improve the surface condition are still necessary to realize enhancement-mode GaN MOSFETs on AlGaN/GaN heterostructure.

The dielectric, structural, and kinetic properties of $(GaN)_{54}$ $(SiO_2)_{50}$ nanoparticles of the following two types are studied in Chapter 7 using the method of molecular dynamics: particles of GaN covered by a SiO_2 layer and particles of SiO_2 with a GaN layer deposited onto it. The absorption and reflection spectra of these nanoparticles are considerably different in the respective frequency ranges $0 \leq \omega \leq 1500$ cm^{-1} and $\omega > 700$ cm^{-1}. The Raman spectrum of $(SiO_2)_{86}$ had three broad bands only, whereas the Raman spectrum of $(GaN)_{129}$ contained a large number of overlapping bands. The shape of Raman spectra for four-component particles depends strongly on the way the GaN-, and SiO_2-components are located in the nanoparticle. The nanoparticle with a GaN core had a continuous fairly smooth spectrum over the frequency range $0 \leq \omega \leq 600$ cm^{-1}, whereas the spectrum of the nanoparticle with a SiO_2 core contained well-defined bands caused by vibrations of groups of atoms of different kinds and atoms of the same kind. The single-centered functions of the radial distribution of the nanoparticles differ in their peak positions and intensities, including the "tail" part for $r > 0.9$ nm. The structure of nanoparticles with the SiO_2 core is completely tetrahedral, while the nearest six-atom environment is dominant in the structure of a particle with the GaN core. The mobility of gallium atoms in the nanoparticles is lower by two orders than that of other elements (N, Si, and O).

In Chapter 8, the authors fabricated a Multi-finger lateral-type AlGaN/GaN Schottky barrier diode (SBD) fabricated on the AlGaN/GaN heterostructure grown on sapphire substrate and also present the electrical properties of the SBD measured by varying temperature. The multi-finger patterned large-area SBD with the bonding pad over active (BPOA) structure was fabricated. The size, the length of the anode perimeter, and the anode-to-cathode distance for both devices are 3000 μm × 3000 μm, 270 mm, and 20 μm, respectively. The fabricated SBD with BPOA structure exhibited excellent device performances, such as forward current of 4.5 A at 1.5 V, leakage current of 6 μA at - 600V, reverse-recovery time (Trr) of 19.4 ns, breakdown voltage of 747 V, and low specific on-resistance (Ron) of 12 mΩ·cm2, resulting in the figure of merit (VBR2/Ron) as high as 47 MW/cm2, which demonstrates that the AlGaN/GaN SBD with BPOA structure has a great potential application to the high power electronics.

Recently, high brightness GaN-based light emitting diodes (LEDs) have attracted a lot of interest owing to their excellent device characteristics. The GaN-based LEDs are widely used for a variety of applications such as general lighting, liquid crystal display and traffic signal. In order to achieve high performance of these devices, it is urgently required to enhance the optoelectronic properties of LEDs. It is well-known that the poor heat dissipation of sapphire substrate is the main limitation to fabricate the high-efficiency LEDs. For a conventional LED, the heat is easily accumulated in the device, resulting in the performance deterioration. In Chapter 9, the improvement heat dissipation of GaN-based LEDs is proposed and discussed.

In this chapter, the authors have proposed several techniques including the employment of cup-shaped copper-diamond heat spreader, the combination of beveled sapphire and cup-shaped copper sheet, and the addition of diamond like carbon layer into LED structure. To develop these techniques, many methods consisting of copper electroplating, chemical mechanical polishing, metal mirror coating, wafer bonding, laser lift-off, and so on. After fabricating the LEDs by using these techniques, the surface temperature, thermal resistance, and light output power of LED devices are investigated in detail. With the aids of above-mentioned techniques, the results reveal that not only the heat dissipation but also the light extraction of LEDs can be improved efficiently.

In Chapter 10 the authors review the recent achievements of experimental and theoretical developments in understanding the magnetic, optical and electronic properties of transition-metal-doped Gallium Nitride semiconductors, which show ferromagnetic order at relatively high temperature and good structural compatibility with existing III–V devices. Electronic structure, magnetic and optical properties of GaN:TM (TM: Cr, V, Mn, Fe, Co, Ni), based dilute magnetic semiconductors (DMS), are investigated using first-principles calculations for bulk and nanofilms. The authors show that density functional theory is a very powerful tool for 1-studying the effects of local doping, defects effect, size and disorder effect on the magnetic and optical properties of these materials, and predicting properties of new materials.

Due to their technological application, a precise prediction for the optical properties of III-nitride alloys is significantly attractive. Numerical simulation based on first-principles calculations is applied to study the structural characteristics and band-energy properties of the zincblende $Ga_{1-x}In_xN$. The purpose of Chapter 11 is to explore theoretically the electronic and optical characteristics of the parent InN and GaN compounds and their ternary alloys by utilizing KKR-CPA approach. The authors show that density functional theory is a very powerful tool for 1-studying the local doping effects, defects effect, size and disorder effect on optical and electronic properties of these materials, and predicting properties of new materials.

In: Gallium Nitride
Editor: Kiera Olivia Peak

ISBN: 978-1-63321-387-6

Chapter 1

EFFECTS OF RADIATION DAMAGE IN GaN AND RELATED MATERIALS

S. J. Pearton[1], Fan Ren[2], Y. H. Hwang[2], Shun Li[2], Yueh-Ling Hsieh[2], Alexander Y. Polyakov[3] and Jihyun Kim[4]

[1]Department of Materials Science and Engineering,
University of Florida, Gainesville, Florida, US
[2]Department of Chemical Engineering, University of Florida,
Gainesville, Florida, US
[3]Institute of Rare Metals, Moscow, Russia
[4]Department of Chemical and Biological Engineering,
Korea University, Sungbuk-gu, Seoul, Korea

Abstract

Due to their relatively high average bond strengths, GaN and related materials are inherently radiation hard when compared to the more conventional Si and GaAs. For example, experimental evidence suggests that GaN is several orders of magnitude more resistant to radiation damage than GaAs, ie. it can withstand radiation doses at least two orders of magnitude higher than those degrading GaAs of similar doping level. We review the literature on different types of ionizing radiation on GaN, InGaN and AlGaN materials and devices, including high electron mobility transistors and light-emitting diodes. A comparison of theoretical and experimental threshold displacement energies is given, along with a summary of energy levels introduced by different forms of radiation, carrier removal rates and role of existing defects. In terms of heterostructures, experimental data shows the radiation hardness decreases in the order AlN/GaN >AlGaN/GaN > InAlN/GaN, consistent with the average bond strengths in the Al-based materials. Future areas for research are identified, including the strong asymmetry in carrier removal rates in n- and p-type GaN and the interaction of radiation defects in doped nitrides with the high dislocation in this material.

I. Introduction

GaN based devices have applications in lighting, displays and power electronics. In AlGaN light emitting diodes (LEDs) and laser diodes (LDs) are the predominant semiconductor light sources for the UV/green/blue range. AlGaN/GaN High Electron Mobility Transistors (HEMTs) and GaN diodes are well-suited to high-power/high-frequency/high-temperature electronics applications, while AlGaN/GaN UV photodetectors have matured in recent years. Many of these devices are used in satellite and military systems where radiation tolerance is critical. The Earth's magnetosphere is bombarded by a nearly isotropic flux of energetic charged particles-85% protons, 14% α-particles, and 1% heavier ions covering the full range of elements. There are also the van Allen belts, an inner belt extending to 2.5 Earth radii and comprising energetic protons up to 600 MeV together with electrons up to several MeV, and an outer belt comprising mainly electrons extending out to 10 Earth radii. In the years around solar maximum, the sun is an additional sporadic source of lower energy particles accelerated during certain solar flares and/or in the subsequent coronal mass ejections. These solar particle events last for up to several days at a time and comprise both protons and heavier ions with variable composition from event to event. Energies typically range up to several hundred MeV and have most influence on high inclination or high altitude systems. Irradiation also provides a versatile method for controllably introducing defects in GaN for fundamental studies. The demand for radiation-resistant electronics devices spans multiple disciplines such as aerospace, nuclear reactor design, weaponry and weather forecasting systems.

In this review we discuss radiation damage effects in GaN and the most common devices fabricated in this materials system. To summarize briefly, results for chosen types of materials, devices and radiation sources are discussed. For devices, the bulk of research was performed for GaN LEDs and AlGaN/GaN HEMTs, the most popular radiation sources were protons and electrons with some work also reported for γ-irradiation and neutron irradiation.

II. Fundamental Studies of Radiation Defects in GaN and Related Materials

II.1. Threshold Displacement Energy: Theory and Experiment

Early studies of displacement threshold energy for GaN involved monitoring of the changes in luminescence spectra of GaN LEDs as a function of electron irradiation with energies 300-1400 keV[1]. Generation of a band attributed to the formation of Ga vacancy started from electron energy of 440 keV, which corresponds to the Ga displacement energy of 19± 2eV. Comparison of defect production efficiency for 2 MeV protons and 2.5 MeV electrons showed that the protons were 1000 times more effective in creating defects, whereas calculations of the number of displaced atoms based on the measured threshold energy predicted a ratio of 250. The discrepancy was attributed to the difference in defects self-annealing rates depending on particle type and energy. The measured displacement threshold in GaN was much higher than in Si and GaAs, similar to SiC, but lower than in diamond. This reflects the difference in the bond strength and can serve as an indication of relative radiation hardness of respective materials. There was no clearcut threshold energy for the nitrogen displacement. However, measurements of the changes in electrical properties of undoped n-GaN films irradiated with electrons with energies 700-1000 keV suggested to Look et al. [2] that nitrogen vacancies were introduced with a rate close to 1 cm^{-1} at these electron energies. Molecular dynamics calculations of displacement effects in GaN [3] might explain why discrepancies exist when comparing effectiveness of radiation defects introduction by various particles based on singular values of threshold defect formation. It was found that a) there exists a wide distribution of threshold energies for both Ga and N sublattices and b) that effects of recombination induced by self-annealing caused by a thermal local energy transfer are very important. The minimal energies of defect formation were found to be 18±1 eV for Ga and 22± 1 for nitrogen, but the average displacement energy was much higher, 45±1 eV (Ga) and 109± 2(N). This modeling predicts about 5 times higher radiation tolerance of GaN compared to GaAs.

Experimental measurements of defect accumulation in GaN produced by various high energy ion species at different temperatures were performed with Rutherford backscattering and transmission electron microscopy [4]. Radiation defects in GaN were mobile even at low temperatures and the doses of ions producing amorphization were more than an order of magnitude higher than for GaAs. For light ions, trapping of primary radiation defects by the implanted species was observed. For hydrogen irradiation at doses exceeding ~10^{16} cm^{-1}, formation of hydrogen bubbles was significant. For heavy ions (starting with Si) the main factor in defect production was the ion mass and energy and hence the number of primary displacements initiated. Recombination of radiation defects during irradiation was very prominent even at low temperature and the rate of recombination increased with irradiation temperature. At room temperature the doses of irradiation producing amorphization increased from ~ 10^{14} cm^{-2} for heavy ions to ~10^{16} cm^{-2} for light ions and increasing the implantation temperature increased the dose of amorphization by several orders of magnitude. For heavily amorphized material the formation of nitrogen bubbles was observed. These experimental observations are in good agreement with the results of theoretical modeling [3] which shows that the main features of radiation defect formation process in GaN are reasonably well understood. Figure 1 shows the energy levels of the most commonly produced radiation-induced defects in GaN.

II.2. Radiation Defects in GaN: Defects Levels, Effects on Charge Carriers Concentration, Mobility, Lifetime of Charge Carriers, Thermal Stability of Defects

II.2.1. Levels of Radiation Defects in GaN

The most basic primary defects produced in GaN are Frenkel pairs in the Ga and N sublattices. Theoretical estimates predict that nitrogen vacancies in GaN are electronic resonances with levels in conduction band. By capturing electrons they are turned into effective-mass-like (EM) shallow donors [5]. The electrical properties of undoped n-AlGaN as a function of Al composition suggest that such EM V_N donors have ionization energies of 40-60 meV in GaN and the Al concentration at which respective resonance levels emerge in the forbidden gap is ~ 20% [6].

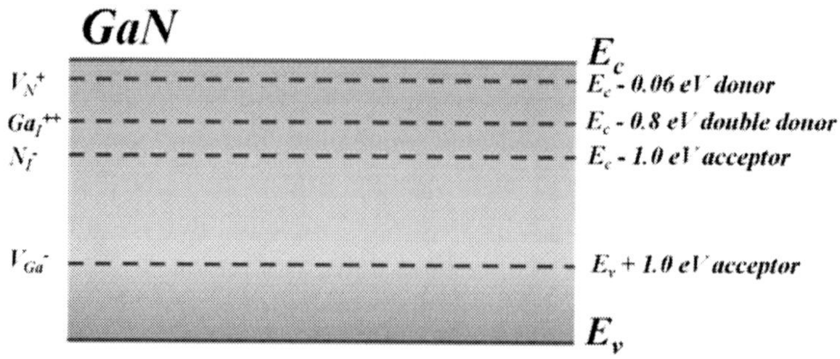

Figure 1. Simple model for radiation defects created in GaN by protons and other ionizing radiation.

For nitrogen interstitials, N_i, theory predicts the existence of a deep acceptor near 1 eV from the conduction band edge [7, 8]. Ga vacancies V_{Ga} in the doubly charged state are supposed to produce an acceptor state near E_v+1 eV, whilst Ga interstitials form negative-U type donors whose +/0 transition level is close to the conduction band edge and 3+/2+ charge transition level is located near E_v+2.6 eV (E_c-0.8 eV) [7, 8]. Most of these states are indeed observed in radiation effects experiments performed for GaN.

Look et al. [2] showed that electron irradiation with energies 0.7-1 MeV introduced new donors with ionization energy of ~ 0.06 eV with introduction rate of 1 cm^{-1}. The net electron concentration hardly changed while the mobility of electrons decreased with increasing dose. The analysis of the temperature dependence of mobility suggested that acceptor centers were introduced at a rate similar to the rate of the 0.06 eV donors. The observed effects were explained by the formation of Frenkel pairs in the nitrogen sublattice, with the nitrogen vacancies V_N responsible for the 0.06 eV donors and nitrogen interstitials N_i associated with the compensating acceptors. Deep level transient spectroscopy (DLTS) measurements performed on 1 MeV electron irradiated n-GaN showed the presence of deep electron traps with an activation energy of 0.9 eV attributed to N_i acceptors [9]. DLTS spectra of n-GaN irradiated with ^{60}Co γ-rays also showed the presence of electron traps G1 with activation energy 80 meV [10] close to that observed for V_N donors in Ref. [2]. In undoped n-GaN samples irradiated with 10 MeV electrons we observed deep electron traps with activation energy 1 eV that were acceptors and that were attributed to N_i acceptors [11]. The authors of Ref. [12] observed that 2.5 MeV electron irradiation at 4.2K produced in GaN a strong defect photoluminescence PL band centered near 0.95 eV, for which optically detected electron paramagnetic resonance ODEPR spectra could be obtained. The PL line was attributed to Ga vacancy V_{Ga} with a level near E_v+1 eV and the ODEPR process was interpreted as interaction with two different Ga_i interstitial centers with levels close to E_v+2.6 eV. The quenching of the ODEPR signal for annealing to room temperature was then attributed to moving of the Ga_i defect away from V_{Ga}.

If, however, only Frenkel pairs were produced by irradiation of GaN, then for n-GaN, the carrier removal rate at the initial stage of irradiation, when the density of radiation defects is lower than the concentration of dopant donors, will be equal to the production rate of V_{Ga} and N_i, whilst for high irradiation doses the Fermi level will be pinned by the shallower of the two EM-like native donors, V_N or $V_{Ga}{}^{1+}$. In p-type GaN the initial carrier removal rate should be close to that in n-type until the aggregate density of V_N and Ga_i donors exceeds the density of acceptor dopants and the Fermi level should jump to the level of V_{Ga} near E_v+1 eV and be pinned there.

This generally is not the case because primary defects recombine, form complexes with each other, with dopants and with extended defects and because at high energy of irradiating particles the energy of the primary recoils becomes so high that they produce collision cascades and form heavily disordered regions with a very high defect density in the core. Figure 2 (top) shows a schematic of the type of simple point defects created in a lattice by protons and electrons, while the bottom of the figure shows extended defects typical of those produced by neutron irradiation.

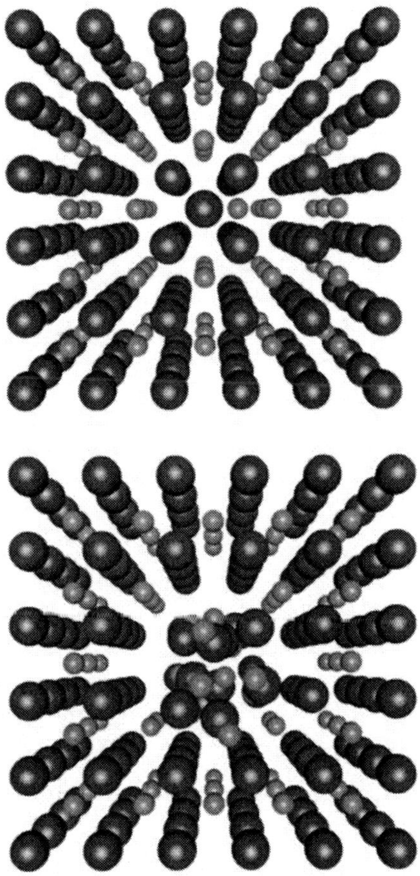

Figure 2. (top) Schematic of point defects created by proton or electron irradiation of crystalline lattice, (bottom) disordered zone created by neutron irradiation.

DLTS measurements on n-GaN irradiated with electrons, γ-rays and protons with MeV energies that will predominantly create point radiation defects showed formation of electron traps with activation energies 0.13 eV, 0.16 eV and 0.18-0.2 eV [10, 13-18](ER1, ER2, ER3 according to [15]). These traps produce a broad feature in DLTS due to the proximity of emission rates of components, but this broad peak can be deconvoluted into separate defect contribution due to the large difference in electron capture cross sections [9, 10, 13-16]. For the ER3 trap, the activation energy is the sum of the trap ionization energy which is 0.06 eV, i.e., close to that of V_N and the capture activation energy of about 0.14 eV[13]. Based on this observation it was suggested that the ER3 traps are complexes of nitrogen vacancies with some other native defects, such as V_N-N_i or V_N-N_{Ga}-Ga_N-N_i [9]. The shallower defects ER1 and ER2 have been observed in a variety of surface treatments bound to produce nitrogen vacancies and are also thought to be V_N-related [9]. It was assumed that the 0.13 and 0.2 eV electron traps were donors as well as the 0.06 eV V_N donors [9]. The carrier removal in irradiated n-GaN should then be regulated by the balance between relatively shallow V_N-related donors, deep N_i acceptors, deep Ga_i donors and V_{Ga} acceptors. However, this simple picture is considerably spoiled by experimental measurements of the ER3 traps ionization energy dependence on applied electric field which indicate these traps are acceptors [14]. An alternative suggested identity of these traps offered in Ref. [14-16] associates them with $V_{Ga}N_i^{2-}$ complexes. DLTS spectra of n-GaN films irradiated with light particles (γ-rays, electrons with several MeV energy, protons with several MeV energy) are dominated by the ER1-ER3 relatively shallow electron traps. With rare exceptions, only for higher electron energies [11], heavier ions (He, N) [15, 19], neutron irradiation [20] or higher density of defects in proton implanted samples (as for 150 keV protons with the dose over 5×10^{14} cm^{-2} [21]) are deeper traps commonly detected. Electron traps that are generally observed in these cases show the activation energies of 0.75-0.8 eV, 0.95-1.2 eV [11,15, 19, 20, 21]. The 0.8 eV traps show decrease of the ionization energy with increasing electric field whereas the 1 eV traps energy does not vary with electric field. The first type of behavior is expected for donors because of the Poole-Frenkel effect [22], whilst the second type of behavior is characteristic of acceptors that are neutral when they emit an electron [22]. The attribution based mostly on theoretical calculations tends to ascribe the first to Ga_i^{2+} deep donors and the second to N_i^{2+} deep acceptors [9, 11,15, 20]. In addition to these deep centers, implantation of n-GaN with 150 keV protons to doses higher than 5×10^{14} cm^{-2} creates deep electron traps whose energies increased from 0.2 eV at low dose to 0.25 eV, 0.32 eV and 0.45 eV at higher dose suggesting that these centers could be larger complexes produced by addition of new radiation defects to the more simple radiation defects formed at low doses [21]. Measurements on neutron transmutation doped and annealed n-GaN [23] indicate that the 0.2 eV and the 0.45 eV traps could incorporate donor doping atoms.

The knowledge of hole trap behavior in irradiated n-GaN is less satisfactory compared to electron traps. DLTS studies on n-GaN have mostly been done on Schottky diodes where probing of the traps in the lower half of the bandgap can only be achieved by optical injection of holes (deep level optical spectroscopy DLOS [24] or optical deep level transient spectroscopy ODLTS [25]). These measurements depend on the ability to provide complete recharging of deep traps within the space charge region. When the lifetime strongly decreases after irradiation the ability to fully recharge hole traps becomes an issue and could lead to erroneous conclusions. A good example is the measurement of concentration for the 0.9-1 eV hole trap attributed to V_{Ga} acceptors. These traps produce a prominent hole traps peak in ODLTS with activation energy close to 0.9 eV [26]. The 0.95 eV defect PL band in irradiated GaN can be assigned to recombination on these hole traps [12, 27-29]. Also, these traps are believed to be related to the yellow recombination band in GaN attributed to a donor-acceptor pairs DAP transition involving a shallow donor and the E_v+0.9 eV hole trap [6, 30-33]. Electron irradiation at 4.2K strongly increases the intensity of the 0.95 eV PL band [12]. These defects are stable up to 500°C, even though the ODEPR signal vanishes because of the increased spatial separation of V_{Ga} and Ga_i [12]. Thus, one would expect that irradiation will increase the magnitude of the ODLTS peak E_v+0.9 eV and enhance the yellow band intensity. The former is seldom the case. The most obvious way to explain the results would be to assume that the peak observed is a superposition of one or several positive and negative peaks due respectively to electron and hole traps. The effective activation energy deduced from the Arrhenius plot of $1/\tau T^2$ versus 1000/T for the peak will depend on contributions from both types of traps (τ is the characteristic emission time of the trap, T is the temperature position of the peak with the chosen value of spectrometer time window [25]). For 10 MeV electron irradiation the electron trap interfering with ODLTS measurements of 0.9 eV hole traps seems to the 1 eV N_i acceptor rather than the 0.9 eV Ga_i donor. These results indicate that, for 10 MeV electron irradiation, the introduction rate for the 0.9 eV hole traps is close to, but lower than the introduction rate of the 1 eV N_i electron trap.

Measurements of deep trap spectra for irradiated p-GaN are very scarce. Part of the problem is that standard DLTS measurements on Schottky diodes on p-GaN are very difficult because of the problems with high series resistance of the films and strong freeze-out of relatively deep Mg acceptors even at moderately low temperatures. Studies of radiation defects in p-GaN were carried out for 100 keV protons in Ref. [34] and for fast reactor neutrons in Ref. [35]. For proton irradiation conductivity versus temperature, low frequency capacitance-voltage C-V measurements, admittance spectra measurements, photoinduced transient current PICTS [36] and current deep traps transient CDLTS [37] spectroscopy indicated that deep electron traps near E_c-0.5-0.6 eV and deep hole traps with levels near E_v+0.3 eV and E_v+0.85 eV were introduced. The electron traps at E_c-0.5-0.6 eV are believed to be complexes of Mg acceptors with native defects and were shown to also give rise to intense blue defect luminescence band centered at 2.9 eV [38, 39]. The E_v+0.85 eV traps are probably the same as the V_{Ga}-related hole traps in n-type films. Their in-grown concentration in p-type GaN should be very low because of the high formation energy [7]. However, irradiation being a very nonequilibrium process, it can introduce such defects and as a result produce the yellow luminescence band in heavily irradiated p-GaN films while yellow luminescence is manifestly absent in virgin p GaN [34].

Neutron irradiation of p-GaN films grown either by by hydride vapor phase epitaxy (HVPE) or (MBE) [35] showed differences in the temperature dependence of conductivity, lower mobility and the presence of additional acceptors with activation energy of 0.12 eV, shallower than the ordinary Mg acceptors with activation energy of 0.15 eV as well as their luminescence spectra and photosensitivity. In HVPE samples the luminescence was dominated by blue luminescence band due to transitions involving the E_c-0.5-0.6 eV Mg-related centers discussed above. In MBE films the concentration of these centers was much lower. The photosensitivity of HVPE samples was higher due to trapping of electrons on the E_c-0.5-0.6 eV level. Irradiation of both types of samples compensated acceptors and increased the activation energy of major acceptors from 0.15 eV to 0.18 eV. In luminescence spectra of the MBE samples it increased the contribution of the blue luminescence band. Both types of samples remained p-type up to the dose of 1.7×10^{17} cm^{-2} and in both we observed the emergence of the yellow luminescence band due to the formation of the E_v+(0.8-0.9) eV V_{Ga}-related acceptors after high doses. At the fluence of 10^{18} cm^{-2} both samples converted to high-resistivity n-type with the Fermi level pinned near E_c-(0.9-1) eV. Clearly, save for the E_c-(0.5-0.6) eV Mg-related centers, the radiation defects introduced in p-type GaN are the same as in n-type, but the carrier removal rate was about 20 times higher than in n-GaN [35].

II.2.2. Carrier Removal and Deep Traps Introduction Rates; Disordered Regions

Basic questions of radiation physics include what determines the changes of electrical properties of irradiated material and where the Fermi level is stabilized after high doses of radiation. Figure 3 shows a compilation of reported carrier removal rates for n- and p-GaN films as well as different types of HEMT devices.

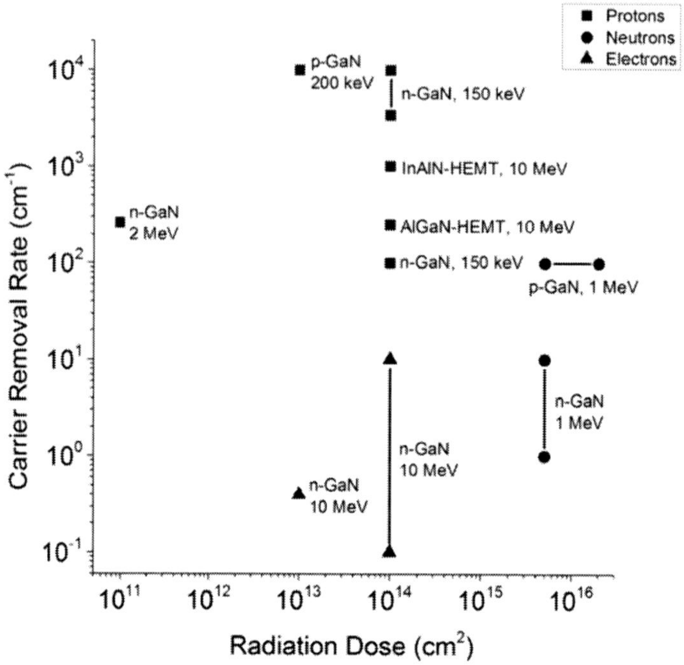

Figure 3. Compilation of carrier removal rates as a function of dose in GaN films and HEMT devices by different types of radiation.(reprinted with permission from American Institute of Physics, Pearton et al., Journal of Vacuum Science and Technology A31, 2013, 050801).

In general, p-type GaN is more susceptible to carrier loss than n-GaN and InAlN/GaN HEMTs are less radiation hard than their AlGaN/GaN counterparts. Protons remove more carriers per ion than electrons. Some examples of carrier removal as a function of neutron dose are shown in Figure 4 for n-GaN films and in Figure 5 for ELOG materials versus conventional MOCVD grown films. The growth method has a clear effect beyond the carrier type and radiation source.

There seem to exist several hole traps in irradiated GaN with a level near E_V+0.8-0.9 eV, but with vastly different hole capture cross section. Even after annealing to 1000°C a hole trap with the activation energy close to that of V_{Ga}, but a lower hole capture cross section is present, as shown in Figure 6. It is tempting to attribute the new trap to the same family of defects as V_{Ga}, which would mean even after very high temperature annealing such traps can survive. The difference between the hole traps in question could be due to some of them being trapped at dislocations. Photoluminescence studies on yellow band related defects in GaN also suggest the presence of several types of such bands [32] which was attributed to complexes of Ga vacancies with impurities, such as hydrogen. Those are most likely to be the members of the 0.9 eV hole traps family annealed at moderate temperatures.

The defect introduction rates measured for the major deep traps detected by DLTS, the 0.18 eV electron traps and the 1 eV electron traps is 0.4 cm^{-1}, the 0.18 eV traps and the 1 eV traps introduction rates are respectively 0.2 -0.8 cm^{-1}. The contribution from the other acceptors, the V_{Ga} related E_v+0.9 eV hole traps in ODLTS spectra of irradiated samples, is underestimated by the interference of the 1 eV electron traps due to N_i acceptors. Deconvolution of the 0.9 eV hole trap ODLTS feature gives the upper limit of the V_{Ga} introduction rate as about 0.4 cm^{-1}.

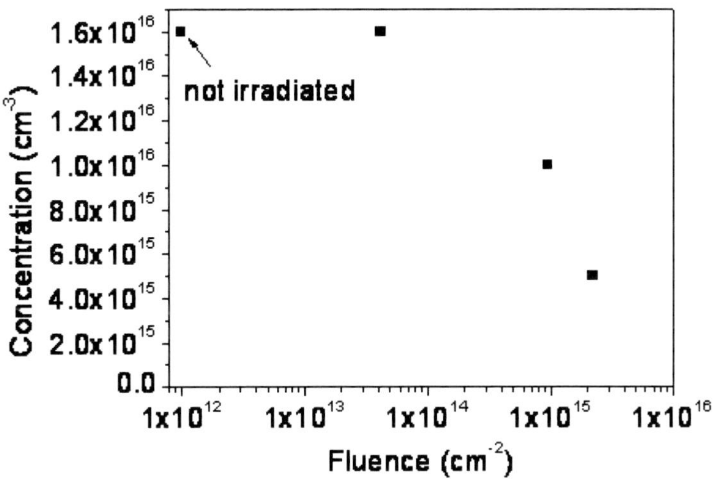

Figure 4. Decrease in electron concentration in n-GaN films of different doping concentration irradiated with fast neutrons to different doses.

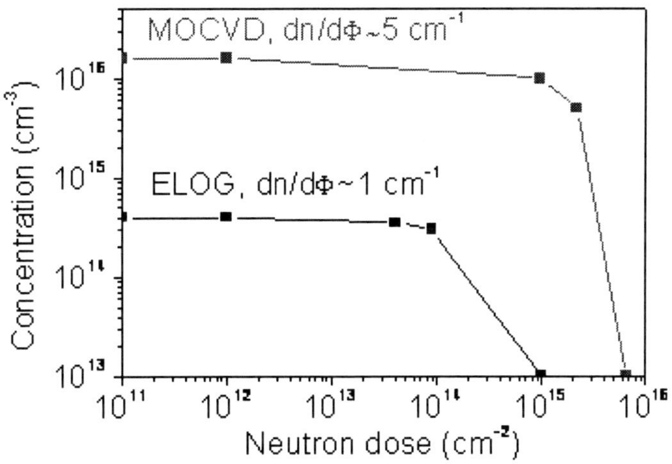

Figure 5. Decrease in carrier concentration in n-GaN films grown by MOCVD or the ELOG process as a function of fast neutron dose.(reprinted with permission from American Institute of Physics, Pearton et al., Journal of Vacuum Science and Technology A31, 2013, 050801).

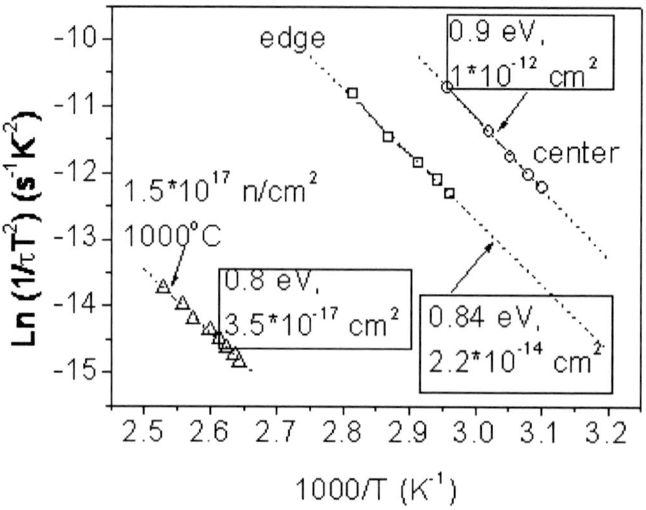

Figure 6. Arrhenius dependences of the dominant hole traps in undoped n-GaN measured by ODLTS: measurements in the center of the wafer, at the periphery of the wafer, in the center of the wafer after irradiation with $1.5x10^{17}$ cm^{-2} fast neutrons and annealing at 1000°C.

Thus, if we assume that the initial carrier removal rate, 0.4 cm^{-1}, comes from the difference in introduction rates of all these acceptors and the introduction rate of the 0.06 eV V_N donors, the latter should be close to 1.4 cm^{-1}. A similar relation between the concentrations of these traps was determined from electron concentration and mobility fitting and DLTS measurements for n-GaN irradiated with 1 MeV electrons [40] (1 cm^{-1} for the V_N centers and 0.2 cm^{-1} for the 0.18 eV centers). Hence, the data on electron removal by relatively low energy electrons can be consistently explained by the introduction rates of well documented radiation point defects.

This is not the case for fast neutron irradiated GaN. The main deep traps that can be associated with such irradiation are the 0.18 eV ER3 electron traps and the 0.8 eV Ga_i electron traps. The introduction rate of the shallower ER3 traps is very low, much lower than for electron irradiation, the introduction rate of the 0.8 eV traps is below 1 cm^{-1} and is much lower than the electron removal rate of 5 cm^{-1}. Since these traps are deep donors they cannot contribute to carrier removal. The removal rate observed in neutron irradiated n-GaN can be explained by the formation of disordered regions DRs of the type described by Gossick [41] and shown schematically in Figure 2(bottom). Irradiation produces a broad hole-trap-like feature at temperatures 100-300K. The activation energy of this peak is 0.6-0.7 eV. If this were a true hole trap or band of hole traps they would be located near E_v+0.6 eV and their ODLTS signal should not be produced by optical excitation with the photon energy lower than 2.8 eV. The observed PPC phenomena and the appearance of the quasi-hole-trap signal in ODLTS stem from the existence of regions in which the bands are bent upwards by about 1 eV so that electrons released from deep centers inside these regions are swept out by the built-in electric field of the region and have to overcome the barrier of about 1 eV to be recaptured by their host traps. It is natural to associate these regions with Gossick-like DRs [41], i.e., heavily disordered core regions surrounded by the space charge region with a strong band bending. For very high neutron irradiation doses the outer regions of the DR's overlap and the Fermi level pinning position in such material gives some idea of the Fermi level position in the core of the DR. Our measurements on heavily neutron irradiated GaN show that, irrespective of the starting conductivity type and doping, the Fermi level in such samples is pinned near E_c-(0.9-1) eV [42]. We pointed to an obvious correlation between this Fermi level pinning position and the Fermi level pinning at the surface of n-GaN Schottky barriers. This is linked to the Fermi level stabilization F_s or charge neutrality CNL concept introduced long time ago to explain similar correlation in many other III-V materials [43, 44]. Several theoretical models have been offered to account for the observed F_s position in various materials. In one class of models the Fermi level is believed to be trapped between the levels of the major native defects (see e.g., [43] and references therein). In GaN it is located between the levels of N_i acceptors and of Ga_i donors. Moreover, lattice parameter measurements in heavily neutron irradiated GaN show a measurable increase indicating that the dominant defects could be interstitials [20, 35, 42]. Rutherford backscattering experiments on neutron irradiated GaN also point to a very high density of interstitials, predominantly Ga_i [45]. In the other set of models the F_s position is determined by the gap-induced states and can be estimated based on the known band structure [44]. Theoretical estimates made along these lines place the F_s in GaN near E_c-0.8 eV [46].

Although any of these models will qualitatively explain the behavior of neutron irradiated n- and p-GaN, much needs to be done to build the quantitative model. For example, the dependence of the carrier removal rate on starting donor density is not easily explained by the classical Gossick model. The carrier removal rate decreases by about 5 times when the donor concentration thus decreases. The carrier removal rate in neutron irradiated p-GaN is about 20 times higher than for n-GaN despite the much higher concentration of acceptors in p-GaN than donors in n-GaN that should have overpowered the effect of increased barrier height in p-GaN DRs [20, 35]. The same asymmetry of carrier removal rates was observed for proton implanted p-GaN and n-GaN samples [21, 34] and suggests that interaction of primary defects with Mg ions present in very high concentrations could be an additional factor in both cases.

The dependence of electron removal rate was also observed for variously doped n-GaN films irradiated with 100 keV protons [47] and ^{60}Co γ-rays [48].

II.2.3. Effect of Irradiation on Lifetime of Nonequilibrium Charge Carriers

It is a common practice in radiation effects community to characterize the radiation damage for lifetime by lifetime degradation constant K_τ (see e.g., [49]):

$$\tau_0/\tau=1+K_\tau F, \tag{1}$$

where τ_0 is the lifetime value before irradiation, τ is the lifetime after irradiation with the fluence F. Direct measurements of lifetimes in GaN of either type before and after irradiation are rather scarce. The diffusion length in n-GaN before proton irradiation with 150 keV protons was 1.2 μm which converts to the lifetime of 10 ns if we assume the hole mobility of 100 cm^2/Vs. Proton irradiation with dose of 5×10^{14} cm^{-2} decreased the diffusion length to 0.6 μm (estimated lifetime of 1.4 ns, the hole mobility was also assumed to be 100 cm^2/Vs). From Eq. (1) the lifetime degradation coefficient K_τ(protons) is found to be 1.6×10^{-14} cm^2 for such irradiation.

Systematic data of this kind has not as yet been collected for GaN. Some idea of the lifetime damage constant could be obtained from photoluminescence intensity measurements because the PL (or in our case microcathodoluminescence MCL) intensity is also related to lifetime. In these measurements the lifetime degradation constant can be obtained from the well-known expression [49]:

$$I_o/I=1+K_\tau F, \tag{2}$$

where I_o is the intensity before irradiation and I is the intensity after irradiation with fluence F. Measurements of the bandedge MCL intensity in neutron irradiated undoped n-GaN were performed in Ref. [20]. Strong changes in intensity occurred after the neutron fluence of 2×10^{15} cm^{-2}. Calculations according to Eq.(2) yielded in this case the K_τ(neutrons) value of about 3×10^{-16} cm^2.

Similar estimates for HVPE p-GaN and MBE p-GaN samples irradiated with fast neutrons yield for lifetime damage constants the values of:

K_τ(neutrons, HVPE p-GaN)=(7-8)$\times10^{-17}$ cm^2, K_τ(neutrons, MBE p-GaN)=4×10^{-16} cm^2.

The nature of defects responsible for the lifetime degradation is not clear. For proton implanted n-GaN we noted that strong decrease in the measured lifetime and in the bandedge luminescence intensity started after doses at which deep electron traps 0.8 eV presumably associated with Ga_i were introduced [21]. In GaN, when the dislocation density is fairly high, one has to consider possible effects of dislocations on diffusion length or lifetime measurements.

II.2.4. Effects of Dislocation Density, Comparison of Irradiation of Standard GaN and Low-Dislocation-Density Material Grown by Epitaxial Lateral Overgrowth

GaN films are mainly grown on lattice mismatched substrates: sapphire, SiC, and contain a very high density of dislocations. In standard GaN films grown on sapphire using low temperature GaN or AlN buffers the dislocation density is on the order of 10^9 cm^{-2}. By optimizating growth conditions, the density of dislocations can be reduced to some 10^8 cm^{-2} for growth on sapphire and to low 10^8 cm^{-2} for growth on SiC. The dislocation density can be decreased to about 10^6 cm^{-2} for epitaxially laterally overgrown ELOG GaN films prepared on sapphire by overgrowing a regular pattern prepared on a GaN template by deposition of a dielectric mask (usually, SiO_2 or Si_3N_4) [50]. A simplest example is a pattern consisting of SiO_2 stripes. Growth over the stripes proceeds predominantly in lateral direction which leads to an efficient filtering out of dislocations in these ELOG wing regions while in between the stripes growth proceeds mostly in the vertical direction and inherits the dislocation density of the starting GaN template[50]. Typically, ELOG films are prepared by metalorganic chemical vapor deposition MOCVD and the dislocation density in the ELOG wing regions is about 10^6-10^7 cm^{-2} whereas in the ELOG window regions it is on the order of some 10^8 cm^{-2} [51]. Thick (over 200 μm) GaN films grown on sapphire by hydride vapor phase epitaxy HVPE have the dislocation density below 10^7 cm^{-2} and can be separated from sapphire by various means to form an improved GaN substrate for epitaxial growth of GaN-based device structures [52]. Therefore, interaction of radiation defects with dislocations is something to be considered for structures prepared by ordinary epitaxy. At the same time, advanced devices, such as high-power LEDs and LDs tend to be grown either on low-dislocation-density GaN substrates or using some version of ELOG technique. This could have an impact on the type of defects created in GaN-based devices.

The most understood manifestation of dislocations is the effect of spatial correlation of electron capture by deep traps decorating dislocations. Correlated capture has been theoretically and experimentally treated [53, 54]. The most obvious effect is the logarithmic dependence of the DLTS peak amplitude of the trap on the injection pulse length t_p [53, 54]. Such behavior has been reported for the 1 eV N_i-related acceptors. Similar behavior was reported for 0.78 eV and 0.95 eV electron traps introduced by 2 MeV proton irradiation [15,55]. These traps are due to Ga_i deep donors and N_i deep acceptors. This shows that during irradiation some of the interstitial defects can travel to dislocation boundaries and decorate them-this needs to be considered in radiation defects models. It might also make us reconsider how we treat the carrier removal effects even in the cases when only point defects are produced because of the possible contribution of capture in the space charge regions surrounding dislocation cores.

Another factor is the impact of dislocations on mobility of charge carriers. A credible theory describing the individual dislocations contribution to scattering of electrons in n-GaN was described in Ref. [9]. The theory predicts a strong increase of the electron mobility with decreasing the dislocation density and increasing the electron concentration (the latter due to enhanced screening of dislocations). However, the effect of dislocations on the carrier mobility is not reduced to the sum of contributions from individual dislocations. For dislocation densities exceeding $\sim 10^8$ cm^{-2} in GaN dislocations form a cellular structure with dislocation boundaries formed mostly by the threading edge dislocations and the characteristic grain size determined by the dislocation density. The carriers travel over the grain boundaries involves in general overcoming a more or less high potential barrier. When the electron concentration is high tunneling through the barriers is efficient and the material behaves as though electrical nonuniformities were not present. With increased doses of radiation the electron concentration within the grains becomes lower and the potential barriers at the grains gain higher importance. For p-GaN tunneling of holes is difficult even at high acceptor densities and hole mobility seldom shows "theoretical" temperature or concentration dependence [56].

The cellular structure of GaN films with high dislocation density should also have a profound effect on diffusion length of nonequilibrium charge carriers. Chernyak et al. [57] postulated that the diffusion length in n-GaN is determined by the distance between the low-angle dislocation boundaries and hence by the dislocation density. The rule worked very reasonably well for HVPE grown n-GaN films with thickness varying from below 1 μm to 80 μm and the dislocation density decreasing respectively from about 10^{10} to about 10^8 cm^{-2} [57]. This is not always the case (we've observed relatively high diffusion lengths on the order of 1 μm for material with dislocation density of about 5×10^8 cm^{-2} [21]; in the model one has to account for the difference in recombination efficiency of dislocation boundaries in different materials [56]). The model is not valid for low-dislocation density material where the dislocations are distributed randomly. There the diffusion length and the lifetime of charge carriers are determined mostly by other defects than dislocations.

Comparison of electrical characteristics, deep traps spectra, diffusion length and MCL intensity of the low dislocation density (~5×10^{6} cm^{-2}) ELOG wing regions and high dislocation density (~10^{8} cm^{-2}) ELOG window regions was carried out by combined DLTS, ODLTS, MCL, EBIC measurements [51, 58-60]. For undoped n-GaN ELOG films with the ELOG layer thickness of 12 μm, the residual donor concentration was much higher in the ELOG window region than in the ELOG wing region (~10^{15} cm^{-3} versus ~3×10^{14} cm^{-3}), that the deep traps density in the ELOG wing region was about an order of magnitude lower than for standard MOCVD material, that the diffusion length in the low-dislocation-density ELOG wing was close to 0.3 μm versus 0.17 μm in the high-dislocation- density window (the lifetime difference of 0.4 ns versus 0.1 ns) and that the bandedge intensity was about 5 times higher in the ELOG windows [58]. Doping of ELOG GaN with Si up to about 10^{18} cm^{-3} resulted in the donor concentration being about 3 times higher in the window region [59]. Decreasing the ELOG layer thickness to 6 μm instead of 12 μm resulted in the appearance of relatively high density of dislocation loops and greatly increased density of deep traps with activation energies of 0.25 eV, 0.6 eV and 0.85 eV. Doping with Si suppressed the formation of the 0.6 eV traps, but enhanced the formation of the 0.85 eV traps with the result of the diffusion length L_d being relatively constant. It was slightly lower for thinner ELOG layers compared to thicker ELOG layers, 0.25 μm versus 0.3 μm [58, 59]. From these results it was inferred that the centers most likely to be the lifetime killers were the 0.6 eV and the 0.85 eV electron traps. The ELOG wing region was also the region of strong band bending which resulted in anomalously long propagation of nonequilibrium charge carriers along the ELOG stripes even far outside the Schottky barrier [58, 59]. Only for relatively heavy donor doping this band bending effect was suppressed, most likely due to enhanced tunneling.

Neutron irradiation of undoped ELOG GaN films resulted in a much lower effective removal rate than for standard MOCVD material, 1 cm^{-1} versus 5 cm^{-1} [61]. The Fermi level position in heavily irradiated material is also the same suggesting that no strong changes in the structure of the core regions of DRs occur with changing the dislocation density.

DLTS spectra of neutron irradiated ELOG samples differed from their MOCVD opposite numbers in that the 1 eV N_i-related acceptor state could be clearly seen. However, as with MOCVD, the introduction rates of all traps were several times lower than the electron removal rate again suggesting the dominant role of DRs in carrier removal. There was a much lower introduction rates of all traps compared to the MOCVD case.

EBIC profiling of the irradiated samples showed that the strongest carrier removal occurred in the high-dislocation-density ELOG window regions and the effect slowly propagated into the ELOG wing region with lower donor concentration and low dislocation density. Estimates of the lifetime degradation constant from changes in bandedge MCL intensity give the values of 3×10^{-16} cm^{2} for undoped ELOG material and 1.2×10^{-16} cm^{2} for lightly doped (2×10^{16} cm^{-3}) material. All in all the impact of the difference in dislocation density on the lifetime degradation does not seem to be strong for neutron irradiation.

It is interesting to check whether the difference in behavior between MOCVD and ELOG and between the wing and window regions of ELOG is related to the difference in dislocation density? The main differences in the carrier removal rates of ELOG wing and ELOG window regions stem from the difference in doping level. However, gradual penetration of the damage from the boundary with high-dislocation-density window into the wing indicates that primary defects interaction with dislocations in the window region is also a factor. We would also like to note that the deep traps introduction rate increased for the doped ELOG samples.

For ELOG material we observed considerable broadening of triple-crystal x-ray rocking curves for the (0006) symmetric reflection and (11-20) asymmetric reflection from 165" to 205" (0006) and from 230" to 270" (11-20) respectively as a result of neutron fluence irradiation of 10^{18} cm^{-2} [55]. This broadening is a manifestation of increased density of extended defects which is confirmed by direct etch pits density EPD measurements that show EPD to increase by about 5×10^{7} cm^{-2} in the high dislocation density regions and the appearance of new irradiation induced inclined dislocation bands in the low dislocation density wings [55].

II.2.5. Thermal Stability of Radiation Defects in GaN

Annealing of defects introduced by light irradiating particles (2 MeV protons, 0.2-2.4 MeV electrons) was performed in [15]. The shallow radiation defects ER1, ER2, ER3 start annealing at 540K and the annealing is complete after 620K. Deeper electron traps ER5 associated with N_i were reported to start annealing also at 540K, but a higher annealing temperature of 660K was needed for complete removal of these defects [15]. For Ga_i deep donors it was found in Ref. [12] that they start moving at room temperature although the fate of these centers was not uncovered in the paper. The V_{Ga} centers responsible for the 0.95 eV PL band were stable up to 500°C.

In GaN samples with a high density of radiation defects (high doses of ~100 keV implanted hydrogen or heavier ions, neutron irradiated material) the thermal stability of radiation damage was much higher. After heavy proton implantation the bandedge luminescence intensity could not be restored to the pre-irradiation value even after furnace annealing of 800°C [21]. For activation of ion implanted donors (Si) or acceptors (C) annealing to temperatures exceeding 1000°C was necessary and even so the degree of impurity activation was relatively low [62]. Figure presents the evolution of the sheet resistivity of undoped GaN sample irradiated with fast and thermal neutrons to fast neutron fluence of 1.5×10^{17} cm^{-2} (the ratio of fast and thermal fluences 1:1) [23]. The as-irradiated resistivity was quite high, it measurably decreased at the 150-250°C stage, increased very strongly at the 250-450°C and then gradually decreased in a very broad stage 500-1000°C [23]. The first stage corresponds to reconstruction of the ER3 and ER5 acceptors as in Ref. [15] which perhaps explains the decrease in resistivity. The reverse annealing stage at 250-450°C is most likely due to movement of the N_i, Ga_i centers with forming new deep compensating centers. The onset of the third stage of recovery at 500°C correlates with the V_{Ga} acceptors annealing stage in Ref. [12] which explains the decrease of the resistivity. Still, even after annealing at 800°C, the pre-irradiation resistivity was not reached, the Fermi level was pinned at relatively deep centers with activation energy 0.45 eV, the sample's series resistance was quite high which resulted in the appearance of DLTS peaks of the wrong sign. The most prominent electron traps were the 0.9 eV and the 1 eV traps that are very likely related to the Ga_i donors and the N_i acceptors, but with a very high binding energy, possibly trapped within disordered regions. The ODLTS spectra were dominated by the hole traps with activation energy close to the V_{Ga} 0.9 eV centers, but with a much lower capture cross section. After 1000°C annealing the Fermi level was pinned near E_c-0.2 eV, DLTS spectra were dominated by the 0.6 eV and 0.9 eV traps in high concentration. The total concentrations of the 0.45 eV traps pinning the Fermi level after 800°C annealing and of the 0.2 eV traps dominant in the 1000°C annealing are close to each other and equal to the number of donor Ge atoms converted from Ga by interaction with thermal neutrons (2×10^{16} cm^{-3}). Hence, there is reason to believe that these relatively deep traps could be complexes of radiation defects with donor atoms [23]. These results show that, even after such not too high doses of neutron irradiation, it is hard to break down the disordered regions and to restore the virgin conductivity.

For very high neutron fluences the resistivity of GaN passes through a maximum related to the onset of hopping conductivity. The activation energy for the temperature dependence of resistivity for doses before the maximum showed the usual value of 0.9-1 eV. After the fluence corresponding to maximum the temperature dependence was much weaker. Annealing of such heavily irradiated samples showed a strong reverse annealing stage up to 300°C where the density of radiation defects decreased and the activation energy returned to the 0.9 eV value. Complete recovery could not be attained even after annealing at 1000°C.

III. Radiation Effects in Other III-Nitrides

Radiation effects studies for III-nitrides other than GaN are scarce. For InN, Hall effect measurements on 150 keV protons and ^{60}Co γ-rays irradiated undoped n-InN films showed that, in contrast to n-GaN, irradiation increases electron concentration in InN. This difference seems to be closely linked to the difference in the position of the Fermi stabilization level in GaN and InN that in turn determines the band offsets in respective heterojunction and the difference in the Schottky barrier height between GaN and InN [63]. Based on calculations carried out in Ref. [63], with increasing the In composition in $In_xGa_{1-x}N$ solid solutions the Fermi stabilization level moves upwards to the conduction band edge. The cross-over point is close to x=0.34 and this composition separates solid solutions in which the electron concentration decreases with irradiation from those in which irradiation increases the electron concentration. Blue LEDs are built on GaN/InGaN QW structures with In mole fraction in the QW close to 0.2 and their behavior should be reasonably close to GaN. For LEDs working in the longer wavelength the In composition is higher and care is needed when extrapolating GaN results to these green or red LEDs.

For AlGaN there are only a few papers describing the effects of proton and neutron irradiation. In Ref. [18] n-AlGaN films with Al mole fraction of x=0.12 and free electron concentration of 10^{17} cm^{-3} were irradiated at room temperature and at 300°C with 2 MeV protons. It was found that the carrier removal rate in AlGaN was about twice as high as for GaN and decreased approximately by two times for high temperature irradiation. A broad feature near 200K was observed in DLTS spectra after irradiation, but the authors failed to get the values of activation energy, capture cross section and concentration for this center because of the strong concentration freeze-out effects. For HEMTs, AlN/GaN devices were more radiation hard than their AlGaN/GaN counterparts, as shown in Figure 7. For undoped n-AlGaN films with Al mole fraction of x=0.4 the effects of 100 keV proton irradiation and fast reactor neutron irradiation were studied in Ref. [64, 65]. Before irradiation electrical properties of such samples were determined by the presence of a high (2.5×10^{18} cm^{-3})

relatively deep donors with ionization energy of 0.25 eV as follows from admittance spectra and low-frequency capacitance-voltage C-V measurements.

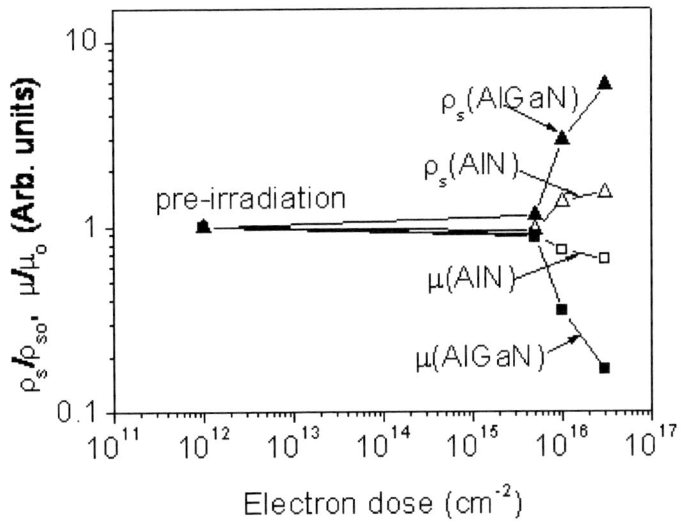

Figure 7. Relative changes of 2DEG resistivity and mobility after 10 MeV electron irradiation of AlGaN/GaN and AlN/GaN HEMT structures.(reprinted with permission from American Institute of Physics, Polyakov et al., Appl.Phys. Lett.. 2008, 93,152101).

The traps in question are believed to be due to defects that form resonance states within conduction band of GaN and emerge in the bandgap of AlGaN at compositions x~0.2-0.3 [6]. These traps were also detected by photoinduced current transients spectroscopy PICTS. Even deeper states giving rise to defect bands peaked at 3.7 eV and 2.2-2.5 eV were detected in microcathodoluminescence MCL spectra. Neutron irradiation led to compensation of the 0.25 eV traps and introduced instead deeper states with activation energy of 0.28 eV at neutron fluences of 10^{15} cm^{-2} to 2.5×10^{16} cm^{-2}. For higher neutron fluences deeper traps with activation energies 0.35 eV and 1 eV were formed according to PICTS. After irradiation with 1.7×10^{17} cm^{-2} neutrons the films became semi-insulating with the Fermi level pinned near 0.35 eV from the conduction band edge. Irradiation with higher neutron fluences increased the sheet resistivity of the layers so that it could not be measured (sheet resistivity in excess of $10^{14}\Omega$/square). If one goes by the value of the Schottky barrier height in such AlGaN films (2 eV [64, 65]) one would expect the Fermi level to be pinned near E_c-2 eV in these samples after high doses of radiation which accounts for very high resistivity (see section II.2.2 devoted to Fermi level pinning in heavily disordered GaN).

Interestingly, the introduction rate for compensating defects for neutron irradiated n-AlGaN was very much higher than for undoped n-GaN (about 500 cm^{-1} versus 5 cm^{-1}) and even higher than in p-GaN (~100 cm^{-1} [21]). Measurable changes of electrical properties started at neutron fluence of 10^{15} cm^{-2}, i.e., similar to undoped n-GaN [20], even though the concentration of centers to be compensated was 2 orders of magnitude higher for n-AlGaN.

The effects of proton implantation were similar to the effects of neutron implantation, but the 100 keV protons started to change electrical properties of such AlGaN layers after the dose of 10^{12} cm^{-2}, again two orders of magnitude lower than for undoped n-GaN, even despite a much higher donor density in n-AlGaN [21]. As for neutron irradiation the resistivity of the samples rapidly rose with increasing the proton fluence and after irradiation with 10^{14} cm^{-2} of protons the sheet resistivity was $10^{13}\Omega$/square. We also observed that irradiation led to a decrease of the overall intensity of the two major defect bands and increase of the contribution of the 2.3 eV defect band at the expense of the 3.7 eV band [64].

For p-AlGaN with Al mole fraction of x=0.12, the effect of 100 keV proton irradiation was studied in Ref. [66]. In the virgin state the electrical properties were determined by Mg acceptors with activation energy of 0.17 eV and concentration of 3×10^{18} cm^{-3} as follows from admittance spectra and C-V measurements. The concentration started to decrease after irradiation with a very very low proton fluence of 10^{12} cm^{-3}. After irradiation with 10^{13} cm^{-2} the apparent activation energy of the dominant acceptors increased to 0.2 eV while the concentration further decreased. Irradiation with the fluence of 10^{14} cm^{-2} totally compensated the p-AlGaN film down to the depth corresponding to the range of 100 keV protons (see C-V concentration profiles in Figure 17). Proton irradiation also led to a strong increase of the series resistance in the forward current-voltage I-V characteristics and decreased the reverse current of Au/p-AlGaN Schottky diodes. The Mg-related MCL band intensity decreased by about 10 times after irradiation with 10^{14} cm^{-2} protons and after irradiation with higher proton doses we saw, alongside with decrease of the intensity of this band, the emergence of the yellow band. The observed changes, both qualitatively and quantitatively are similar to proton irradiation effects in p-GaN which is not altogether astonishing given the low Al composition of p-AlGaN.

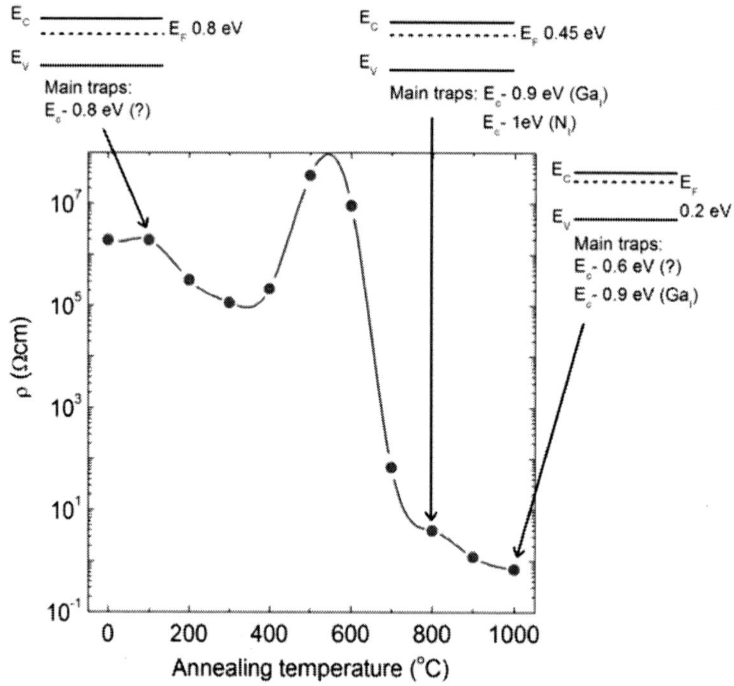

Figure 8. Sheet resistivity as a function of annealing temperature for undoped GaN sample irradiated with fast and thermal neutrons to a fluence of 1.5×10^{17} cm^{-2}.The position of the Fermi level at different stages is also shown.

In all these cases, it is important to realize that the resistivity and Fermi level position as changing with annealing-an example is shown in Figure 8 for GaN irradiated with neutrons to a dose of 101^7 cm^{-2} and annealed at different temperatures.

IV. Radiation Effects in GaN Schottky Diodes, in AlGaN/GaN and GaN/InGaN Heterojunctions and Quantum Wells

The simplest device is a Schottky diode. It can be a stand-alone device as in a Schottky diode photodetector or power rectifier or a part of a more complex device, such as a high electron mobility transistor HEMT. The authors of [10] studied the effects occurring in Ni/n-GaN Schottky diodes upon ^{60}Co γ-rays irradiation and found a considerable, about 0.15 eV, increase in the Schottky barrier height after exposure to the dose of 20 MRad (Si). The effect was explained by the formation of a negatively charged defect interfacial layer near the Ni/GaN boundary. Similar increase by about 0.1 eV in the Ni/GaN Schottky barrier height was reported for the Schottky gates of AlGaN/GaN HEMTs irradiated with 1.8 MeV protons for protons doses in excess of 10^{14} cm^{-2} [67]. We observed similar slight Schottky barrier height increase for Au and Ni n-GaN films after irradiation with fast reactor neutrons and 10 MeV electrons. The increase of the specific ohmic contacts resistance reported in several papers (see e.g., Ref. [67, 68]) for proton irradiated AlGaN/GaN HEMTs is most likely also due to the increase of the Schottky barrier height upon irradiation: it is well known that good ohmic contacts are Schottky diodes with a low barrier height and strong tunneling of electrons provided by various means.

In most cases low and moderate doses of radiation decrease the reverse current of GaN Schottky diodes (see e.g., the example for neutron irradiated n-GaN in Figure 9).

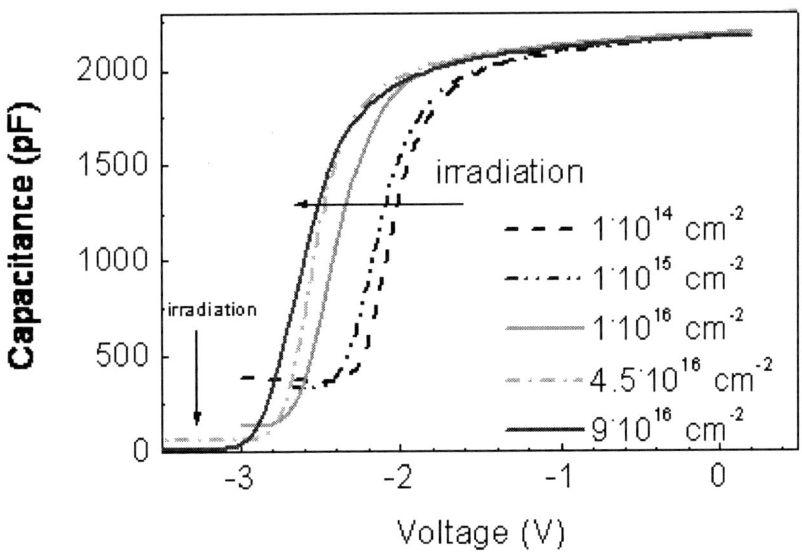

Figure 9. C-V characteristics as a function of 10 MeV electron dose for AlGaN/GaN Schottky diodes.

AlGaN/GaN heterojunction is the heart of any HEMT structure. Because of the strong electrical polarization field the two-dimensional electron gas 2DEG density in such structures is on the order of 10^{13} cm^{-2} even without selective doping of the AlGaN barrier which makes such devices preferable to their AlGaAs/GaAs opposite numbers for high-frequency/high-power applications (see e.g., Ref. [69]). Typically the structure consists of a substrate (sapphire, SiC, Si), a buffer layer (low temperature GaN or AlN for sapphire, AlN for SiC or Si), ~1 μm of GaN that has to be highly resistive, ~20-30-nm-thick AlGaN barrier with Al mole fraction typically in the x=0.2-0.4 range and ~2-nm-thick GaN cap. Multiple variations to this basic structure are possible. Selective donor doping of the AlGaN barrier to ~10^{18} cm^{-3} Si donors is supposed to increase the 2DEG concentration. This selectively doped layer can be followed by undoped AlGaN spacer to suppress the scattering effect of donors in the barrier and to increase the 2DEG mobility. The barrier can be made of AlN to increase the 2DEG density or a composite AlGaN/AlN barrier can be used for the same purpose (see e.g., a recent review [70] and references therein). Gaudreau et al. [71] irradiated AlGaN(undoped, 13 nm, x=0.3)/GaN(200 nm)/SI-GaN(C) (2 μm)/AlN/sapphire with 2 MeV protons and reported that substantial changes of 2DEG concentration occurred after the proton fluence of 10^{15} cm^{-2} while 2DEG mobility changes started after 10^{14} cm^{-2}. The results were explained by compensation of both the GaN channel layer and the AlGaN barrier layer and disordering of the AlGaN/GaN boundary. Perhaps the most revealing studies of the matter were performed in Ref. [67]. The authors irradiated the simplest HEMT structure consisting (from top to bottom) of 2 nm undoped GaN cap, 23 nm Si doped AlGaN barrier (x=0.3) and 2 mm undoped GaN (all films grown by MBE on sapphire) with 1.8 MeV protons at room temperature. It was found that the 2 DEG density started measurably decreasing after the proton fluence of 10^{14} cm^{-2} while the 2DEG mobility started decreasing after the fluence of 10^{13} cm^{-2}. Secondary ions mass spectrometry SIMS measurements and MCL spectra measurements probing very thin layers of material near the AlGaN/GaN interface indicated that proton irradiation leads to mixing of AlGaN and GaN near the boundary which should decrease the strain in the structure and hence the polarization-field-induced charge in the 2DEG. The authors also postulated a strong decrease of donor doping in AlGaN that should contribute to the decrease of 2DEG density after the highest doses of irradiation. The mobility degradation was ascribed to the effects of AlGaN/GaN mixing at the boundary. Since the mobility is more sensitive to the irregularities of the boundary and to the interface roughness than the 2DEG concentration the mobility changes occur earlier than concentration changes. Radiation defects introduced into the GaN channel also contribute to the mobility degradation although these effects should be strongly screened in the high density 2DEG. Very similar changes of the 2DEG mobility and concentration were observed in Ref. [68] for more advanced HEMT structure with a combined barrier consisting of (top to bottom) a much thicker (250 nm) undoped GaN cap, 10 nm Si doped graded AlGaN (x increases from 0 to 0.22), 40 nm undoped AlGaN (x=0.22) and 7 nm AlN (the structure was grown by MOCVD on sapphire).

Effects of neutron irradiation on 2DEG concentration and mobility in a simple AlGaN/GaN HEMT structure were studied by us in Ref. [72]. As for proton irradiation, we observed that the decrease of the 2DEG mobility started from the neutron fluence of 10^{14} cm^{-2} for mobility measured at 77K and after irradiation with 10^{15} cm^{-2} of fast neutrons for the 300K 2DEG mobility. The degradation of the 2DEG concentration occurred only after the fluence of 2.5×10^{16} cm^{-2}. ODLTS and PICTS spectra measurements suggest that hole traps with activation energies 0.2 eV and 0.7 eV were produced by neutrons.

The results of 10 MeV electrons irradiation of AlGaN/GaN HEMT structures grown by MBE were described by us in Ref. [73]. A simple structure with a combined SI-GaN(Fe)(0.7 mm)/undoped GaN (2.2 μm) buffer/undoped AlGaN (x=0.28, 20 nm)/GaN cap(2 nm) grown by MBE was irradiated at room temperature. The 2DEG mobility started to decrease after irradiation with electron fluence of 5×10^{15} cm^{-2} (decrease of about 14%). After irradiation with the fluence of 3×10^{16} cm^{-2} the mobility decreased by 6 times. Since the 2DEG concentration was not affected by these doses of radiation the change in the sheet resistivity of the 2DEG channel was totally reflecting the changes in 2DEG mobility. An interesting observation also reported in Ref. [73] concerns the electron irradiation results for HEMT structures in which the barrier was 3-nm-thick AlN instead of 20 nm AlGaN. The 2DEG concentration in such AlN-barrier HEMT structures was 1.5 times higher than for their AlGaN-barrier opposite numbers (3×10^{13} cm^{-2} instead of 2×10^{13} cm^{-2}) with the room temperature mobility virtually the same (1360 cm^2/Vs for AlN versus 1450 for AlGaN). (Detailed description of electrical properties of these two types of heterojunctions can be found in Ref. [74].) As seen from Figure, for the HEMT structure with AlN barrier, the changes in mobility were much slower than in the structure with AlGaN barrier: even after the fluence of 3×10^{16} cm^{-2} the mobility was still 50% of its pre-irradiation value. This almost an order of magnitude higher radiation tolerance of the AlN/GaN HEMTs compared to AlGaN/GaN HEMTs is due to the fact that the AlN barrier is 7 times thinner than its AlGaN counterpart with correspondingly lower energy deposited by 10 MeVelectrons into the formation of defects in the barrier and near the barrier/channel interface. Similar considerations were invoked in Ref. [75] to explain an increased radiation tolerance of AlGaN/GaN HEMTs to proton irradiation as the energy of protons increased from 2 MeV to 108 MeV. In addition, one expects the displacement energy in AlN to be considerably higher than in AlGaN.

One more characteristic that is important for AlGaN/GaN heterojunctions developed for HEMTs is the threshold voltage, i.e., the voltage you have to apply to your Schottky barrier gate to totally pinch off the 2DEG channel near the AlGaN/GaN interface. In simple theory this voltage V_T is determined by the magnitude of the electrical polarization field, the Schottky barrier height ϕ_B and the donor doping level of the AlGaN channel N_d as [76]:

$$V_T=\phi_B-(qN_dt_{AlGaN}^2/2\varepsilon)-(\Delta E_c/q)-(\sigma_{pol}t_{AlGaN}/\varepsilon), \tag{3}$$

where q is the electronic charge, ε is the dielectric permittivity, ΔE_c is the conduction band discontinuity, σ_{pol} is the polarization charge, t_{AlGaN} is the AlGaN barrier thickness. As discussed above in relation with proton induced changes of 2DEG mobility and concentration there is reason to believe that the second and fourth terms in Eq. (3) could decrease after bombardment with high energy particles which should lead to increasing the threshold voltage. This indeed happens in many cases after proton irradiation (see Ref. [67,68]). Typically the proton fluence after which the threshold voltage starts to grow is close to 10^{14} cm^{-2} for 2 MeV protons.

However, the situation could be more complicated. For example, in Ref. [73, 77] we reported the changes in C-V characteristics of 10 MeV irradiated AlGaN/GaN structures. The capacitance of Schottky dides made on these heterojunctions was high at low negative bias on the diode because of the very high charge accumulated in the 2DEG region. As the negative bias increased the states in the AlGaN/GaN quantum well were emptied which resulted in decreased capacitance. When all electrons were driven out of the AlGaN/GaN QW the capacitance was the capacitance of the GaN buffer and its change with voltage was determined by the donor concentration in GaN. The voltage at which the capacitance becomes the capacitance of the GaN is, of course, the same as the threshold voltage discussed above for HEMT structures. This was shown earlier in Figure 9, where the C-V characteristics measured at 400K for AlGaN/GaN Schottky diode before and after irradiation with increasing fluences of 10 MeV electrons show changes in threshold. In Ref. [77] we argue that the low-dose negative shift of threshold voltage could be due to radiation induced positive charge accumulation in the AlGaN barrier that is then superseded by the above described processes of radiation induced interface disordering. We observed similar behavior for the initial stages of neutron irradiation of AlGaN/GaN HEMT structures. The shift of the threshold voltage to more negative values was also reported for γ-irradiated AlGaN/GaN HEMTs [78]. This positive trapped charge could also increase the 2DEG concentration and we indeed reported some increase of the 2DEG concentration in neutron irradiated AlGaN/GaN heterojunctions for low neutron fluences [72].

For GaN LED structures perhaps the most important part for operation of the device is the GaN/InGaN quantum well QW or multiple quantum well MQW region where recombination of injected charge carriers should occur. Processes induced in MQWs by 10 MeV electron irradiation at room temperature showed a strong influence of tunneling. C-V profiling at 85K revealed the presence of 4 QW related peaks due to the 4 lowermost QWs). At 400K the profiles were strongly smeared and only a kink due to the lowest QW could be seen. The observed profiles were due to interface traps decorating GaN/InGaN QW interfaces. Irradiation compensated the MQW region and increased the ionization energy of the traps decorating the QW interfaces from 20-60 meV to 90 meV after irradiation with 5×10^{16} cm^{-2} electrons and to 190 meV after irradiation with 10^{17} cm^{-2} electrons. As a result at low temperature the whole MQW region behaved as dielectric layer while at 400K the QW related profile became more pronounced and showed a much enhanced concentration of interface traps. The traps in question are quite similar to the electron traps introduced by electrons in n-GaN which probably points to the traps being located in the GaN barriers. DLTS spectra measured on irradiated samples indicated, in addition to these interface traps, a deeper trap with activation energy 1.1 eV reminiscent of the N_i related acceptors in n-GaN. These traps were also located in the GaN barriers or at the GaN/InGaN interfaces on the GaN side of them [79].

Among other effects important for device applications we note a measurable increase of the series resistance of the MQW Schottky diodes and a strong decrease of the QW-related recombination. The latter indicates a lifetime degradation constant close to 6×10^{-16} cm^2 [79].

V. Radiation Effects in GaN-Based Devices

AlGaN/GaN HEMTs irradiated with 1.8 MeV protons showed decrease of transconductance, threshold voltage and drain saturation current occurred after proton fluences of 10^{14} cm- [67, 68, 80]. This dose is about two orders of magnitude higher than for

AlGaAs/GaAs HEMTs [81]. Annealing at 800°C was shown to be efficient in partially restoring the electrical characteristics [80]. For higher proton energies of 40 MeV both DC characteristics (transconductance, threshold voltage, drain saturation current) and AC characteristics were little affected by proton fluences of up to 5×10^{10} cm^{-2} [83-85]. Proton irradiation of AlGaN/GaN HEMTs with energies 1.8 MeV, 15 MeV, 40 MeV, 105 MeV and proton fluences up to 10^{13} cm^{-2} showed the strongest changes for the lowest proton energy and the effect was explained by the decrease of the energy transferred to Al, Ga and N atoms in elastic collisions occurring within the active region of devices as the range of protons increased with increasing energy. Gamma irradiation of AlGaN/GaN HEMTs up to a dose of 600 MRad did not substantially change the characteristics [86].

The carrier removal rate was about four times higher in InAlN/GaN HEMTs compared to the AlGaN/GaN devices shown in Figure 10. Although all the proton-irradiated HEMTs exhibited good pinch-off characteristics, the amount the saturation drain currents were degraded was dependent on the irradiation energy. For the 10 MeV irradiated HEMTs, the reduction of saturation drain current at V_G = 0V was 23.6%. Much larger saturation drain current reduction, 46.4%, was observed for the HEMTs irradiated with proton energy at 5 MeV and only 11.5% drain current reduction exhibited for the HEMTs irradiated with proton energy at 15 MeV.

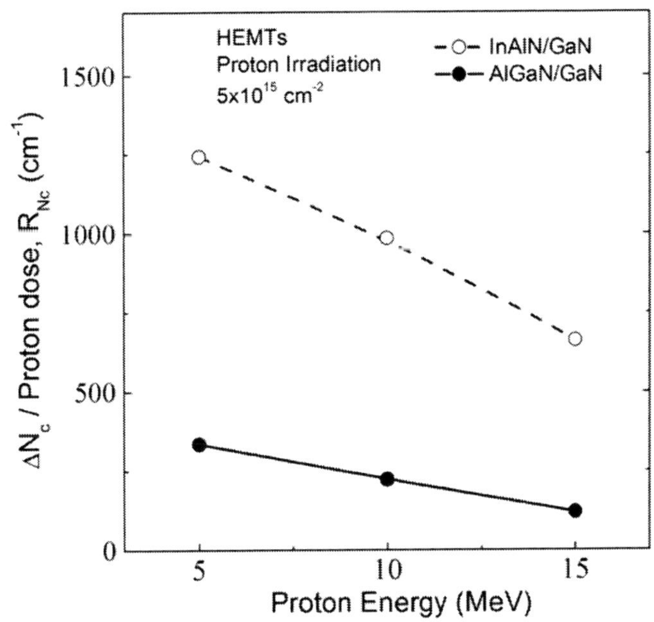

Figure 10. Carrier removal rates by protons in InAlN/GaN and AlGaN/GaN HEMTs.

One more important thing to be noted in conjunction with AlGaN/GaN HEMTs performance as affected by radiation is the current collapse phenomena in these devices. This phenomenon is a strong dependence of AC characteristics of transistor on frequency: a substantial loss of current that can be passed through a device at high frequency compared to DC characteristics, a substantial lag in the switching performance when rapidly changing the gate voltage, long-term drift of parameters after driving the device at high current [87-91]. The most important among those is the loss of current transfer characteristics at high frequencies believed to occur because of the trapping of charge carriers in the AlGaN barrier and the formation of a virtual gate with a much increased area compared to the actual metallic gate [87-91]. The effect can be mitigated by deposition of dielectric layers (Si_3N_4, Gd_2O_3, MgO) on top of the barrier [93-97] due to a decrease of the surface trap density [98]. For Gd_2O_3 and MgO passivating layers, their beneficial effect is not affected by proton and γ-ray irradiation [84-86].

Another important class of GaN-based devices is LEDs. For double heterostructure blue GaN/InGaN LEDs degradation of the light output started after irradiation with 10^{14} cm^{-2} neutrons [99]. For proton irradiated AlGaN/GaN QW LEDs, the threshold dose for the start of degradation was two orders of magnitude higher than for AlGaAs/GaAs QW LEDs (10^{12} cm^{-2} versus 10^{10} cm^{-2} for 3 MeV protons). Increasing the proton energy from 3 MeV to 5 MeV measurably increased the dose necessary for the onset of light output degradation, most likely due to a lower energy going into elastic collisions within the active region of devices. Even higher proton doses were found necessary for changing the characteristics of proton-irradiated blue GaN/InGaN LEDs [100]. For green GaN/InGaN LEDs [101-105] it was reported that 2 MeV protons produce about 40% light output decrease after a fluence of 1.7×10^{12} cm^{-2}.

VI. Enhancement of Off-State Breakdown Voltage in Proton Irradiated HEMTs

As we have seen above, typically, proton irradiated AlGaN/GaN HEMTs exhibited degradation of Schottky barrier height, gate leakage, saturation drain current and extrinsic transconductance (g_m), whose magnitude depends on the proton energy and dose. However, it was recently reported that the off-state drain breakdown voltage and the critical voltage during the off-state drain-voltage step-stress improved in the proton irradiated HEMTs [106,107]. The defects created by the high energy proton irradiation were evenly distributed throughout the entire HEMT structure. Since the effect of the proton irradiation created defects in the GaN buffer layer or in the 2DEG could not be separated, it is very difficult to identify the cause of drain breakdown voltage and step-stress critical voltage improvement through the proton irradiation. It is also difficult to separate the effects of the high dislocation density in HEMT structures, as shown in the TEM cross section of Figure 11.

Recent results with proton irradiation performed from the backside of the samples through via holes fabricated on the Si substrate allowed the effect of defects precisely placed at specific locations inside the AlGaN/GaN HEMT structure.

There was no degradation observed for the proton irradiated AlGaN/GaN HEMTs with proton energy of 225 or 275 keV, in which the defects created by the proton irradiations were intentionally placed in the GaN buffer. On the contrary, the irradiated HEMTs with the defects placed in the 2DEG channel region and AlGaN barrier using 330 and 340 KeV protons showed degradations on drain current and extrinsic transconductance. However, the off-state drain breakdown voltage of these degraded AlGaN/GaN HEMTs significantly increased. Finite-element simulations showed that the sharp increase of the off-state drain breakdown voltage resulted from the formation of a virtual gate at the 2DEG region of the AlGaN/GaN structure to reduce the peak electric field around the gate edges, as shown in Figure 12.

Figure 11. Cross-sectional TEM image of AlGaN/GaN HEMT structure. (reprinted with permission from American Institute of Physics, Pearton et al., Journal of Vacuum Science and Technology A31, 2013, 050801).

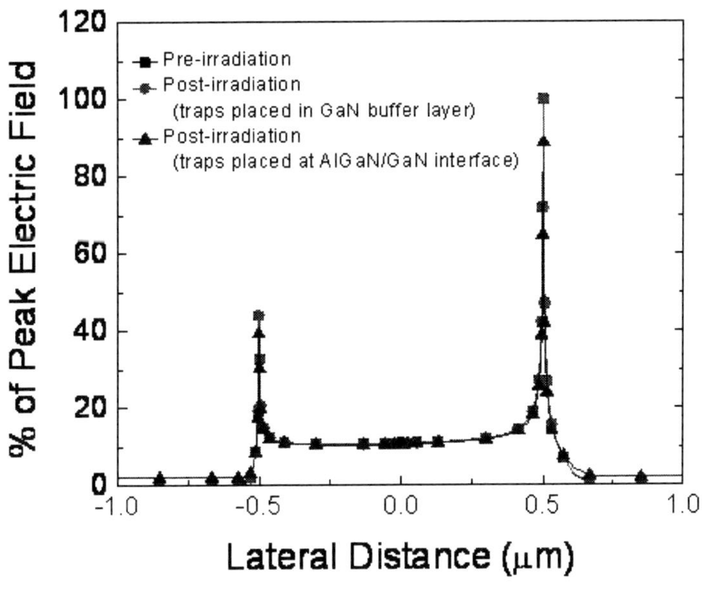

Figure 12. Comparison of the magnitude of pre-irradiation electric field with that of post irradiation as a function of lateral distance taken 5 nm under the AlGaN surface with the defects placed inside the GaN buffer layer or at the AlGaN/GaN interface.

Summary and Conclusion

The model for radiation effects in GaN shows the main radiation defects are due to shallow V_N and deep Ga_i donors and deep V_{Ga} and N_i acceptors. This picture places the V_N donors near E_c-0.06 eV, the Ga_i doubly charged donors near E_c-0.8 eV, the V_{Ga} acceptors near

E_v+1 eV and the N_i acceptors near E_c-1 eV. A schematic of the levels in the bandgap for this model is shown in Figure 10. Other prominent defects in n-GaN, relatively shallow ER1-ER3 traps, seem to be complexes of these primary defects, mostly of V_N, with unidentified species, possibly with donors in the case of ER3 if one considers the results of neutron doping and annealing experiments. In p-GaN there is evidence of formation of defects near E_c-0.5 eV that seem to be Mg acceptors complexes with native defects and also of defects of unidentified nature with a level near E_v+0.3 eV. The carrier removal rate in GaN for light particles is well accounted for by the introduction of these simple defects.

For particles such as fast neutrons that produce large recoil cascades, the data suggests carrier removal by disordered regions in which the Fermi level in the core is pinned between the Ga_i donor level and the N_i acceptor level. In heavily irradiated samples these disordered regions pin the Fermi level near E_c-1 eV irrespective of initial type of conductivity and doping level. Charge trapping in DRs contributes to strong persistent photoconductivity effects in neutron irradiated GaN. Both theoretical modeling and experiment show a much higher radiation tolerance of GaN compared to Si or GaAs. When comparing the results for different types of particles and different energies of particles, modeling provides a reasonable guide for predicting the relative effectiveness of different radiation types in changing the electrical properties of GaN.

Many issues still have to be addressed, including: 1) strong asymmetry in carrier removal rates in n- and p-type materials and interaction of radiation defects with Mg acceptors; 2) scarcity of knowledge of radiation effects in III-nitrides other than GaN; 3) poor state of understanding of radiation defects in nitrides with dislocations present in the as-grown material; 4) radiation defects interaction with dopants and impurities; 5) better understanding of recombination processes in irradiated nitrides; 6) electrical and recombination effects in homojunctions, GaN/InGaN QW heterojunctions, effects at the AlGaN/GaN, GaN/InGaN interfaces for various compositions of HJs and QWs. There has been little study of the electronic collisions of high-energy ions which can create damage tracks in materials, including GaN [102-105].Some results, such as lower radiation stability of n-AlGaN films compared to GaN, do not fit theoretical predictions and suggest checking effects of crystalline quality on performance. More has to be done to understand defect transformation upon increasing irradiation temperature and upon annealing.

Acknowledgments

We thank Prof. In-Hwan Lee of Chonbuk University in Korea, Dr. Nikolai Smirnov and Dr. Anatoliy Govorkov at the Institute of Rare Metals, Dr. Nikolai Kolin at the Institute of Physical Chemistry in Obninsk for their help, support and fruitful discussions. We also acknowledge Dr. Erin Patrick and Prof. Marl Law for their collaborations on electric filed distributions.The work was supported in part by International Science and Technology Center ISTC grant #3870. The work at UF is supported by DTRA awardHDTRA1-08-10-BRCWMD-BAA.

References

[1] A. Ionascut-Nedelcsescu, C. Carlone, A. Houdayer, H.J. von Bardelleben, J.L. Cantin, S. Raymond, *IEEE Trans. Nucl. Sci.* Vol. 49, p. 2733, 2002.

[2] D.C. Look, D.C. Reynolds, J.W. Hemsky, J.R. Sizelove, R.L. Jones, R.J. Molnar. *Phys. Rev. Lett.* Vol.79 p. 2273, 1997.

[3] J. Nord, K. Nordlund, J. Keininen, *Phys. Rev. B* Vol 68, p.184104, 2003.

[4] S.O. Kucheyev, J.S. Williams, C. Jagadish, *Phys. Rev. B* Vol.64, p. 035202, 2001.

[5] D.W. Jenkins, J.D. Dow, M.-H. Tsai, *J. Appl. Phys.*, Vol. 72, p. 4130, 1992.

[6] A.Y. Polyakov, M. Shin, J.A. Freitas, M. Skowronski, D.W. Greve, R.G. Wilson, *J. Appl. Phys*, Vol. 80, p.6349, 1996.

[7] J. Neugebauer, C.G. Van de Walle, *Phys. Rev. B* Vol. 50, p.8067, 1994.

[8] P. Boguslawski, E.I. Briggs, J. Bernholc, *Phys. Rev. B*, Vol. 51, p.17255, 1995.

[9] D.C. Look, Z. Fang, L. Polenta, *MRS Internet J. Nitride Semicond. Res.*, Vol.5S1, p W10.5, 2000.

[10] G.A. Umana-Membreno, J.M. Dell, G. Parish, B.N. Nener, L. Faraone, U.K. Mishra, *IEEE Trans. Electron Dev.*, Vol. 50, p. 2326, 2003.

[11] A.Y. Polyakov, N.B. Smirnov, A.V. Govorkov, A.V. Markov, C.R. Lee, I-H. Lee, N.G. Kolin, D.I. Merkurisov, V.M. Boiko, J.S. Wright, S.J. Pearton, *Abstracts of the Fall MRS 2006 Meeting: Symposium I: Advances in III-Nitrides Semiconductor Materials and Devices* (MRS, Warrington, 2006) paper I7.46.

[12] K.H. Chow, L.S. Vlasenko, P. Johannesen, C. Bozdog, G.D. Watkins, *Phys. Rev. B* Vol. 69, p. 045207, 2004.

[13] L. Polenta, Z-Q. Fang, D.C. Look, *Appl. Phys. Lett.* Vol. 76, p 2086, 2000.

[14] F.D. Auret, S.A. Goodman, F.K. Koschnick, J.-M. Spaeth, B. Beaumont, P. Gibart, *Appl. Phys. Lett.* Vol.74, p.407, 1999.

[15] S.A. Goodman, F.D. Auret, F.K. Koschnick, J.-M. Spaeth, B. Beaumont, P. Gibart, *Mat. Sci. & Engineering* B, Vol.71, p.100, 2000.

[16] S.A. Goodman, F.D. Auret, F.K. Koschnick, J.-M. Spaeth, B. Beaumont, P. Gibart, *Appl. Phys. Lett.* Vol. 74, p. 809, 1999.

[17] F.D. Auret, S.A. Goodman, F.K. Koschnick, J.-M. Spaeth, B. Beaumont, P. Gibart, *Appl. Phys. Lett.* Vol. 73, p. 3745, 1998.

[18] M. Hayes, F.D. Auret, L. Wu, W.E. Meyer, J.M. Nel, M.J. Legodi, *Physica B*, Vol. 340-342, p. 421, 2003.

[19] P. Hacke, T. Detchprohm, K. Hiramatsu, N. Sawaki, *Appl. Phys. Lett.* Vol. 63, p. 2676, 1993

[20] A.Y. Polyakov, N.B. Smirnov, A.V. Govorkov, A.V. Markov, S.J. Pearton, N.G. Kolin, D.I. Merkurisov, V.M. Boiko, C.-R. Lee, In-Hwan Lee, *J. Vac. Sci. Technol. B* Vol. 25, p. 436, 2007.

[21] A.Y. Polyakov, A.S. Usikov, B.Theys, N.B. Smirnov, A.V. Govorkov, F. Jomard, N.M. Shmidt, W.V. Lundin, *Solid-State Electronics*, Vol. 44, p. 1971, 2000.

[22] J. Bourgoin, M. Lannoo, "Point defects in semiconductors II, *Experimental aspects* (Springer Verlag, Berlin, 1983) chapter 6.

[23] A.Y. Polyakov, N.B. Smirnov, A.V. Govorkov, N.G. Kolin, D.I. Merkurisov, V.M. Boiko, A.V. Korulin, S.J. Pearton, *J.Vac. Sci. Technol.* B Vol. 28, p. 608, 2010.

[24] A. Chantre, G. Vincent, D. Bois, *Phys. Rev. B* Vol 23, p. 5335, 1981.

[25] G.M. Martin, A. Mitonneau, D. Pons, A. Mircea and D.W. Woodard, *J. Phys. C* Vol. 13, p.3855, 1980.

[26] A.Y. Polyakov, N.B. Smirnov, A.V. Govorkov, Z.-Q. Fang, D.C. Look, R.J. Molnar, A.V. Osinsky, *J. Appl. Phys.* Vol.91, p. 6580, 2002.

[27] M. Linde, S.J. Uftring, G.D. Watkins, *Phys. Rev.* B Vol. 55, p. R10177, 1997.

[28] I.A. Buyanova, M. Wagner, W.C. Chen, B. Monemar, J.I. Lindstrom, H. Amano, I. Akasaki, *Appl. Phys. Lett.* Vol. 73, p. 2968, 1998.

[29] K.H. Chow, G.D. Watkins, A. Usui, M. Mizuta, *Phys. Rev. Lett.* Vol. 85, p. 2761, 2000.

[30] F.J. Sanchez, D. Basak, M.A. Sanchez-Garcia, E. Calleja, E. Munoz, I. Izpura, F. Calle, J.M.G. Tijero, B. Beaumont, P. Lorenzini, P. Gibart, T.S. Cheng, C.T. Foxon and J.W. Orton, *MRS Internet Journal-Nitride Semiconductor Research* Vol. 1, article 7, 1996.

[31] A.Y. Polyakov, N.B. Smirnov, A.S. Usikov, A.V. Govorkov, B.V. Pushniy, *Solid State Electronics,* Vol. 42, p. 1959, 1998.

[32] M.A. Reschikov, H. Morkoc, *J. Appl. Phys.* Vol. 97, p. 061301, 2005.

[33] T. Ogino, M. Aoki, *J. Appl. Phys.* Vol. 19, p. 2395, 1980.

[34] A.Y. Polyakov, N.B. Smirnov, A.V. Govorkov, S.J. Pearton, J.M. Zavada, R.G. Wilson, *J. Appl. Phys.* Vol. 94, p. 3069, 2003.

[35] A.Y. Polyakov, N.B. Smirnov, A.V. Govorkov, and A.V. Markov, N.G. Kolin, D.I. Merkurisov, V.M. Boiko, K.D. Shcherbatchev, V.T. Bublik, M.I. Voronova, S.J. Pearton, A.Dabiran, A.V. Osinsky, *J. Vac. Sci. Technol.* B Vol. 24, 2256, 2006.

[36] A.Y. Polyakov, N.B. Smirnov, A.V. Govorkov, J.M. Redwing, *Solid State Electron,* Vol. 42, p. 831, 1998.

[37] L.S. Berman, Purity control of semiconductors by the method of capacitance *spectroscopy* (Electronic Integral Systems, St. Petersburg, 1995) 180 pp.

[38] P. Hacke, H. Nakayama, T. Detchprom, K. Hiramatsu and N. Sawaki, *Appl. Phys. Lett.* Vol. 68, p.1362, 1996.

[39] G. Popovici and H. Morkoc, Growth and Doping of Defects in III-Nitrides, in *GaN and Related Materials* II, ed. S.J. Pearton (Gordon and Breach Science Publishers, the Netherlands, 1999) pp. 93-172.

[40] Z-Q Fang, J.W. Hemsky, D.C. Look, M.P. Mack, *Appl. Phys. Lett.* Vol. 72, p. 448, 1998

[41] B.R. Gossick, *J. Appl. Phys.* Vol. 30, p. 1214, 1959.

[42] A.Y. Polyakov, N.B. Smirnov, A.V. Govorkov, A.V. Markov, N.G. Kolin, D.I. Merkurisov, V.M. Boiko, K.D. Shcherbatchev, V.T. Bublik, M.I. Voronova, I-H. Lee, C.R. Lee, *J. Appl. Phys.* Vol. 100, p. 093715, 2006.

[43] W. Walukiewicz, *J. Vac. Sci. Technol.* B Vol.5, p.1062, 1987.

[44] J. Tersoff, *J. Vac. Sci. Technol.* B Vol. 4, p. 1066, 1986.

[45] K. Kuriyama, T. Tokumasu, Jun Takahashi, H. Kondo, and M. Okada, *Appl. Phys. Lett.* Vol. 80, p. 3328, 2002.

[46] V.N. Brudnyi, A.V. Kosobutskiy, N.G. Kolin, *Sov. Phys. Semicond.* Vol. 43, p. 1312, 2009 (in Russian).

[47] V.V. Emtsev, V.Yu. Davydov, E.E. Haller, A.A. Klochikhin, V.V. Kozlovskii, G.A. Oganesyan, D.S. Poloskin, N.M. Shmidt, V.A. Vekshin, A.S. Usikov, *Physica B* Vol. 308-310, p. 58, 2001.

[48] N.M. Shmidt, D.V. Davydov, V.V. Emtsev, I.L. Krestnikov, A.A. Lebedev, W.V. Lundin, D.S. Poloskin, A.V. Sakharov, A.S. Usikov, A.V. Osinsky, *Phys. Stat. Sol. B*, Vol. 216, p.533, 1999.
[49] B. H. Rose, C.E. Barnes, *J. Appl. Phys.* Vol. 53, p.1772, 1982.
[50] C. A. Usui, H. Sunakawa, A. Sakai, and A. A. Yamaguchi, *Jpn. J. Appl. Phys.* Vol. 36, p. L899, 1997.
[51] In-Hwan Lee, A.Y. Polyakov, N.B. Smirnov, A.V. Govorkov, A.V. Markov, S.J. Pearton, *Phys. Stat. Sol.* C, Vol..3, p. 2087, 2006.
[52] K. Fujito, Kiyomi, T. Mochizuki, H. Oota, H. Namita, S. Nagao, and I. Fujimura, *Physica Status Solidi A* Vol. 205, p.1056, 2008.
[53] T. Wosinsky, *J. Appl. Phys.* Vol.65, p.1566, 1988.
[54] Y. Tokuda, Y. Matuoka, K. Yoshida, H. Ueda, O. Ishiguro, N. Soejima, and T. Kachi, *Phys. Stat. Sol.*C Vol. 4, p. 2568, 2007.
[55] A.Y. Polyakov, N.B. Smirnov, A.V. Govorkov, A.V. Markov, T.G. Yugova, C.R. Lee, I-H. Lee, N.G. Kolin, I.D. Merkurisov, V.M. Boiko, K.D. Scherbatchev, V.T. Bublik, M.I. Voronova, S.J. Pearton, Neutron Irradiation Effects on GaN films Prepared by Lateral Overgrowth, *Abstracts of the European Workshop on III-Nitride Semiconductor Materials,* Heraklion, Crete, Greece, September 2006 (Heraklion University, Heraklion, 2006), pp. 91-92.
[56] A.Y. Polyakov, Structural and Electronic Properties of AlGaN, in *GaN and related Materials* II, ed. S. Pearton (Gordon and Breach, NY, 1999) pp. 173-233.
[57] L. Chernyak, A. Osinsky, G. Nootz, A. Schulye, J. Jasinski, M. Benamara, Z. Liliental-Weber, D.C. Look, and R.J. Molnar, *Appl. Phys. Lett.* Vol. 77, p. 2695, 2000.
[58] E.B. Yakimov, P.S. Vergeles, A.Y. Polyakov, N.B. Smirnov, A.V. Govorkov, In-Hwan Lee, Cheul Ro Lee, S.J. Pearton, *Appl. Phys. Lett.* Vol.90, p.152114, 2007.
[59] E.B. Yakimov, P.S. Vergeles, A.Y. Polyakov, N.B. Smirnov, A.V. Govorkov, In-Hwan Lee, Cheul Ro Lee, S.J. Pearton, *Appl. Phys. Lett.* Vol.94, p.042118, 2008.
[60] A.Y. Polyakov, N.B. Smirnov, A.V. Govorkov, A.V. Markov, E.B. Yakimov, P.S. Vergeles, In-Hwan Lee, Cheul Ro Lee, S.J. Pearton, *J.Vac. Sci. Technol. B*, Vol. 26, p. 990, 2008.
[61] A.Y. Polyakov, N.B. Smirnov, A.V. Govorkov, A.V. Markov, E.B. Yakimov, P.S. Vergeles, N.G. Kolin, D.I. Merkurisov, V.M. Boiko, In-Hwan Lee, Cheul-Ro Lee, and S.J. Pearton, *J. Electron. Mater.* Vol. 36, p. 1320, 2007.
[62] S.J. Pearton, "GaN device processing", in *GaN and related materials*, ed. S.J. Pearton (Gordon and Breach, NY, 1999) pp. 475-540.
[63] J.W. Ager, R.E. Jones, D.M. Yamaguchi, K.M. Yu, W. Walukiewicz, S.W. Li, E.E. Haller, H. Lu, and W.J. Schaff, *Phys. Stat. Sol. B*, Vol. 244, p. 1820, 2007.
[64] A.Y. Polyakov, N.B. Smirnov, A.V. Govorkov, N.V. Pashkova, S.J.Pearton, J.M. Zavada, R.G. Wilson, *J. Vac. Sci. Technol. B* Vol. 21, p.2500, 2004.
[65] A.Y. Polyakov, N.B. Smirnov, A.V. Govorkov, A.V. Markov, N.G. Kolin, V.M. Boiko, D.I. Merkurisov, S.J. Pearton, *J. Vac. Sci. Technol.* B Vol. 24, p. 1094,2006.
[66] A.Y. Polyakov, N.B. Smirnov, A.V. Govorkov, K.H. Baik, S.J. Pearton and J.M. Zavada, *J. Vac. Sci. Technol.* B Vol. 22, 2291, 2004.
[67] B.D. White, M. Bataiev, S.H. Gross, X. Hu, A. Karmarkar, D.M. Fleetwood, R.D. Schrimpf, W.J. Schaff, L.J. Brillson, *IEEE Trans. Nucl. Sci.* Vol. 50, p.1934, 2003.

[68] X. Hu, A. Karmarkar, B. Jun, D.M. Fleetwood, R.D. Schrimpf, R.D. Geil, R.A. Weller, B.D. White, M. Bataiev, L.J. Brillson, U.K. Mishra, *IEEE Trans. Nucl. Sci.* Vol. 50, p.1791, 2003.
[69] M.S. Shur and M. A. Khan, in *GaN and Related Materials* II, ed. S.J. Pearton (Gordon and Breach Science Publishers, New York, 1999) pp.47-92.
[70] H. Miyamoto, *Phys. Stat. Sol.* C, Vol. 3, p. 2254, 2006.
[71] F. Gaudreau, P. Fournier, C. Carlone, S.M. Khanna, H. Tang, J. Webb, A. Houdayer, *IEEE Trans. Nucl. Sci.* Vol. 49, p. 2702, 2002.
[72] A.Y. Polyakov, N.B. Smirnov, A.V. Govorkov, A.V. Markov, S.J. Pearton, N.G. Kolin, D.I. Merkurisov, V.M. Boiko, *J. Appl. Phys.* Vol. 98, p. 033529, 2005.
[73] A.Y. Polyakov, N.B. Smirnov, A.V. Govorkov, A.V. Markov, S.J. Pearton, A.M. Dabiran, A. M. Wowchak, B. Cui, A.V. Osinsky, P.P. Chow, N.G. Kolin, V.M. Boiko, and D.I. Merkurisov, *Appl. Phys. Lett.* Vol. 93, p. 152101, 2008.
[74] A.Y. Polyakov, N.B. Smirnov, A.V. Govorkov, A.V. Markov, A.M. Dabiran, A. M. Wowchak, B. Cui, A.V. Osinski, P.P. Chow, S.J. Pearton, *J. Appl. Phys.* Vol. 104, p. 053702, 2008.
[75] X. Hu, B.X. Choi, H.J. Barnaby, D.M. Fleetwood, R.D. Schrimpf, S. Lee, S. Shojah-Ardalan, R. Wilkins, U. Mishra, R.W. Dettmer, *IEEE Trans. Nucl. Sci.* Vol. 51, p.293, 2004.
[76] A.Rashmi, S. Kranti, S. Haldar, P.S. Gupta, *Microelectron. J.* Vol. 33, p. 205, 2002.
[77] A.Y. Polyakov, N.B. Smirnov, A.V. Govorkov, N.G. Kolin, V.M. Boiko, D.I. Merkurisov, S.J. Pearton, :Effects of electron irradiation on GaN-based transistor structures", in the extended abstracts of the *6th Russian Conference "Nitrides of gallium, indium and aluminum: structures and devices"* S-Petersburg, Ioffe Physico-Technical Institute, 2008.
[78] B. Luo, J.W. Johnson, F. Ren, K.K. Alums, C.R. Abernathy, S.J. Pearton, A.M. Dabiran, A.M. Wowchak, C.J. Polley, P.P. Chow, D. Shoenfeld, A.G. Baca, *Appl. Phys. Lett.* Vol. 80, p. 604, 2002.
[79] A.Y. Polyakov, N.B. Smirnov, A.V. Govorkov, In-Hwan Lee, Jong Hyeob Baek, N.G. Kolin, V.M. Boiko, and D.I. Merkurisov, S.J. Pearton, *J. Electrochemical Soc.* Vol. 155, p. H31, 2008.
[80] A.P. Karmarkar, B. Jun, D.M. Fleetwood, D.D. Schrimpf, R.A. Weller, B.D. White, L.S. Brillson, U.K. Mishra, *IEEE Trans. Nucl. Sci.* 51, p. 3801, 2004.
[81] S.J. Cai, J.S. Tang, R. Li, J. Wei, L. Wong, J.L. Chen, K/L/ Wang, M. Chen, Y.F. Zhao, R.D. Schrimpf, J.C. Keay, K.F. Galloway, *IEEE Trans. Electron Devices* Vol. 47, p.304, 2000.
[82] Q. Wang, H.Q. Xu, P. Omling, C. Jang, G. Malmqvist, *Nucl. Instrum. Methods Phys. Res.* B Vol. 160,p. 33, 2000.
[83] B. Luo, J.W. Johnson, F. Ren, K.K. Allums, C.R. Abernathy, S.J. Pearton, R. Dwivedi, T.N. Fogarty, R. Wilkins, A.M. Dabiran, A.M. Wowchak, C.J. Polley, P.P. Chow, A.G. Baca, *Appl. Phys. Lett.* Vol. 79, p. 2196, 2001.
[84] B. Luo, J. Kim, F. Ren, J.K. Gillespie, R.C. Fitch, J. Sewell, R. Dettmer, G.D. Via, A. Crespo, T.J. Jenkins, B.P. Gila, A.H. Onstine, K.K. Allums, C.R. Abernathy, S.J. Pearton, R. Dwidevi, T.N. Fogarty, R. Wilkins, *Appl. Phys. Lett.* Vol. 82, p.1428, 2003.

[85] B. Luo, F. Ren, K.K. Allums, B.P. Gila, A.H. Onstine, C.R. Abernathy, S.J. Pearton, R. Dwivedi, T.N. Fogarty, R. Wilkins, R.C. Fitch, J.K. Gillespie, T.J. Jenkins, R. Dettmer, J. Sewell, G.D. Via, A. Crespo, A.G. Baca, R.J. Shul, *Solid-State Electron.* Vol. 47, p.1015, 2003.
[86] B. Luo, J.M. Johnson, F. Ren, K.K. Allums, C.R. Abernathy, S.J. Pearton, A.M. Dabiran, A.M. Wowchak, C.J. Polley, P.P. Chow, D. Schoenfeld, A.G. Baca, *Appl. Phys. Lett.* Vol. 80, p.604, 2002.
[87] V. Tilak, B. Green, V. Kaper, H. Kim, T. Prunty, J. Smart, J. Shealy, and L.F. Eastman, *IEEE Electron Dev. Lett.*, Vol. 22, p.504, 2001.
[88] Daumiller, D. Theron, C. Gaquiere, A. Vescan, R. Dietrich, A. Wieszt, H. Leier, R. Vetury, U. K. Mishra, I. P. Smorchkova, S. Keller, C. Nguyen, and E. Kohn, *IEEE Electron. Dev. Lett.*, Vol. 22, p. 62, 2001.
[89] U.K. Mishra, P. Parikh and Y.-F. Wu, *Proc. IEEE* Vol. 90, p. 1022, 2002.
[90] C. Lee, H. Wang, J. Yang, L. Witkowski, M. Muir, M. Asif Khan and P. Saunier, *Electron. Lett.* Vol. 38, p. 924, 2002.
[91] O. Mitrofanov, M. Manfra and N.G. Weimann, *Appl. Phys. Lett.* Vol. 82, p. 4361, 2003.
[92] S. C. Binari, K. Ikossi, J. A. Roussos, W. Kruppa, D. Park, H. B. Dietrich, D. D. Koleske, A. E. Wickenden, and R. L. Henry, *IEEE Trans. Electron. Dev.*, Vol. 48, p. 465, 2001.
[93] B. M. Green, K. K. Chu, E. M. Chumbes, J. A. Smart, J. R. Shealy, and L. F. Eastman, *IEEE Electron. Dev. Lett.*, Vol. 21, p. 268, 2000.
[94] L. F. Eastman, V. Tilak, J. Smart, B. M. Green, E. M. Chumbes, R. Dimitrov, H. Kim; O. S. Ambacher, N. Weimann, T. Prunty, M. Murphy, W. J. Schaff, and J. R. Shealy, *IEEE Trans. Electron Dev.,* Vol. 48, p. 479, 2001.
[95] X. Hu, A. Koudymov, G. Simin, J. Yang, M. Asif Khan, A. Tarakji, M. S. Shur, and R. Gaska, *Appl. Phys. Lett.* Vol. 79, p. 2832, 2001.
[96] B. Luo, J. W. Johnson, J. Kim, R. M. Mehandru, F. Ren, B. P. Gila, A. H. Onstine, C. R. Abernathy, S. J. Pearton, A. G. Baca, R. D. Briggs, R. J. Shul, C. Monier, and J. Han, *Appl. Phys. Lett.*, Vol. 80, p. 1661, 2002.
[97] A.Y. Polyakov, N.B. Smirnov and A.V. Govorkov,V.N. Danilin,T. A. Zhukova, B. Luo and F. Ren, B. P. Gila, A. H. Onstine, C. R. Abernathy and S. J. Pearton, *Appl.Phys. Lett.,* Vol. 83, p. 2608, 2003.
[98] A.Y. Polyakov, N.B. Smirnov, B.P. Gila, M. Hlad, A.P. Gerger, C.R. Abernathy, S.J. Pearton, *J. Electrochem. Soc.,* Vol. 154, p. H115, 2007.
[99] Shyam M. Khanna, Diego Estan, Alain Houdayer, Hui C. Liu and Richard Dudek, *IEEE Trans. Nucl. Sci.* Vol. 51, p. 3585, 2004.
[100] F Gaudraeau, C. Cardone, A. Noudayer and S.M. Khamna, *IEEE Trans. Nucl Sci.* Vol. 48, 1778, 2001.
[101] M. Osinsky, P. Perlin, H. Schone, A.H. Paxtone and E.W. Taylor, *Electron. Lett.* Vol. 33, 1252, 1997.
[102] J. Grant, W. Cunningham, A. Blue, V. O'Shea, J. Vaitkus, E. Gaubas, and M. Rahman, *Nucl. Instr. Methods in Physics Research*, A Vol. 546, p. 213, 2005.
[103] J. Grant, R. Bates, W. Cunningham, A. Blue, J. Melone, F. McEwan, J. Vaitkus, E. Gaubas, and V. O'Shea, *Nucl. Instr. Methods in Physics Research*, A Vol. 576, p. 60, 2007.

[104] A.Y. Polyakov, N.B. Smirnov, A.V. Govorkov, A.V. Markov, E.A. Kozhukhova, I.M. Gazizov, N.G. Kolin, D.I. Merkurisov, V.M. Boiko, A.V. Korulin, V.M. Zalyetin, S.J. Pearton, I.-H. Lee, A.M. Dabiran, P.P. Chow, *J. Appl. Phys.* Vol. 106, p. 103708, 2009.

[105] R.G. Wilson, S.J. Pearton, C.R. Abernathy, J.M. Zavada, *Appl. Phys. Lett.* Vol. 66, p. 2238, 1995.

[106] L. Liu, C. Velez Cuervo, Y. Y. Xi, and F. Ren, S. J. Pearton, H. Y. Kim, J. Kim and I. I. Kravchenko, *J. Vac. Sci. Technol. B* 31, 042202-1-6 (2013).

[107] L. Liu, C.-F. Lo, Y. Xi, Y. Wang, F. Ren, S. J. Pearton, H.-Y. Kim, J. Kim, R. C. Fitch, D. E. Walker, Jr., K. D. Chabak, J. K. Gillespie, S. E. Tetlak, G. D. Via, A. Crespo and Ivan I. Kravchenko, *J. Vac. Sci. Technol. B* 31, 022201-1 (2013).

In: Gallium Nitride
Editor: Kiera Olivia Peak
ISBN: 978-1-63321-387-6

Chapter 2

MECHANISMS DETERMINING RELIABILITY IN ELECTRICALLY STRESSED AlGaN/GaN HIGH ELECTRON MOBILITY TRANSISTORS

Fan Ren, Y. H. Hwang[1], Shun Li[1], Yueh-Ling Hsieh[1], P. G. Whiting[2], M. R. Holzworth[2], S. J. Pearton[2] and K. S. Jones[2]
[1]Department of Chemical Engineering, University of Florida, Gainesville, Florida, US
[2]Department of Materials Science and Engineering, University of Florida, Gainesville, Florida, US

Abstract

AlGaN/GaN high electron mobility transistors are gaining commercial acceptance for use in high power and high frequency applications, but the degradation mechanisms that drive failure in the field are not completely understood. Since some of these mechanisms are current or field driven, reliability studies must go beyond the typical Arrhenius accelerated life tests. In this chapter we summarize recent work on electric field or current driven degradation in devices with different gate metallization, device dimensions, electric field mitigation techniques (such as source field plates), and the effect of device fabrication processes for both dc and RF stress conditions. Some of the mechanisms now established include gate or Ohmic metal reaction, hot carrier trap generation, stress-induced crack generation and oxidation reactions involving the AlGaN layer.

Introduction

Gallium Nitride based high electron mobility transistors (HEMTs) have become attractive for use in high power and high frequency applications. GaN transistors offer both higher breakdown voltages as well as 2DEG density compared to their GaAs counterparts [1-30]. As AlGaN/GaN HEMTs are introduced in commercial and military applications, it has become critical for reliability studies to extend beyond the normal temperature-accelerated life tests. It is common to determine the failure rate at operating temperatures by extrapolating from the failure rate under thermally accelerated stress conditions through the Arrhenius equation but this does not capture failure mechanisms that are not accelerated by elevated temperatures.

Employment of GaN HEMTs for high power radar systems requires devices to be subjected to large-signal RF while being driven into saturation, resulting in devices experiencing high electric fields and high current densities. Impressive mean-time-to-failure values of greater than 10^7 hours have been reported at operating temperatures below 200 °C, with activation energies ranging from 0.18 eV to 2 eV [1-4]. However, these studies neglect the effect of high electric field and current on device lifetime. As the device design and material processing technology for AlGaN/GaN HEMTs has matured over the years, several failure mechanisms that limited device lifetime have been addressed and improved. However, there are still numerous degradation mechanisms that have been reported to plague AlGaN/GaN HEMTs.

These mechanisms can be grouped together into three main categories that affect lifetime: contact degradation, hot electron effect, and inverse piezoelectric effect. Both Schottky and Ohmic contacts have shown excellent stability below 300 °C. Piazza et al. have reported an increase in contact resistance and passivation cracking due to Ga out-diffusion and Au inter-diffusion after a 100 h thermal storage test stress at 340 °C [8]. Nickel based Schottky contacts have been shown to form nickel nitrides on GaN at annealing temperatures as low as 200 °C, resulting in a significant decrease in Schottky barrier height [8]. The observed current collapse and gate lag under high voltage and high current operation have been attributed to hot electrons. Hot electrons are electrons that have been accelerated in a large electric field, resulting in high kinetic energy, which can result in trap formation. Creation of traps can occur in both the AlGaN layer and the buffer, leading to reversible degradation of transconductance and saturated drain current [2, 5, 9].

GaN is a piezoelectric material and under high bias conditions, the electric field induces additional tensile stress to the already strained AlGaN layer. Upon reaching a "critical voltage," irreversible damage to the device occurs resulting in defect formation through which electron leakage can occur [9-12].

The HEMT Device Structure and Processing

Let us begin by considering the HEMT layer structure, since this is instructive in determining likely failure mechanisms. The AlGaN barrier layer Al content typically is in the range of 15 - 30% mole fraction with a thickness of nominally 15 - 30 nm. The 2DEG characteristics follow the trends of (1) higher Al content → higher charge density, and (2) thicker barrier → higher (absolute value) pinch-off voltage. There are limits, however. There is a lower limit on thickness below which the 2DEG will not form due to the above mentioned conduction band bending polarization effects being insufficient to create the well. There is an upper limit on thickness above which the lattice mismatch stress will exceed the maximum yield strength causing the AlGaN to relax through the formation of cracks. Both limits are functions of the Al content. With regards to Al content, while a higher mole percentage produces more charge, this typically comes at the expense of device leakage issues and crystal quality above ~ 30 %. However, in the ranges noted, the 2DEG will have representative sheet charge density of 0.8 - 1.1 x 10^{13} cm^{-2} with mobility of 1300 - 2000 cm^2 / V-s.

Conventional GaN HEMT processing schemes involve (i) mesa isolation, (ii) Ohmic metallization and anneal, (iii) gate and contact pad metallization, and (iv) passivation. However, in this sequence, the channel region of the HEMT is exposed during all pre-passivation process steps to various photoresist coats, developer solutions, solvents, and plasma descums, as well as effects from the ambient. The surface of GaN HEMTs is sensitive to disruption, passivating as early as possible during the device fabrication sequence is necessary. This practice effectively 'locks-in' the qualities of the surface / passivant interface, eliminating unwanted contributions from subsequent process steps. Process sequences with early passivation and dielectrically-defined T-gates are common, borrowing from similar progression of GaAs pHEMT processing to protect the surface and imnprove device performance.

The process flow proceeds through Ohmic metallization and anneal, passivation, implant isolation, gate and contact pad etching and metallization. For a typical device, subsequent fabrication steps include formation of a source-connected field plate over an additional encapsulant dielectric, frontside electroplating and/or interconnect, and backside processing (e.g., thinning, via etch, backside metal).

Most GaN HEMT Ohmic contact schemes employ Ti/Al/x/Au, where x may be Ti, Ni, Pt, Mo, or Pd. Cross-sectional transmission electron microscopy (XTEM) has shown the existence of Au at the GaN surface of Ti/Al/Pt/Au contacts, suggesting significant Au diffusion into the contact during annealing. In fact, Ni, Pd, and Pt have all been shown to be permeable to Au during typical contact processing. One issue unique to GaN HEMT contact formation is the extremely high annealing temperatures required to minimize contact resistance. Typical annealing conditions are ~850°C, 30 sec. in N_2. This very high annealing temperature is well above the melting point of Al (661°C), and typically leads to rough Ohmic contact morphology. The edge definition of the patterned contact is also affected by the roughening, and edge acuity may become an issue for very small device features, such as the channel of an HEMT. For the traditional Ti/Al-based scheme, a tradeoff between electrical properties and surface morphology seems to exist, with smoother morphologies available at lower annealing temperatures while sacrificing Rc.

Most AlGaN/GaN HEMTs employ Ni/Au as the gate contact due to comparable electrical properties and improved adhesion to the (Al)GaN surface relative to Pt. For Ni gates, it has been shown that an interfacial layer of NiO_x can lead to increased HEMT parametric drift but can be partially mitigated by thermal treatment. Recently, the use of Pt has become more prevalent in GaN HEMTs and state-of-the-art enhancement and depletion mode devices have been demonstrated using a Pt/Au or Pt/Ti/Au Schottky contacts. Pt may be preferable to Ni for reduced leakage and improved device stability under reliability stress. The large barrier height of Pt, combined by the propensity of Ni to oxidize, favor Pt as the Schottky barrier metal. Future devices may employ thin gate insulator layers and optimization of metal scheme for these structures has yet to emerge.

High density chlorine-based plasmas have become the most common method to etch GaN due to high plasma uniformity over large areas, lower ion energy (and thus lattice damage). However, despite the success of high density plasma etch techniques, dry etch-induced lattice damage can severely degrade material properties and HEMT performance. The damage may occur as lattice defects, unintentional passivation, or preferential N_2 loss producing surface non-stoichiometry and/or conductivity. Such damage can be manifested as extrinsic leakage currents in isolated areas of GaN HEMTs, decreasing breakdown voltage and limiting power performance.

Ion implantation can be used to produce interdevice isolation without disrupting device planarity. This may be advantageous in certain processing sequences where etching or step coverage of metalization layers off mesa sidewalls is undesirable. Implantation into semiconductors creates damage-related energy states that trap free carriers or create recombination sites, leading to high-resistivity material.

Although most energy levels created by implant damage in (Al,Ga)N are not near midgap, they are still typically sufficient to produce insulating material with sheet resistances of $\sim 10^9$ - 10^{12} Ω/sq.

Polarization effects in GaN HEMTs lead to surface states that can have a significant detrimental impact on device performance. When unpassivated, these positively charged surface donor states can trap negative charge and lead to a "virtual gate" phenomenon that depletes the 2DEG of carriers, reducing drain current and power performance . SiN_x is still the most widely employed material for terminating the surface of traditional AlGaN/GaN heterostructures. Such SiN_x layers are typically deposited by PECVD at a baseplate temperature of ~300°C using SiH_4 and NH_3 precursors combined with inerts such as Ar, N_2 and/or He. Stress and fixed charge in these SiN_x films can be tailored by modifying growth parameters and can have an impact on device performance. The primary results of appropriate passivation are 1) an increase in sheet electron concentration due to termination of surface states and 2) a reduction in trap-related dispersion in fabricated devices. Both serve to greatly increase drain current and power density of RF GaN HEMTs relative to unpassivated structures. The SiN_x passivant can also be used to create dielectrically-defined T-gates. The dielectrically-defined T-gate geometry has both advantages and disadvantages at the device level.

Overlap of gate metallization onto the surface of the passivant can be optimized to create a gate field plate and reduce peak electric field in the semiconductor, increasing breakdown voltage. The presence of high-k SiN_x surrounding the gate also leads to parasitic loading and increases the fringing capacitance, degrading gain of RF GaN HEMTs.

Additionally, for aggressively scaled HEMTs with gate length <100nm, the aspect ratio of SiN_x thickness to gate length may complicate accurate patter transfer required for gate length definition. For such devices, it is advantageous to make the dielectric very thin.

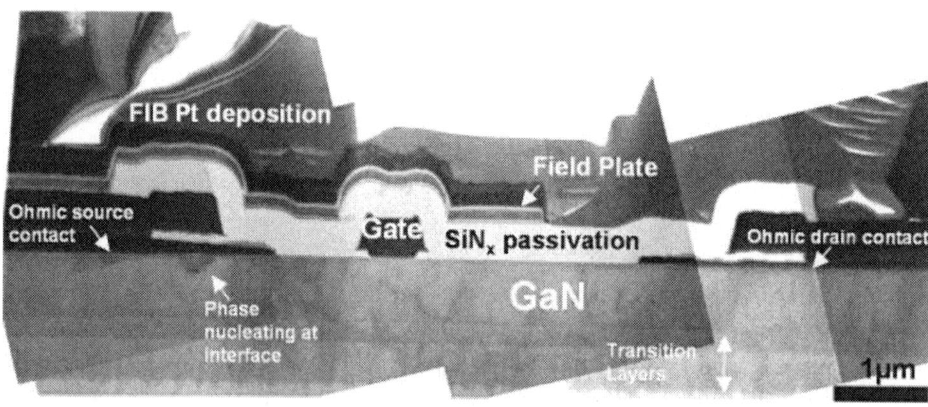

Figure 1. Cross section TEM of a typical AlGaN/GaN HEMT. Note that there can be reactions that have occurred under the Ohmic contacts even prior to reliability testing.

In addition to the field plate formed in the gate electrode, most GaN HEMTs employ an additional dielectric encapsulation layer over the drift region that serves to protect the device and act as the capacitor dielectric for deposition of a 2nd field plate. This field plate is typically tied to the grounded source electrode (source field plate, or SFP) and is located above the high field region at the drain edge of the gate. Optimization of SFP geometry can lead to simultaneous reduction in peak field and feedback capacitance. Reduction in peak field has been demonstrated to increase breakdown voltage and also increase output power density of RF HEMTs. Reduction in feedback capacitance increases microwave gain relative to devices without such field plate(s).

A cross section TEM of a typical AlGaN/GaN HEMT is shown in Figure 1, illustrating the T-gate, Ohmic source and drain contacts and SiN_x passivation layer.

GaN Degradation Mechanisms

(I) Hot-Carriers and Trap Generation

Permanent device degradation after high V_{DG} stress under on-state conditions has been attributed to the presence of hot electrons. In GaAs- based devices, hot electrons generate holes which are accumulated by the gate and result in a negative shift in V_T [1, 4, 14]. Typically, I_G is used to derive the field-acceleration laws for failure. Impact ionization in GaN HEMTs has been reported by some groups although high gate current can hamper its observation [1, 4]. The hot electrons likely lead to trap generation at the AlGaN/GaN interface or at the passivation/GaN cap interface, leading to an increase in the depletion region between the gate and the drain and resulting in an increase in drain resistance and a decrease in I_{DSS}. Under off-state conditions, the degradation is greatly reduced due to the reduction of electrons present in the channel. Sozza et al. showed that AlGaN/GaN HEMTs that underwent a 3000 hour on-state stress resulted in an increase in surface traps with an activation energy of about 0.55 eV [3, 7]. Devices stressed under off-state conditions showed only a very small increase in traps density.

Meneghesso et al. employed electroluminescence (EL) to study the effect of hot-carriers and the dependence on stress conditions [1]. Uniform EL emission due to hot electrons was observed along the channel for devices stressed at V_{GS} = 0 V and V_{DS} = 20 V, with no hot spots or current crowding. Under off- state conditions of V_{GS} = -6 V and V_{DS} = 20 V (resulting in a V_{GD} = -26 V), the EL emission from the channel was not uniform. These hot spots may be due to injection of electrons from the gate into the channel. Due to the high bias conditions, the electrons acquire enough energy to give rise to photon emission. This is a clear correlation between the gate leakage current degradation and the hot spots intensity [31].

(ii) Contact Degradation

Contact degradation and gate sinking are significant degradation mechanisms at elevated temperatures in GaAs and InP based HEMTs. This has not yet proven to be a significant issue with AlGaN/GaN HEMTs at temperatures below 400°C for Pt/Au Schottky contacts and Ti/Al/Pt/Au annealed Ohmic contacts [1, 10]. An increase in Schottky barrier height was observed for Ni/Au Schottky contacts after dc stress at elevated junction temperatures (200°C) [1, 10, 32-35]. This was due to a consumption of an interfacial layer between the Schottky contact and the AlGaN layer, causing a positive shift in Schottky barrier height and pinch-off voltage but a decrease in I_{DSS}. Unstressed devices were subjected to annealing after the Schottky contact was deposited in to decrease the interfacial layer between the gate and semiconductor. Devices that underwent this gate anneal showed 50% less degradation during a 24 hour stress test as opposed to devices that did not receive a gate anneal [1, 9]. Thermal storage tests up to 2000 hours on Ti/Al/Ni/Au Ohmic contacts at and above 290°C showed an increase in contact resistance as well as surface roughness due to growth of Au-rich grains that ultimately led to cracks in passivation [33-41]. The two primary degradation mechanisms were Au inter-diffusion within the metal layers and Ga out-diffusion from the semiconductor into the metallic compounds. Similar degradation was observed after dc stress tests that resulted in junction temperatures equivalent to the thermal storage tests. Due to the high power capability of AlGaN/GaN HEMTs, proper temperature management is crucial in order to optimize device performance under high current and high voltage operation [24-28]. Self-heating of devices can ultimately result in poor device performance through contact degradation.

(iii) Inverse Piezoelectric Effect

Several groups have shown that high reverse bias on the gate results in a defect-generated path of gate current leakage [1, 9, 31-34]. This defect formation mechanism has been ascribed to at four different mechanisms, including (i) inverse piezoelectric effect (ii) oxidation of the surface, with subsequent generation of pits [42] (iii) defect generation/percolation in the AlGaN [43] and (iv) generation of point defects within the AlGaN [44] Since GaN and AlGaN are piezoelectric, the presence of high electric fields results in an increase in stress within these layers. AlGaN is lattice mismatched to GaN, resulting in significant tensile strain, even in the absence of an electric field. If under electric stress the elastic energy within the AlGaN/GaN layers surpasses a critical value, the strained layer will relax through crystallographic defect formation and these defects can be electrically active and result in device degradation [32]. The role of oxygen has been implicated by observing slower degradation rates under vacuum conditions [42]. Negative charge creation in the AlGaN and positive charge at the AlGaN/GaN interface due to defects has been reported during reverse-biasing of HEMTs [43] Permanent degradation occurred with generation of parasitic leakage paths due to a defect percolation process. These were observed during long stress times, where degradation can occurs even below the critical voltage [43]. There was an onset of noise in the gate current prior to the permanent degradation, which is consistent with defect creation and also the time to breakdown strongly depended on the initial defect density in the HEMT layers [43]. Similarly, current deep level transient spectroscopy has observed creation of trap states with thermal activation energies of 0.5 eV from the conduction band in devices subjected to high reverse baises [44].

Joh et. al. have established that I_D and I_G degradation under high reverse gate bias occurs at a critical voltage, typically above 20 V on V_{DG} [34]. This is also correlated with a sharp rise in both source and drain resistance and a positive shift in V_T. However, the critical voltage for devices can deviate substantially within one wafer [33-35, 41, 42]. The density of hot electrons increases exponentially as the field increases and only linearly with current [1, 4]. Though stress experiments in the high power state have shown that increasing $I_{Dstress}$ does not significantly accelerate degradation, the critical voltage for reverse bias stress in which I_{Goff} dramatically increases is dependent upon the $I_{Dstress}$, as V_{crit} increases with increasing $I_{Dstress}$ [37, 42]. The hot electrons are not the driving degradation mechanism for this stress condition. TEM cross sections after stressing with V_{DS} = 40 V and I_{D0} = 250 mA/mm at various base-plate temperatures corresponding to a junction temperature of 250 °C, 285 °C, and 320 °C [9] showed that all stressed devices showed evidence of pit-like defects on the drain side of the gate. The depth of the pit was about 10nm, and remained within the AlGaN layer. Crack-like defects were observed in a few of the stressed devices, and originated at the bottom of the pit defect, extending to the heterointerface of the AlGaN/GaN layer and occasionally into the GaN buffer. As the junction temperature increased, the time after which the crack appeared decreased, developing within 6 hours at a temperature of 320°C. Gate metal was also observed to diffuse ~2nm into the defect crack. The formation of the crack was hypothesized by to originate in the deepest points in the defect pit and spread along the gate width, thus explaining the presence of cracks in very shallow defect pits [10].

Del Alamo et al. [9-11] have postulated that the converse piezoelectric effect is solely an electric field driven degradation mechanism due to the fact that it is the induced mechanical stress that results in the relaxation of the AlGaN layer [11, 37-39] The current should not drive this mechanism, except for indirect self-heating that would accelerate degradation of the device. Device design that affects the profile of the electric field on the drain side of the gate will also, in turn, impact the critical voltage [40-48].

Sarua [13] investigated the effect of piezoelectric strain in AlGaN/GaN FETs under bias with micro-Raman spectroscopy. 2D finite element simulations showed that a pinched-off device at V_{DS} = 20 V resulted in a peak electric field on the drain side of the gate within the AlGaN layer. However, the z component of the electric field extended down into the GaN layer [13]. Fe doped GaN buffers layers raise the acceptor concentration and decreases the depletion width in the GaN buffer, confining the z component of the electric field to the AlGaN/GaN interface [14, 29]. Self-heating will occur under high power stress conditions, which results in a compressive thermal strain/stress [30, 31]. It is possible that due to the mitigation of the piezoelectric stress by the thermal stress, slower device degradation was seen in devices stressed under the high power state as opposed to off-state condition. This is in contrast to the hypothesis of del Alamo et al. that higher temperatures will result in an acceleration of device degradation [11, 32-35].

Other issues can lead to additional compressive and tensile strains on the underlying epitaxial layers, including SiN passivation, which is used extensively to minimize surface traps on the AlGaN surface. Mastro et al. reported the simulated effects of non-uniform strain due to SiN passivation [17]. Typically, SiN has a relatively small magnitude of stress as compared to the tensile strain present in the AlGaN layer due to lattice mismatch. The strain in SiN is highly dependent on processing conditions When deposited on the device, variations and discontinuities can increase the stress fields. For instance, the opening at the edge of the gate metal will result in a force on the AlGaN which will be perpendicular to the gate edge and parallel to the surface of the AlGaN [40]. It was predicted that as the gate length decreases, the magnitude of the strain fields increases.

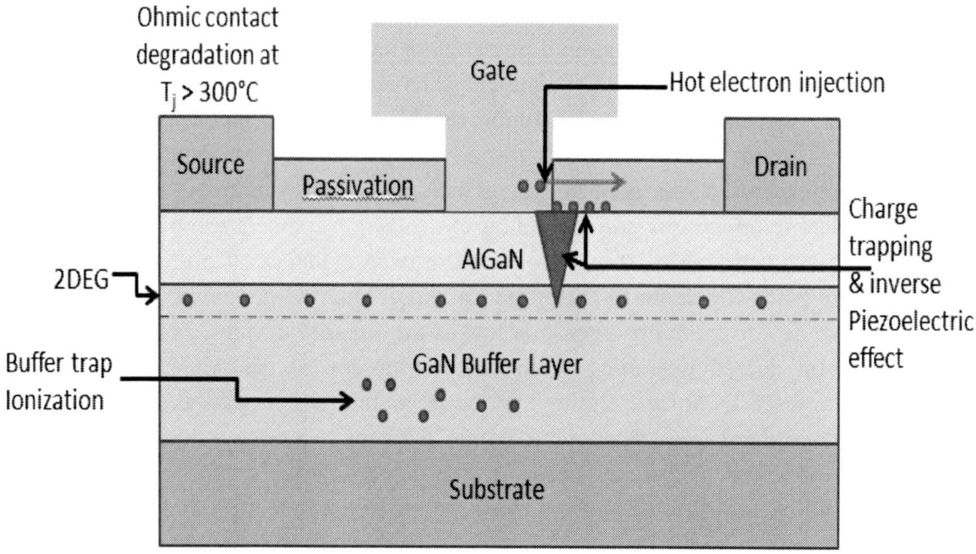

Figure 2. Schematic of failure mechanisms and their location in AlGaN/GaN HEMTs.

Figure 2 summarizes the type and location of reported failure mechanisms in AlGaN/GaN HEMTs. These include reactions of the gate or source/drain contacts with the underlying semiconductor, trap formation through hot electron injection and cracking or pitting at the gate edge due to the inverse piezoelectric effect. These give rise to various signatures, including additional noise, current changes and transients due to trap formation and pitting or other metal reactions.

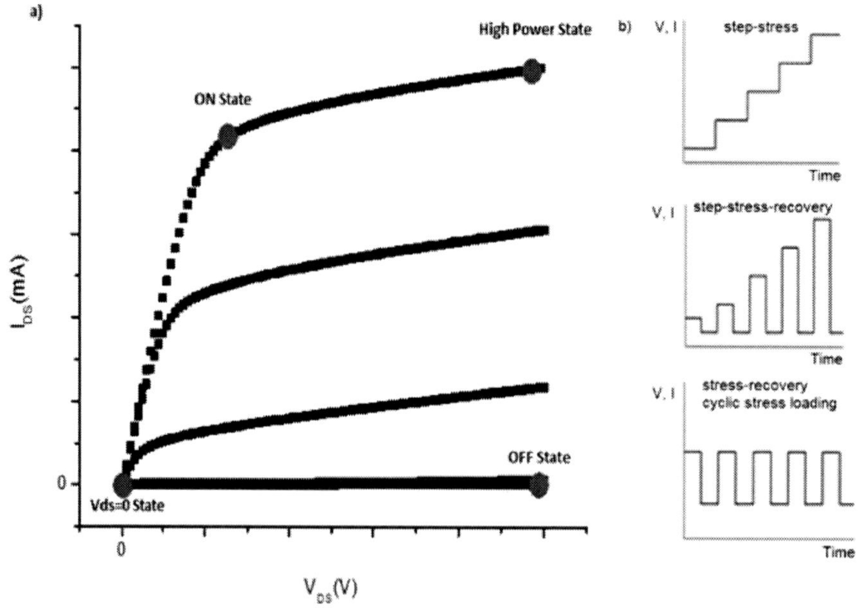

Reprinted from Semiconductor Science and Technology, 28, Cheney et al., Reliability Studies of AlGaN/GaN high electron mobility transistors, 074109, Copyright 2013, with permission from Elsevier.

Figure 3. Schematic of (a) different biasing conditions for AlGaN/GaN HEMTs and (b) bias-stress protocols.

DC Stress Testing

There are four basic stress conditions for HEMTs, as shown in Figure 3 (a). For each of these four conditions (on-state, off-state, high power and V_{DS}= 0 state), the voltage or current can be increased in stepped fashion, a step-stress recovery cycle or step-recovery cycle, as shown in Figure 3(b).

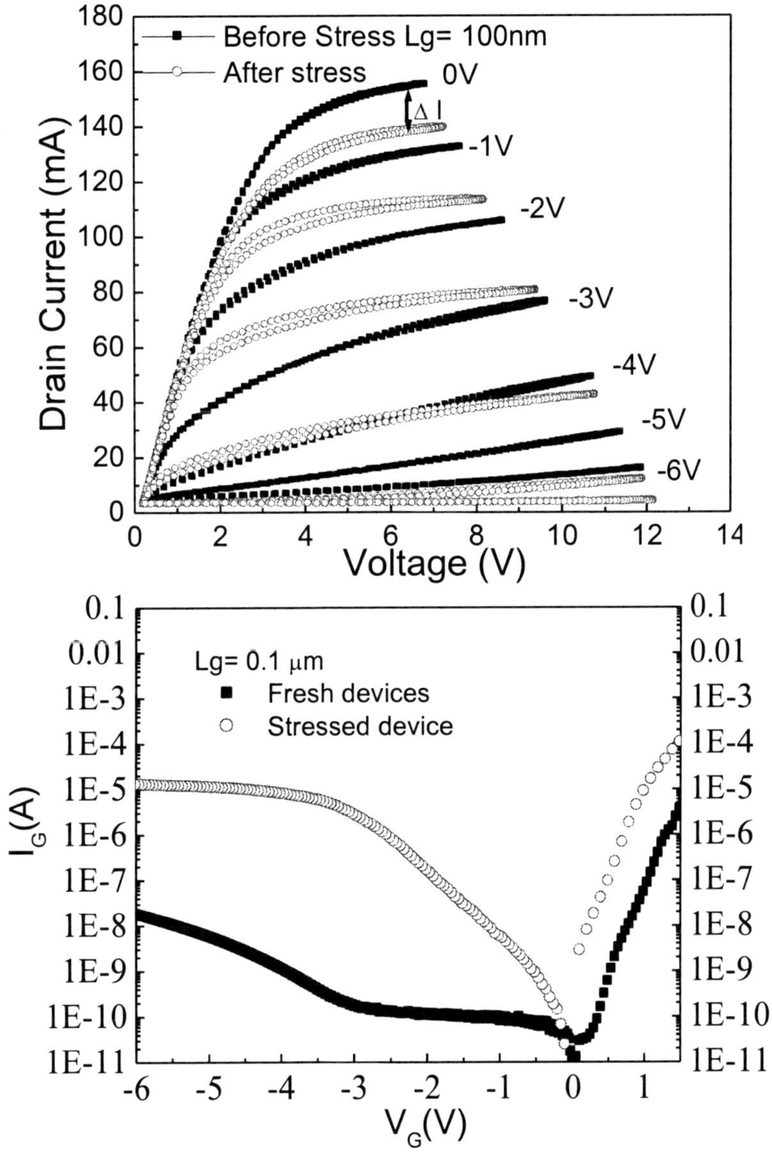

Reprinted from Semiconductor Science and Technology, 28, Cheney et al., Reliability Studies of AlGaN/GaN high electron mobility transistors, 074109, Copyright 2013, with permission from Elsevier.

Figure 4. Drain current (top) and gate current (bottom)- voltage characteristics before and after high reverse gate bias step stress.

These protocols are used in place of increasing temperature to provide a more realistic estimate of the reliability under various operating conditions. The recovery cycle protocols are especially relevant if deep traps are created by the stressing, since that will lead to current transients. This can be manifested as exponential recovery waveforms and increased noise and gate leakage. To study the effect of high electric field on GaN HEMTs, off-state step-stresses were performed in the dark at room temperature with the gate biased up to -100 V reverse gate voltage at various source-drain biases. The effect of gate length, gate to source and gate to drain distance, as well as the effect of applied drain bias on breakdown was investigated on structure A. As the gate voltage is stepped in 1 V increments for 1 minute per step from -10 V to -42 V, gate leakage current in AlGaN/GaN HEMTs with Ni/Au-based gate metallization increased until the critical voltage (V_{CRI}) is reached, resulting in a permanent increase in gate leakage current of several orders of magnitude as well as a reduction in drain current (Figure 4). This sharp rise in current has been attributed to the inverse piezoelectric effect.

Even though a large relative increase in gate current is observed after stress, this is accompanied by a small relative decrease (~20 %) in the drain current, due to an increase in on-state resistance. However, the change of the drain current in absolute value is larger than that of the gate leakage. Similarly, the transfer characteristics typically show a reduction in maximum transconductance of as a result of the gate bias stress cycle and a shift in threshold voltage.

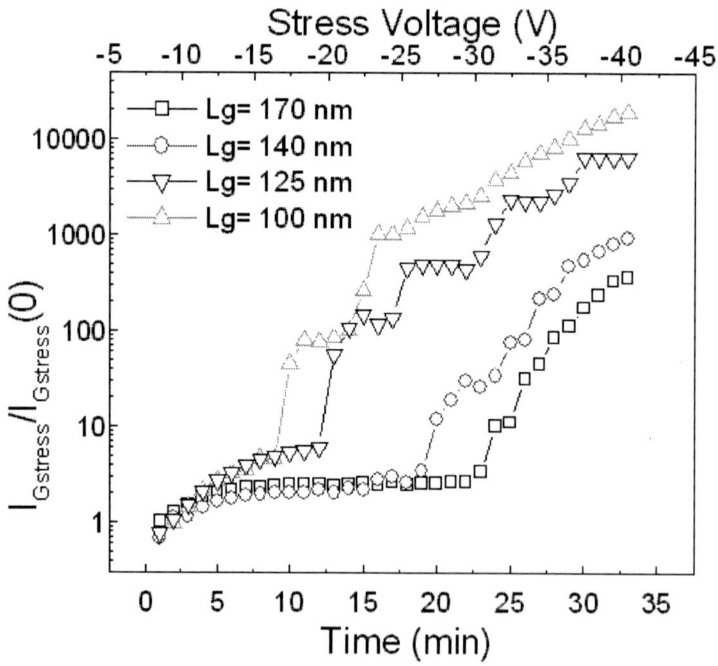

Reprinted from Semiconductor Science and Technology, 28, Cheney et al., Reliability Studies of AlGaN/GaN high electron mobility transistors, 074109, Copyright 2013, with permission from Elsevier.

Figure 5. Gate stress current and stress voltage as a function of time for various gate lengths.

Devices with sub-micron gate lengths ranging from 0.1 to 0.17 μm were step stressed from -10 V to -42 V in the same manner. The bias at which the sudden onset of high gate current was observed was strongly dependent on the gate length, as shown in Figure 5. While a linear dependence of gate length on V_{CRI} is observed, ATLAS/Blaze simulations confirm that degradation occurs at a critical maximum electric field in the channel, ~2.5MV/cm (Figure 6) [41]. This is consistent with previously published results indicating that failure is due to electric field induced tensile strain [41-43]. Photoluminescence (PL) of an ungated, open area of an unstressed HEMT showed that both the GaN peak centered at 366nm and the AlGaN peak centered at 352nm decreased in intensity after stressing. This degradation is indicative of the formation of non-radiative centers in both the GaN and AlGaN layers.

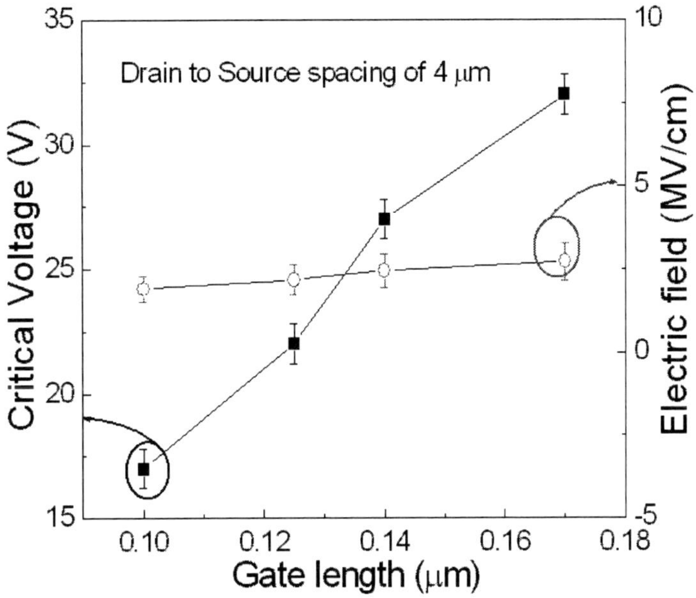

Reprinted from Semiconductor Science and Technology, 28, Cheney et al., Reliability Studies of AlGaN/GaN high electron mobility transistors, 074109, Copyright 2013, with permission from Elsevier.

Figure 6. Dependence of critical voltage and electric field on gate length for AlGaN/GaN HEMTs.

Field Plate

A source field plate over the gate electrode can increase device breakdown voltage by reducing the peak electric field at the drain side of the gate edge. This resulted in a significant increase in V_{CRI} from -40V to -65V [40-48]. Figure 7 shows the typical step-stress results of more than 30 HEMTs with and without the source field plate; the gate current, Ig, has been plotted as a function of the stressed gate voltage. The devices were step stressed in -1 V increments for 1 min at each increment, while grounding the source electrode and maintaining +5 V to the drain. TEM images indicated the presence of Ni from the gate metal stack interacting with the underlying nitride layer in close proximity to a threading dislocation, while simulations using the Florida Object-Oriented Process Simulator (FLOOPS) [47] showed that such metal diffusion can lead to a drop in I_D of at least 40%. A one-to-one correlation between the characteristics of I–V and gate leakage curves of electrically step-stressed AlGaN/GaN HEMTs and the presence of specific microstructural defects in the gate region of each device could not be established for all devices studied. Inverse piezoelectric effects are often cited as the driving force for the evolution of gate-edge "pit" or V-shaped defects which are associated with dramatically increased leakage current but it is possible that gate metal diffusion into the barrier layer, which was observed in all samples, may also have some effect. The possible role of the thin and irregular oxide layer beneath the gate metal contact also remains to be determined [48]. In addition, the threading dislocation density has been found to have a strong influence on the early degradation of HEMTs, with much more significant effects for high dislocation density [49].

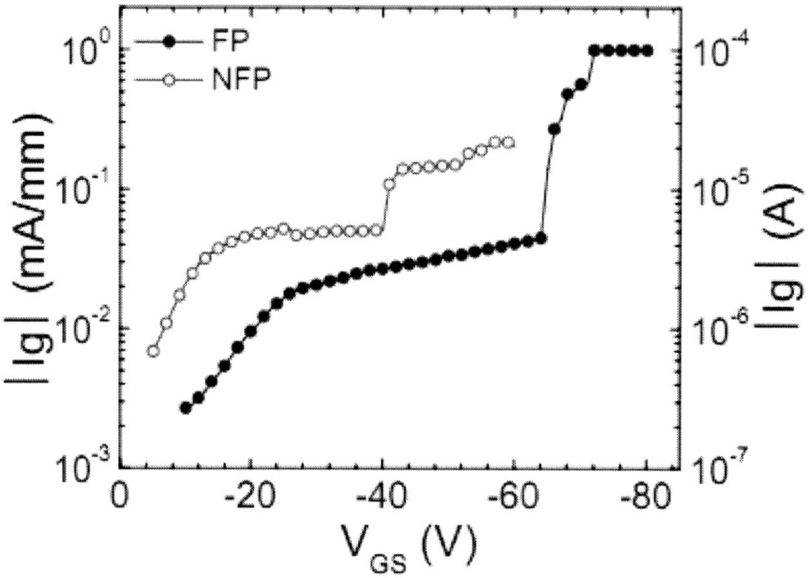

Figure 7. Off-state gate currents as a function of Vgs for HEMT with (FP) and without source field plate (NFP).

Defects Related to Contact Regions

(i) Source-Drain Contacts

The alloying of the source-drain contacts leads to metal inclusions resulting from reaction of the Al,Ti Ni and Au with each other and with N from the GaN. For example, Figure 8 shows SEM images of TiN metal inclusions under a source contact on a HEMT. Cracks nucleate on the facetted corners of the inclusions and can extend into the channel as shown in the image at the right of Figure 8. An image of one of these cracks is shown in Figure 9 and TEM/diffraction analysis shows the crack extending through the AlGaN layer. We speculate that the AlGaN is under compression from the presence of the metal inclusions and enhances crack formation and growth. Once formed in the channel, electrical stressing could cause additional growth through the inverse piezoelectric effect. These cracks may locally short out the AlGaN when the resulting voids fill with gate metal or dielectric passivation material.

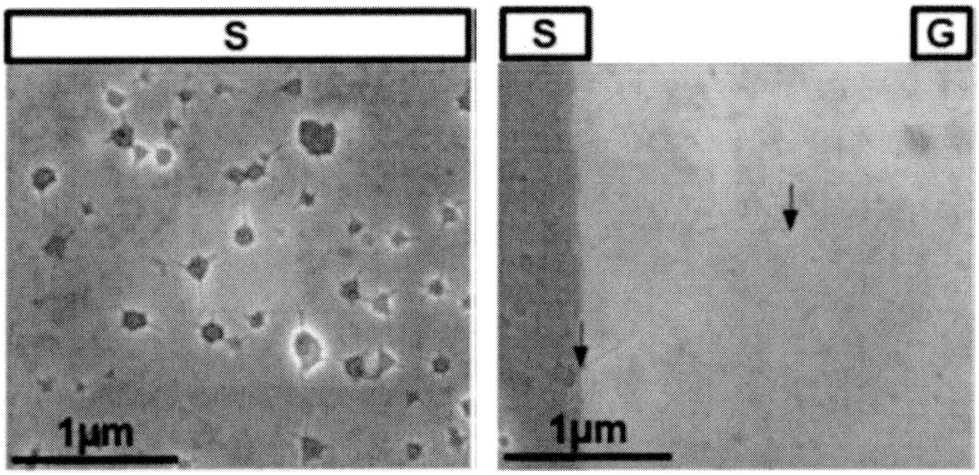

Figure 8. SEM images of alloyed Ohmic contacts on HEMTs. The image at left shows metal inclusions that occur throughout the contact region, while the image at right shows cracks that nucleate on the faceted corners of the metal inclusions. These cracks can extend into the channel region.

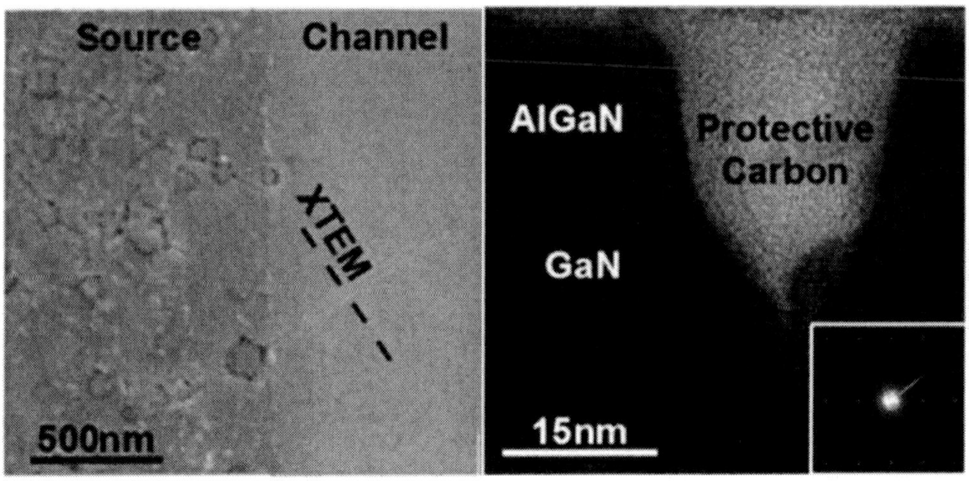

Figure 9. SEM image (right) of a crack extending from the Ohmic contact region into the channel of a HEMT. Cross-sectional TEM was performed in this area to show the crack extended through the AlGaN layer (left).

(ii) Gate Contact

Crack formation on the drain side of the gate is commonly seen under high drain-source voltage electrical stressing. A schematic of this process is shown in Figure 10. The crack eventually becomes filled with other material from the gate or the passivation layer. The presence of such gate degradation is accompanied by a dramatic increase in reverse bias leakage current due to shorting of the gate to the underlying two dimensional electron gas. This phenomenon is quite complex, involving not only the formation and refilling of cracks under the gate, but also metal diffusion down threading dislocations and electrochemical reactions between the gate, the ambient and the AlGaN layer in the HEMT. The type of gate metal makes a large difference in the onset of these processes, with Pt being much more stable than Ni.

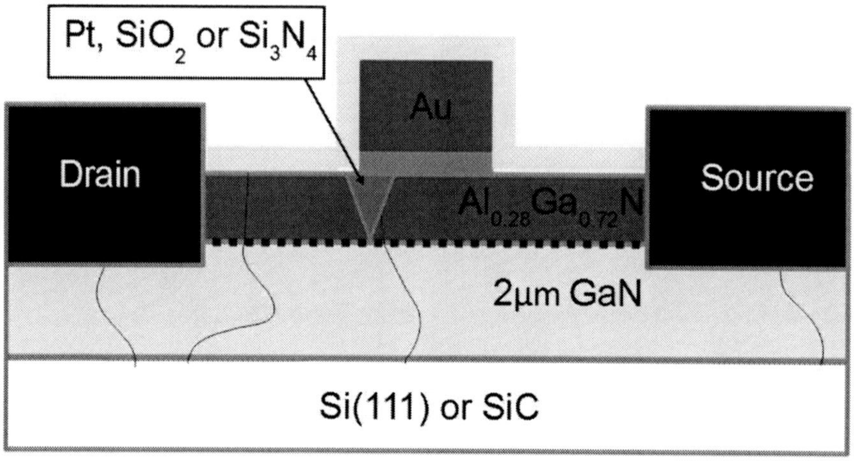

Figure 10. Schematic of crack formation on the drain side of the gate of a HEMT. During electrical stressing, the crack can refill with material from the gate metal, or dielectric passivation layers.

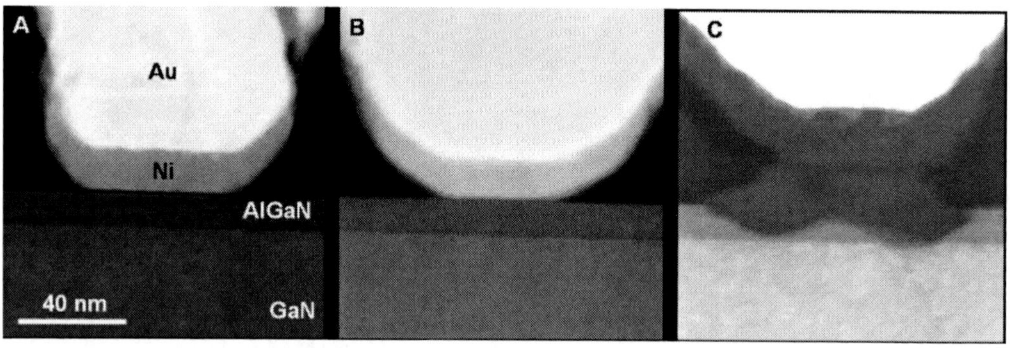

Figure 11. Cross-section TEM images of gates on HEMTs that were electrically stressed and showed slow degradation with V_{DS} =0V(right), slow degradation at V_{DS} =15 V (center) or rapid degradation at V_{DS} =10V where a defect forms by reaction of Ni from the gate contact (right).

Figure 11 shows TEM images of Ni gate HEMTs that displayed slow degradation at zero drain-source bias (a), slow degradation at higher drain-source bias (b) or rapid degradation at this type of bias (c). Note how a large defect forms from the Ni layer and penetrates in the AlGaN layer. Another view of this type of defect is shown in Figure 12-the electrochemical reaction of Ni and the subsequent consumption and pitting of the AlGaN leads to an enhanced leakage current through the gate contact. The defect is not confined to the drain side of the gate contact, pushes the gate electrode out of contact with the AlGaN and spans the full thickness of the AlGaN, directly shorting the gate to the two dimensional electron gas. Chemical analysis shows it is composed of a mixture of Ni and O and the shape is consistent with the electric field distribution. The degradation of devices in the off-mode, on-mode and with V_{DS} =0 V is driven by the vertical field in the AlGaN and not by the current density through the channel as is the case for hot- electron effects that occur in the semi-on mode.

The formation of pitting and cracking is also accompanied by formation of a band of deep traps whose presence can be determined by pulsed I-V measurements.

The kinetics of trap formation correspond to the activation energy for O down dislocation lines in GaN and suggests pitting-related defects form by metal diffusion down pre-existing dislocations under the gate, enhanced by the additional piezoelectric strain induced by the applied electric field during electrical stressing.

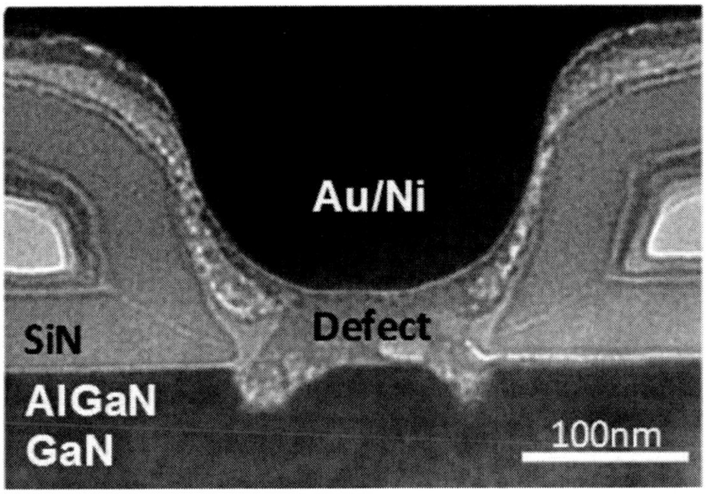

Figure 12. Cross section TEM image of defect formed under the gate during electrical stressing. The electrochemical of Ni leads to consumption and pitting of the underlying AlGaN and enhanced gate leakage current.

Deprocessing of stressed devices reveal pitting of the AlGaN layer on the drain side of the gate where the field is a maximum.

Generally the increase in gate leakage in stressed Pt gate devices stabilizes after the initial jump, whereas in Ni gate devices, the off-state leakage continues to increase. The spike in leakage current can also occur at voltages lower than V_{CRIT} and is strongly time-dependent. Thus, in Pt gate HEMTs, cracking induced by the piezoelectric effect at high fields on the drain side of the gate occurs and refilling of these cracks with passivation material or metal shorts out the gate to the two dimensional electron gas. In Ni gated devices, the same shorting occurs and produces a spike in leakage current at V_{CRIT} but this is not induced by cracking but by formation of large defects consisting of Ni, O and trace concentrations of Al and Ga. The defect formation is correlated with the electric field distribution and by the ambient, with failure occurring faster in oxygen ambients. The presence of a native oxide between the gate and the AlGaN also enhances the defect formation in Ni-gated devices that are electrically stressed.

There are three basic regimes-for operation below $V_{CRIT,}$ no permanent increase in leakage current is observed. At voltages well above V_{CRIT}, the NiO-related defects form. At biases near V_{CRIT}, dark bands form where the gate was located, as shown in Figure 13. Stepped stressing in excess of V_{CRIT} eliminates this banding and produces pitting, as shown in Figure 14. The banding is partially an electrochemical reaction and is enhanced by the presence of oxygen in the gate region, either from the ambient or a native oxide.

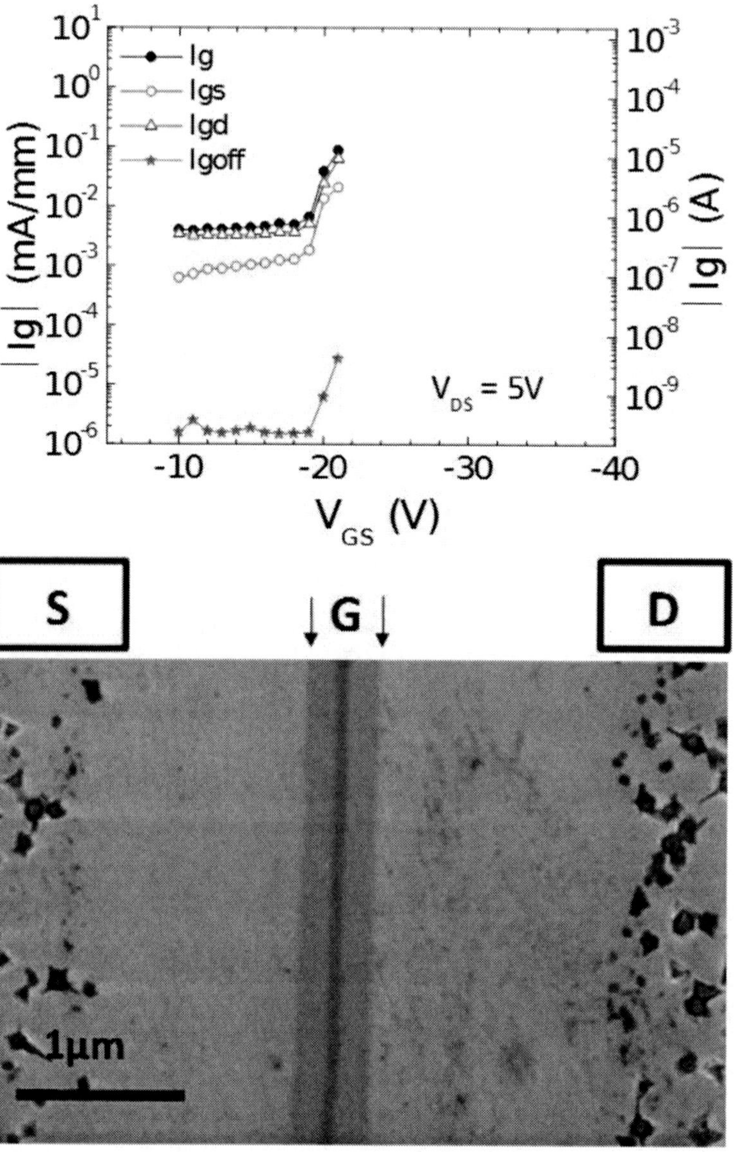

Figure 13. Gate current degradation at the stressing voltage reaches V_{CRIT}. At this value, the gate leakage during both stressing and in the off-state increases by two orders of magnitude (top). SEM image of the deprocessed surface after stressing shows a pair of dark bands where the T-gate was located.

The formation of the Nio-related defects is enhanced in devices where there are cracks or gaps in the SiN_x passivation, providing a direct path for O to react with the Ni in the gate contact. This is illustrated in Figure 15, showing a channel-like void that connects an airgap in the passivation directly to the contact.

This shows the importance of process control and quality in determining the ultimate failure mode of AlGaN/GaN HEMTs.

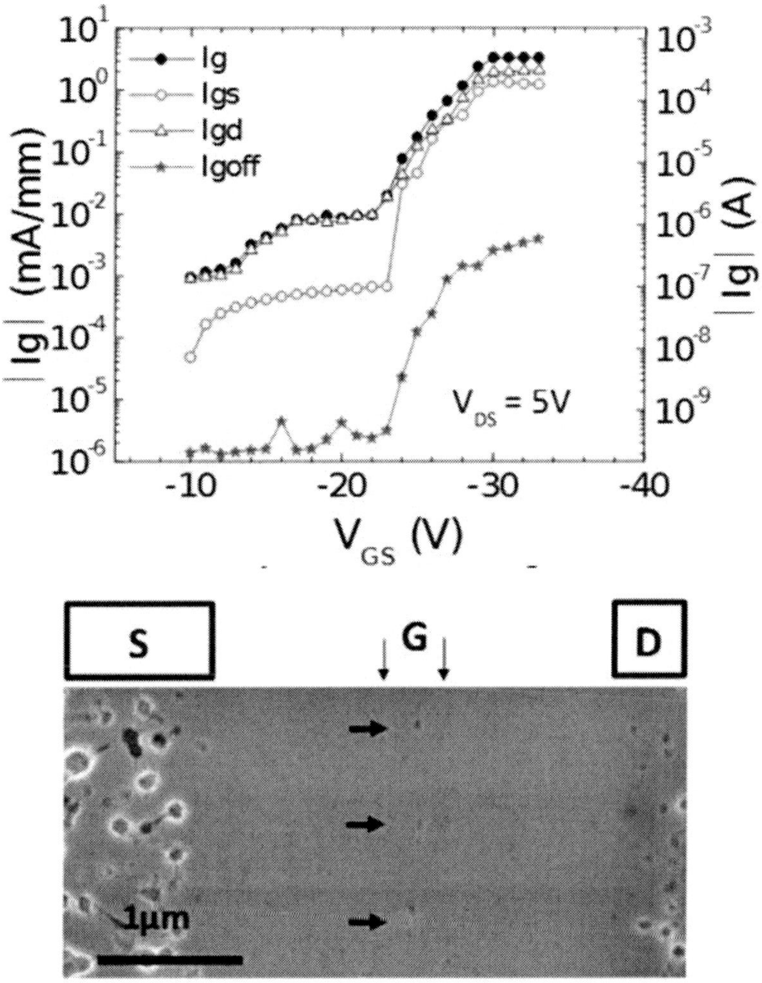

Figure 14. For step-stressing at voltages in excess of V_{CRIT}, there is a more pronounced increase in gate leakage (top). In this case, the banding in the channel is not observed, but pitting of the AlGaN is now observed in the channel where the gate was located.

Temperature Dependence

The interacting effects of temperature and electric field during off-state stress HEMTs with a gate length of 0.14 μm were stepped stressed from -10 V to -42 V in the dark at various temperatures ranging from room temperature to 150°C, regulated by a heated chuck. Both source and drain were held at ground in order to symmetrically stress the gate contact. The voltage at which irreversible degradation occurred was a strong function of temperature, as shown in Figure 16 [44].

V_{CRI}, followed a log relationship of the form, $V_{CRI} = V_0 e^{-E_a/kT}$, where the activation energy is 38 meV. The abrupt and permanent gate degradation at elevated temperatures was observed to occur at similar gate leakage currents regardless of critical voltage and stress temperature (~10^{-7} A). Gate leakage currents immediately after critical voltage was reached also displayed similar values at all stress temperatures (~10^{-6} A).

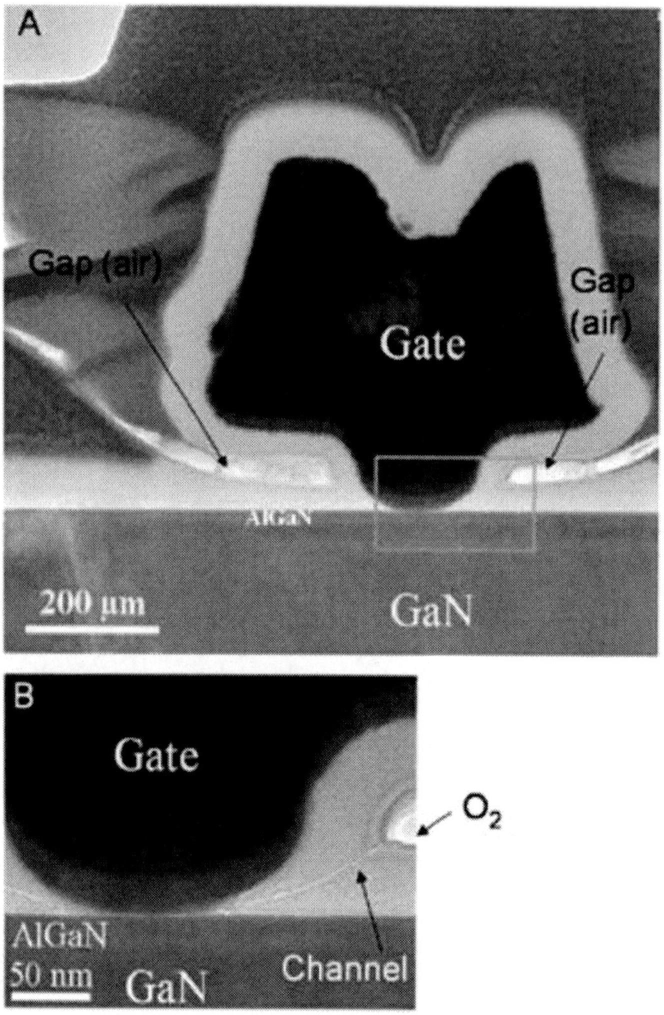

Figure 15. Cross section TEM showing how air gaps in the SiN_x passivation can lead to a pathway for oxidation around the gate region. The image at bottom shows a close-up view of how O_2 might access the gate region to form the large extended oxidized Ni defects observed in some devices.

The critical voltage of these devices at 24 °C was -30 V. As the temperature was increased to 150° C, the critical voltage was observed to decrease. Due to the fact that V_{CRI} occurs at lower voltages as the temperature increases, the maximum electric field present at the edge of the gate at V_{CRI} also decreases.

ATLAS/Blaze electrical simulations indicate that the peak electric field decreases from 3.3 MV/cm at a critical voltage of -28 V with a stress temperature of 28 °C to 2.6 MV/cm at a critical voltage of -18 V at 150 °C. The gate leakage current measured during the step stress of four different devices stressed at four different temperatures ranging from 24 °C to 150 °C further illustrate the negative temperature dependence. This result reveals that the breakdown which results in a sharp increase in gate leakage current does not occur at the same electric field, and therefore does not occur at the same piezoelectric induced stress.

In addition, the gate leakage current was independent of critical voltage, indicating similar leakage paths for all stress temperatures. Devices stressed at room temperature have been reported to display a significant dependence on gate voltage / electric field. Ni/Au metallization schemes for Schottky contacts on GaN have been determined to be thermally unstable above 400 °C, with numerous nickel nitrides reported being formed at temperatures as low as 200 °C [43-46].

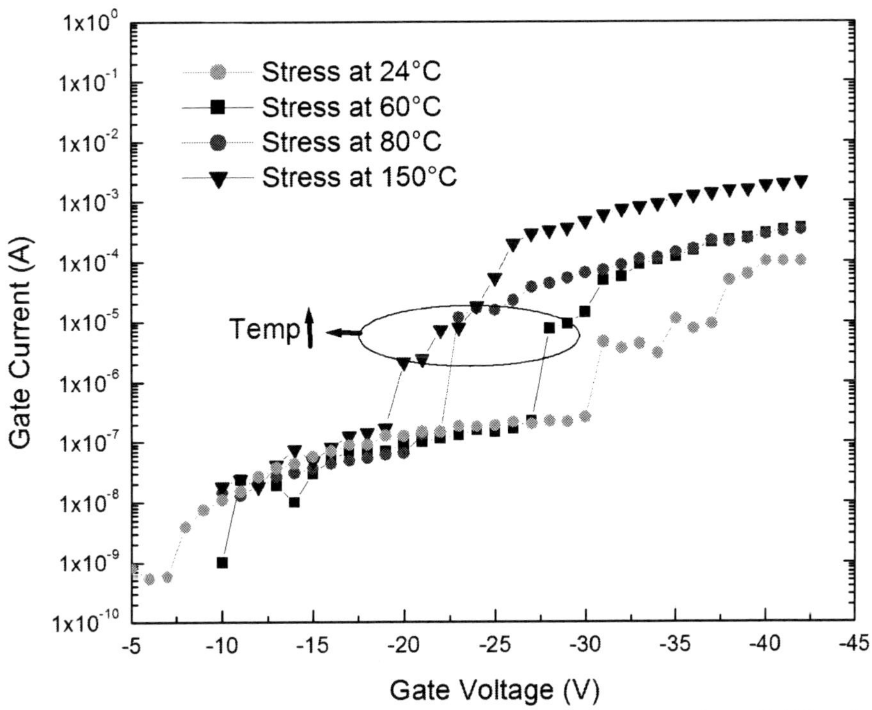

Figure 16. Gate current of four AlGaN/GaN HEMTs step stressed from -10 V to -42 V at 24°C, 60°C, 80°C, and 150°C.

Additionally, thermal instabilities in Ni/Au based Ohmic contacts have been reported due to Ga out-diffusion and Au inter-diffusion at elevated temperatures [48-52].

Electrical Signatures of Degradation with Different Gate Metals

As shown in Figure 17, HEMTs with Ni/Au gate metallization exhibited a critical voltage around -55 V. However, there was no critical voltage observed for the HEMT with the Pt/Ti/Au gate metallization up to -100 V, limited by the instrument. This suggests that the use of Pt based gate metallization could extend the operating bias conditions and improve the device reliability. The Schottky barrier height and ideality of the Pt/Ti/Au were 1.23 and 1.2 V, respectively, which did not change as a result of stressing. By contrast,

HEMTs with Ni/Au gate metallization showed signficantly higher gate reverse bias leakage current and much lower breakdown voltage. The forward gate characteristics of the Ni/Au gate contact appeared very leaky after the stress and the Schottky height reduced from 1.09 to 0.66 V after stress.

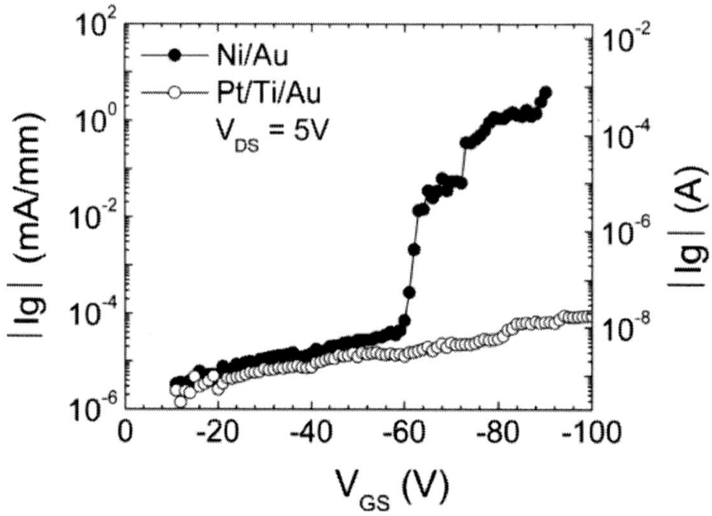

Figure 17. Off-state gate currents as a function of gate for the HEMTs fabricated with Ni/Au or Pt/Ti/Au gate metallization.

The interface between Ni/Au gate metal stack and the underlying semiconductor layer showed an oxide layer present prior to performing off-state stress. Consumption of this oxide layer, and even diffusion of oxygen, has been observed after off-state stress. The stability of Pt as a gate metal is due in part to its low reactivity with oxygen [49-56].

RF Stress Tests

Devices with a 0.125 μm gate length were biased under Class AB conditions with a quiescent drain current (I_{DQ}) of 200 mA/mm at drain bias (V_{DS}) of 10 V, 20 V, and 25 V. RF stressing for up to 350 hours was performed at 10 GHz, driven into 3dB compression and a baseplate temperature of 30 °C [57].Both P_{OUT} and I_{DS} showed little to no degradation for drain bias of 20 V, but rapid permanent degradation occurred at a drain bias of 25V, as shown in Figure 18. A small recovery in both P_{OUT} and I_{DS} is present after pausing at ~25 hours to obtain mid-stress device characteristics. Operating at a drain bias of 25 V is above the threshold necessary for the onset of degradation and is consistent with the observed critical voltage during dc stress. Small degradation in saturated drain current (I_{DSS}), less than 10%, occurred for all conditions after stress and is consistent with prior reports on RF reliability [58].

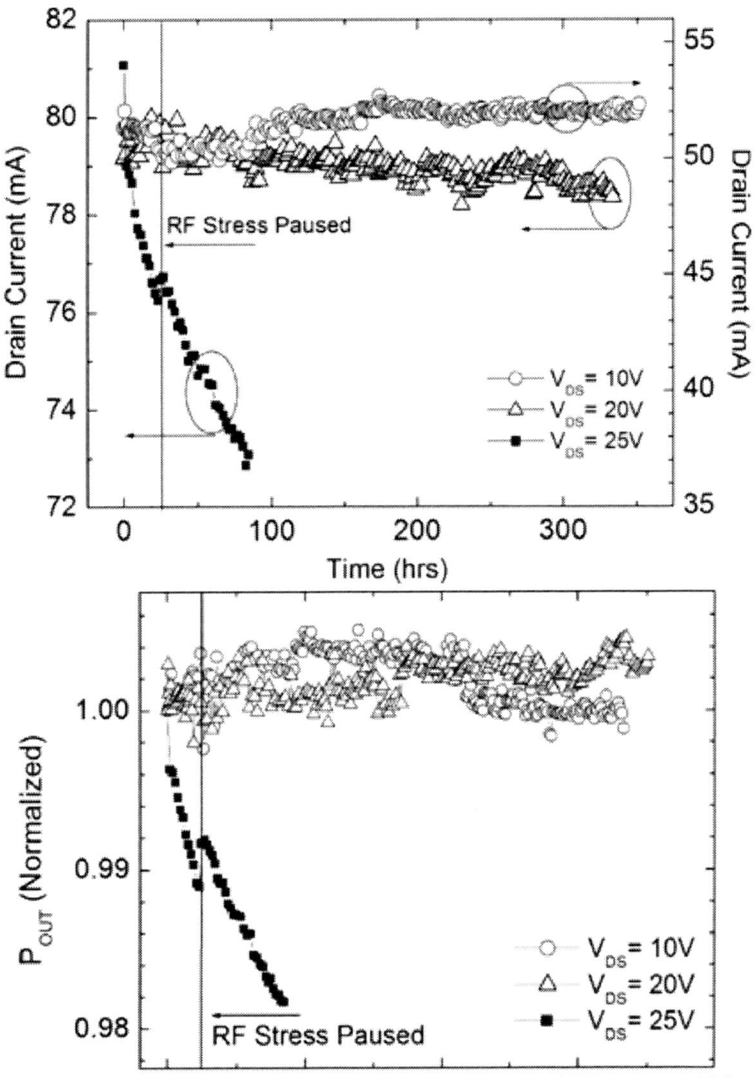

Reprinted from Semiconductor Science and Technology, 28, Cheney et al., Reliability Studies of AlGaN/GaN high electron mobility transistors, 074109, Copyright 2013, with permission from Elsevier.

Figure 18. (top) Drain current and (bottom) output power during RF stress at 10 GHz under 3 dB compression of 0.125 μm gate length AlGaN/GaN HEMT.

For the lowest drain bias of 10 V, a small increase (~3%) in PAE and I_{DS} (~1 mA) was observed during RF stressing for 350 hours, and may indicate that device burn-in is necessary in order to stabilize and improve RF device performance at lower bias conditions. Similarly, minimal degradation in PAE, P_{OUT}, gain, and transconductance occurred for a drain bias of 20 V. An increase inV_{TH} was observed for all stress conditions, with the largest ΔV_{TH} of 0.37 V at a drain bias of 25 V, indicating significant degradation of the Schottky contact [57].The Schottky barrier height reduced from 673 mV to 602 mV after stress at V_{DS} = 25 V. Even though device characteristics for drain bias at or below 20 V do not show significant RF degradation, the gate current-voltage sweeps with V_{DS} = 0 V indicate that the Schottky contact suffered degradation, with almost a 2 order of magnitude increase in gate leakage current.

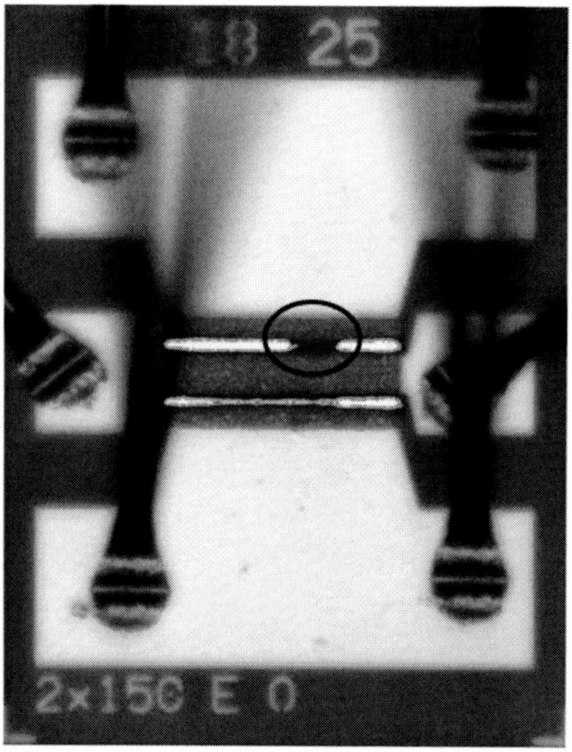

Figure 19. Electroluminescence of AlGaN/GaN HEMT after stress at V_{DS} = 25 V and Class AB operation. Circled region indicates area of channel that has an increase in non-radiative trap formation after stress. Uniform EL emission during forward bias was observed prior to stress. Photoluminescence spectra taken at circled region.

Even though the stress at a drain bias of 25 V was performed for a shorter duration, only ~85 hours, gate leakage current shows the largest increase with almost a 3 order of magnitude rise in current.

Prior to stress, devices exhibited uniform EL emission under forward bias. However, points of gate leakage appeared with gate bias of -10 V. After RF stress, gate leakage paths were not present on the same channel as present prior to stress (Figure 19) [57].It appears from EL emission that degradation due to RF stress may be localized along the channel on the right.

Conclusion

AlGaN/GaN HEMTs show severe degradation after reverse bias step-stressing of drain voltage whereas TLM structures, in which there is no Schottky contact, exhibited excellent stability. Increased baseplate and channel temperatures decreased the voltage at which catastrophic failure occurred. The combination of the Schottky gate material itself and the processing steps preceding the gate deposition is the likely cause for permanent degradation. There are a number of mechanisms theories to explain the reverse degradation of HEMTs, including oxidation of the surface, with subsequent generation of pits [42], defect generation/percolation in the AlGaN [43], generation of point defects within the AlGaN [44] and the inverse piezoelectric effect [37]. Off-state stress showed a linear relationship between critical degradation voltage and gate length. Device degradation is not a threshold process, and may occur even below the so-called critical voltage given enough time under stress. In this view, degradation is supposed to be related to a defect percolation process (activated by the high electric field and/or by the injection of hot carriers), and not unequivocally to the inverse piezoelectric effect.

An oxide layer is present between gate and GaN cap layer before stressing, but consumed after DC stress. Gate leakage current also increases after DC stress. Simulations show constant electric field for breakdown at various critical voltages. Interfacial oxide layers are reactive with the Ni from the Ni/Au gate metal stack after high reverse gate bias stress. In some instances, oxygen was observed diffusing into the underlying AlGaN layer and further through threading dislocations that terminated below the Schottky contact. Both dc and RF high electric field stress indicated a decrease in Schottky barrier height and large increases in gate leakage current. However, a small decrease in saturate drain current was observed in all instances. Devices with Pt/Ti/Au gate metal scheme showed superior stability up to gate voltages of -100 V. These studies indicate that the reliability of AlGaN/GaN HEMTs can be significantly improved for high electric field applications through the use of much more stable gate metals, such as Pt.

Sub-micron gate length HEMTs stressed under Class AB RF conditions with drain bias of 10, 20 and 25 V. AlGaN/GaN HEMTs show excellent reliability up to a drain bias of 20 V, however the Schottky contact properties show substantial degradation in terms of threshold voltage shift, gate leakage current, and Schottky barrier height increase. This indicates the gate contact is the cause for device failure. RF stressing at a drain bias of 25 V is above the threshold necessary for the onset of degradation and resulted in rapid, permanent degradation in P_{OUT} and I_{DS}.

Ni gate and Pt gate HEMTs degrade by different mechanisms, with cracking on the drain side of the gate prevalent for Pt gates at high fields [63-69]. Ni gate devices exhibit formation of band-like defects under the gate near V_{CRIT} and large NiO-related defects at the highest fields in excess of V_{CRIT}.

Acknowledgment

The work performed at UF is supported by an AFOSR MURI monitored by James Hwang.

References

[1] Meneghesso G., Verzellesi G., Danesin F., Rampazzo F., Zanoni F., Tazzoli A., Meneghini M., and Zanoni E., 2008, *IEEE Transactions on Device and Materials Reliability,* 8, 332.

[2] Singhal S., Roberts J. C., Rajagopal P., Li T., Hanson A. W., Therrien R., Johnson J. W., Kizilyalli I. C. and Linthicum K. J., 2006, *44th Annual IEEE International Reliability Physics Symposium Proceedings*, pp. 95-98.
[3] Sozza A., Dua C., Morvan E., Grimber B. and Delage S. L., 2005, *Microelectronics Reliability*, 45, 1617.
[4] Zanoni E., Meneghesso G., Meneghini M., Tazzoli A., Ronchi N., Stocco A., Zanoni F., Chini F., Verzellesi G., Cetronio A., Lanzieri C. and Peroni M., 2009, *European Microwave Integrated Circuits Conf.*, pp. 212-217.
[5] Rivera C., 2009, *Appl. Phys. Lett.*, 94, 053501.
[6] Dammann M., Leuther A., Benkhelifa F., Feltgen T. and Jantz W., 2003, *Phys. Stat. Solidi. (A),* 195, 81.
[7] Sozza A. Dua C., Morvan E. Grimber B. and Delage S. L., 2005, *Microelectron. Rel.,* 45, 1617.
[8] Piazza M., Dua C., Oualli M., Morvan E., Carisetti D. and Wyczisk F., 2009, *Microelectron. Rel.*, 49, 1222.
[9] Chowdhury U., Jimenez J. L., Lee C., Beam E., Saunier P., Balistreri T., Park S. Y., Lee T., Wang J., Kim M. J., Joh J. and del Alamo J. A., 2008, *IEEE Electron. Dev. Lett.,* 29, 1098.
[10] del Alamo J. A. and Joh J., 2009, *Microelectron. Rel.*, 49,1200.
[11] Joh J. and del Alamo J. A., 2006, *IEDM '06*, pp. 1-4.
[12] Faqir M., Verzellesi G., Chini A., Fantini F., Danesin F., Meneghesso G., Zanoni E. and Dua C., 2008, *IEEE Trans. Dev. Mat. Rel.*, 8, 240.
[13] Sarua A., 2006, *Appl. Phys. Lett.*, 88, 103502.
[14] Meneghesso G., 1997, *J. Appl. Phys.* 82, 5547.
[15] Uren M. J., Hayes D. G., Balmer R. S., Wallis D. J., Hilton K. P., Maclean J. O., Martin T., Roff C., McGovern P., Benedikt J. and Tasker P. J., 2006, *1st European Microwave Integrated Circuits Conference*, pp. 65-68.
[16] Buchta M., Beilenhoff K., Blanck H., Thorpe J., Behtash R., Heckmann S., Jung H., Ouarch Z. and Camiade M., 2010, *Int. Microwave and Millimeter Wave Technology Conference*, pp. 488-491.
[17] Mastro, M. A., LaRoche, J. R., Bassim N. D. and Eddy C. R.,2005, *Microelectron. Journal,* 36, 705.
[18] 18 Schroder D. K., Semiconductor material and device characterization, 3rd ed. [Piscataway, NJ] : Hoboken, N.J: IEEE Press; Wiley, 2006.
[19] Oda O., Compound semiconductor bulk materials and characterization , Singapore ; Hackensack, N.J: World Scientific, 2007.
[20] Brillson L., 2001, *J. Vac. Sci. Technol. B,* 19, 1762.
[21] Micovic M., Kurdoghlian A., Hashimoto P., Hu M., Antcliffe M., Willadsen P., Wong W. S., Bowen R., Milosavljevic I., Schmitz A.,Wetzel M. and Chow D. H., International Electron Devices Meeting 2006. IEDM '06., 2006, pp. 1-3.
[22] Arulkumaran S., 2002, *Appl. Phys. Lett.*, 80, 2186.
[23] Vitusevich S. A., Danylyuk S. V., Klein N., Petrychuk M. V., Avksentyev A. Y., Sokolov V. N., Kochelap V. A., Belyaev A. E.,Tilak V., Smart J., Vertiatchikh A. and Eastman L. F., 2003, *Appl. Phys. Lett.*, 82,743.
[24] Vitusevich S. A., Kurakin A. M., Klein N., Petrychuk M. V., Naumov A. V. and Belyaev A. E., 2008, *IEEE Trans. Dev. Mater. Rel.*, 8, 543.

[25] Chou Y. C., Leung D., Smorchkova I., Wojtowicz M., Grundbacher R., Callejo L., Kan Q., Lai R., Liu P. H., Eng D. and Oki A., 2004, *Microelectron. Rel.*, 44, 1033.
[26] Kuball M., 2003, *Appl. Phys. Lett.*, 82, 124.
[27] Batten T., A. Manoi Uren M. J., Martin T. and Kuball M., 2010, *J. Appl. Phys.*, 107, 074502.
[28] Batten T., Pomeroy J. W., Uren M. J., Martin T. and Kuball M., 2009, *J. Appl. Phys.*, 106, 094509.
[29] Kim J., Freitas J. A., Mittereder J., Fitch R., Kang B. S., Pearton S. J. and Ren F., 2006, *Solid-State Electron*, 50, 408.
[30] Chih-Yang C., Dougla E. A., Jinhyung J., Liu L., Lo C. F., Chu B. H., Cheney D. J., Gila B. P., Ren F., Via G. D., Cullen D. A., Lin Z., Smith D. J., Jang S. and Pearton S. J., 2011, *IEEE Trans. Dev. Mater. Rel.*,11,187.
[31] Montes-Bajo, M., Hodges, C., Uren, M. J. and Kuball, M., 2012, *Appl. Phys. Lett.*, 101,033508.
[32] Meneghini, M., Stocco, A., Bertin, M., Marcon, D., Chini, A., Meneghesso, G., and Zanoni E., 2012, *Appl. Phys. Lett.*, 100, 033505.
[33] Marko, P., Alexewicz, A., Hilt, O., Meneghesso, G., Zanoni, E., Würfl, J., Strasser, G. and Pogany, D., 2012, *Appl. Phys. Lett.*, 100, 143507.
[34] Marcon, D., Kauerauf, T., Medjdoub, F., Das, J., Van Hove, M., Srivastava, P., Cheng, K., Leys, M., Mertens, R., Decoutere, S., Meneghesso, G., Zanoni, E., Borghs, G., 2010, *Electron Devices Meeting (IEDM),* 2010 IEEE International Pages: 20.3.1 - 20.3.4.
[35] Coffie R., Chen Y., Smorchkova I. P., Heying B., Gambin V., Sutton W., Chou Y., Luo W. B., Wojtowicz M. and Oki A., 2007, in *Proc. 45th IEEE International Reliability Physics Symp. 2007*, pp. 568-569.
[36] Demirtas S. and del Alamo J. A., 2010, *Microelectronics Reliability*, 50, 758.
[37] Joh J., del Alamo J. A. and Jimenez J., 2008, *IEEE Electron. Dev. Lett.*, 29, 665.
[38] Joh J. and del Alamo J. A., 2008, *IEEE Electron. Dev. Lett.*, 29,. 287.
[39] Joh J., Gao F., Palacios T. and del Alamo J. A., 2010, *Microelectron. Rel.*, 50, 767.
[40] Lo C.-F., Liu L., Kang T.-S., Davies R., Gila B. P., Pearton S. J., Kravchenko I. I., Laboutin O., Cao Y., Johnson J. W. and Ren F., 2011, *Electrochem. Solid-State Lett.*, 14, H264.
[41] Douglas E. A., Chang C. Y., D.J. Cheney, Gila B. P., Lo C. F., Liu L., Holzworth R. Whiting P. Jones K., Via G. D., Kim J., Jang S., Ren F., Pearton S. J., 2011, *Microelectron. Rel.*, 51, 207.
[42] Gao, F., Lu, B., Li, L., Kaun, S., Speck, J. S., Thompson, C. V., and Palacios T., 2011, *Appl. Phys. Lett.*, 99, 223506.
[43] Meneghini, Matteo, Stocco, Antonio, Bertin, Marco, Marcon, Denis, Chini, Alessandro, Meneghesso, Gaudenzio and Zanoni, Enrico, 2012, *Appl. Phys. Lett.*, 100, 033505.
[44] Chini, A., Esposto, M., Meneghesso, G. and Zanoni, E., 2009, *Electron. Lett.*, 45, 426.
[45] Zanoni E., Danesin F., Meneghini M., Cetronio A., Lanzieri C., Peroni M. and Meneghesso G., 2009, *IEEE Electron. Dev. Lett.*, 30, 427.
[46] Ivo P., Glowacki A., Pazirandeh R., Bahat-Treidel E., Lossy R., Wurfl J., Boit C. and Trankle G., 2009, *IEEE International Reliability Physics Symposium*, pp. 71-75.
[47] Beechem T., Christensen A., Graham S. and Green D., 2008, *J. Appl. Phys.*, 103, 124501.

[48] Joh, J., Ling X. and del Alamo J. A., 2007, *International Electron Devices Meeting*, pp. 385-388.

[49] Joh J., del Alamo J. A., Langworthy K., Xie S. and Zheleva T., 2011, *Microelectron. Rel.*, 51, 201.

[50] http://swamp.mse.ufl.edu/floops.html

[51] Johnson, M. R., Cullen, D. A., Liu, L., Kang, T. S., Ren, F., Chang, C. Y., Pearton, S. J., Jang, Johnson, J. W. and Smith, D. J., 2012, *J. Vac. Sci. Technol. B,* 30, 062204.

[52] Tapajna, T., Kaun, S. W., Wong, M. H., Gao, Palacios, T., Mishra, U. K., Speck, J. S., and Kuball, M., 2011, *Appl. Phys. Lett.,* 99,223501.

[53] Guo J. D., Pan, F. M. Feng, M. S., Guo R. J., Chou P. F. and Chang C. Y., 1996, *J. Appl. Phys.,* 80, 1623.

[54] Miura N., Nanjo T., Suita M., Oishi T., Abe Y., Ozeki T., Ishikawa H., Egawa T. and Jimbo T., 2004, *Solid-State Electron.*, 48, 689.

[55] Douglas E. A., Chang C. Y., Gila B. P., Holzworth R., Jones, K. S., Liu L., Kim, J., Jang, S., Via G., Ren F., Pearton S. J., 2012, *Microelectron. Rel.*, 52, 23.

[56] Jung H., Behtash R., Thorpe J. R., Riepe K., Bourgeois F., Blanck H., Chuvilin A. and Kaiser U., 2009, *Physica Status Solidi (c),* 6, S976.

[57] Mattila T. and Nieminen R. M., 1996, *Phys. Rev. B,* 54,16676.

[58] Holzworth M. R., Rudawski N., Pearton S. J., Jones K. S., Lu L., Kang T. S., Ren F. and Johnson J. W., 2011, *Appl. Phys. Lett.*, 98, 122103.

[59] Liu L., Kang T., Cullen D., Zhou L., Kim J., Chang C., Douglas E., Jang S., Smith D. J., Pearton S. J., Johnson J. W. and Ren F., 2011, *J. Vac. Sci. Technol. B,* 29, 032204.

[60] Chini A., Di Lecce V., Esposto M., Meneghesso G. and Zanoni E., 2009, *European Microwave Integrated Circuits Conference*, pp. 132-135.

[61] Douglas E. A., Pearton S. J., Poling B., Via G. D., Liu L. and Ren F., 2011, *Electrochem. Solid-State Lett.,* 14, H464.

[62] Joh J. and del Alamo J. A., 2012, *Microelectron. Rel.*, 52, 33.

[63] Douglas, E. A., Zeenberg D., Maeda M., Gila B. P., Abernathy C.R., Ren F., Pearton S. J., 2013, *ECS Solid State Letters*, 2, Q39.

[64] Holzworth M. R., Rudawski N. G., Whiting P. G., Pearton, S. J., Jones K. S., Lu L., Kang T. S., Ren F., Patrick, E. and Law M. E., 2013, *Appl. Phys. Lett.,* 103, 023503.

[65] Cheney D., Douglas E., Liu L., Lo C. F., Xi Y., Gila B. P., Ren F., Horton D., Law, M. E., Smith D. and Pearton, S. J., 2013, *Semicond. Sci. Technol.* 28, 074019.

[66] Liu L., Lo C. F., Xi Y., Ren F., Pearton S. J., Laboutin O., Cao Y., Johnson J. W., and Kravchenko I., 2013, *J. Vac. Sci. Technol. B,* 31, 011805.

[67] Cheney D. J., E. A. Douglas E. A., Liu L., Lo C. F., B. P. Gila B. P., Ren F. and Pearton S. J., 2012, *Materials,* 5, 2498.

[68] Johnson M., Cullen D., Liu L., Kang T. S., Ren F., Chang C. Y., Pearton S. J., Jang S., Johnson J. W. and Smith D. J., 2012, *J. Vac. Sci. Technol.* B30, 062204.

[69] Cheney D. J., Deist R., Gila B., Navales J., Ren F., Pearton S. J., 2012, *Microelectron. Rel.*, 52, 2884.

In: Gallium Nitride
Editor: Kiera Olivia Peak
ISBN: 978-1-63321-387-6

Chapter 3

LOW-DIMENSIONAL GaN: STRUCTURE, SYNTHESIS AND PHYSICAL PROPERTIES

***Dongwei Xu*[1]*, Haiying He*[2,*] *and Ravindra Pandey*[1,†]**
Department of Physics, Michigan technological University,
Houghton, Michigan, US
Department of Physics and Astronomy, Valparaiso University,
Valparaiso, Indiana, US

Abstract

Recent developments in the low-dimensional GaN materials have revealed their novel structural and physical properties which can be exploited for applications in electronics and optoelectronics. In this paper, we review the work performed on the low-dimensional GaN materials, including clusters, nanotubes, nanowires, nanoribbons and sheets. Specifically, we will review their uniqueness in structure and electronic structure compared to those of the bulk GaN. A special example is discussed about the structure evolution from atomically thin layers to bulk. The stability of these low-dimensional materials is also discussed in light of the available thermodynamic data. It is followed by a summary of synthesis methods used for these GaN nanostructures together with their morphology and structural characterization. We also review the electronic, magnetic and optical properties of various GaN nanostructures ranging from pure to chemically doped or passivated GaN nanostructures. Lastly, their potential applications such as the use of GaN-based devices in light emitting diodes (LED) are discussed. The review ends with a summary of the current achievements, challenges and future directions in the research and development of low-dimensional GaN materials.

1. Introduction

Gallium nitride (GaN), a member of the family of III-V semiconductor compounds, has been investigated intensively and has found extensive applications [1,2]. Recent development has suggested that GaN may be used as an excellent host material for light-emitting devices

[*] E-mail address: haiying.he@valpo.edu (Corresponding author)
[†] E-mail address: pandey@mtu.edu (Corresponding author)

(LEDs) that operate in the blue and ultraviolet region [3], due to its wide and direct band gap (~ 3.4 eV for the bulk). Nano-materials often present novel properties that are different from the corresponding bulk materials due to their unique structures, high specific surface area and quantum confinement effects. For instance, the discovery of graphene (a monolayer of graphite) opens a new horizon in the carbon-based layered materials, where graphene shows drastically different properties and potential applications from those of graphite [4-8]. One would imagine the same enrichment of properties for the low dimensional GaN could be found. In spite that much less zeal has been shown to GaN so far compared to graphene, studies do show the uniqueness of low dimensional GaN as compared to the homogeneous system of carbon materials. Take for example, graphene nanoribbons are nonmagnetic semiconductors, which can be half metallic by applying external transverse electric field [9] or edge modification [10]. However, GaN [11] graphitic nanoribbons are found to be ferromagnetic together with other zigzag MoS_2 [12], SiC [13], and ZnO [14] nanoribbons.

The low-dimensional GaN materials, confined to nanoscale in one, two or three dimensions, have attracted additional attention in recent years due to their outstanding performance in optics, electronics, and magnetics. GaN nanostructures are considered to have great potential application in electro-optical devices owing to their tunability in electronic structure, thereby electronic and optical properties. Unlike bulk GaN with direct band gap, GaN sheets have indirect band gaps, and the band gap smoothly decreases as the thickness increases [15,16]. In addition, the band gap can also been tuned by chemical modulation or electric field [17], where the tunability of the gap is essential in the application of optical devices.

The advances in modern techniques to successfully synthesize GaN nanoscale structures are another catalyst in the ever-increasing research and development of GaN low-dimensional materials. The tubular structure of GaN was firstly observed in epitaxially grown GaN materials [18]. With epitaxial casting techniques, single-crystal GaN nanotubes were successfully synthesized in 2003 [19]. Theoretical studies have shown that the GaN nanotubes have about a 2 eV bandgap, independent of their chirality [20]. Other types of GaN nanomaterials, such as nanobelts [21], nanoribbons [22,23], nanowires [24] were also synthesized.

In this review, Sec. 2 introduces the wide range of low-dimensional GaN structures which have been put on stage, from the two-dimensional sheets to the one-dimensional nanoribbons, nanotubes, nanowires, to the zero-dimensional clusters. Both theoretical predictions and experimental studies are included with a discussion of the thermodynamic stability. Sec. 3 lists the present experimental efforts on the synthesis of these nanoscale GaN materials. Electronic, magnetic and optical properties of these nanostructures are summarized for both computational simulations and experimental measurements in Secs. 4, 5 and 6. Lastly, the future perspectives of applications and challenges are discussed in Sec. 7.

2. Structure and Stability

Even though the bulk GaN has most commonly the wurtzite or zinc blende structure [25], the low dimensional GaN nanostructures may have their unique structures. In this section, we will list the main structures reported in the literatures, from two-dimensional (2D) sheets, to one-dimensional (1D) nanoribbons, nanotubes and nanowires, to zero-dimensional (0D) clusters. In these structures, one, two or three dimensions are reduced to the range of 1-100 nm. As a result, the size confinement, surface effect and quantum confinement are all exaggerated leading to reformation of atoms in the structure.

2.1. Sheets

Theoretical study predicted that wurtzite III–V semiconductors transform into a 2D graphitic structures when they are in the form of an ultrathin (0001) films [26]. These ultrathin films remove the dipole by adopting a layered graphite-like structure in which the atoms are threefold coordinated. This type of structure becomes thermodynamically the most favorable one for GaN ultrathin films [15,27,28]. DFT calculations show that when the number of layers is less than six, planar structure of GaN is preferred [16,26,28]. The planar structure can be removed by chemical modification. For example, the planar GaN sheet becomes buckled after fluorination [29,30].

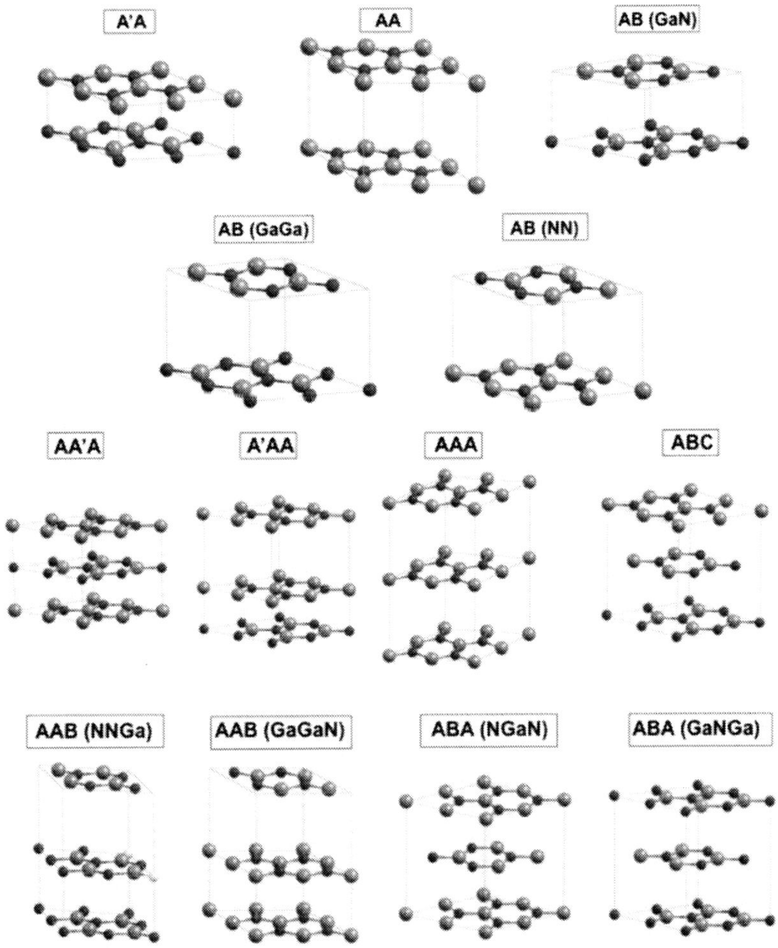

Figure 1. Stacking configurations considered for the bilayer and trilayer of GaN. The large brown (gray) balls represent 'Ga' while the small pink (gray) balls represent 'N'. Reproduced with permission from [15]. Copyright 2013, IOP Publishing Ltd.

The most stable GaN monolayer has the structure similar to graphene. The calculated Ga–N bond length is 1.85 Å at the generalized gradient approximation- density functional theory (GGA–DFT) or van der Waals (vdW) corrected DFT level of theory [15], which is smaller than the bulk value of 1.96 Å. However, the semi-ionic bonding is still predominant in such a single atomic layer as the case with the bulk GaN.

By stacking single layers of GaN together, one can form multilayer GaN. As can be seen in Figure 1, there are five (AA, AA', AB (NN), AB (GaGa), and AB (NGa)) and eight (AA'A, A'AA, AAA, ABC, AAB (NNGa), AAB (GaGaN), ABA (NGaN), ABA (GaNGa) possible stacking sequences for bilayer and trilayer GaN, respectively. It was found from first-principles calculations that the hexagonal A'A and AA'A stacking are the most energetically favorable configurations for the bilayer and the trilayer, respectively [15]. The interlayer binding strength can be attributed to the electrostatic and vdW interactions. In the semi-ionic GaN lattice, the cation (Ga) tends to attract an anion (N), and repel another cation (Ga). Thus, the low-energy stacking in GaN bilayers and trilayers still mimic that of the bulk (the wurtzite phase). The difference at this low dimension, however, is the unbuckled GaN planes. One would expect that the most stable stacking sequence of GaN atomic layers continue to be AA'AA'..., and the GaN layer will become buckled when the number of layers is large enough. The calculated cohesive energy increases with the number of layers from 8.38 eV per GaN pair for monolayer, 8.68 eV per GaN pair for the AA' bilayer, to 8.78 eV per GaN pair for the AA'A trilayer indicating the increasing stability of the multilayer system with increasing the number of layers and approaching the value of wurtzite GaN bulk (9.74 eV per GaN pair).

Surface decoration provides a way to tune the properties with unprecedented potential applications. Some groups explored the semifluorinated or semihydrogenated 2D GaN sheets. Half and full passivated GaN monolayers are shown in Figure 2 and 3, respectively.

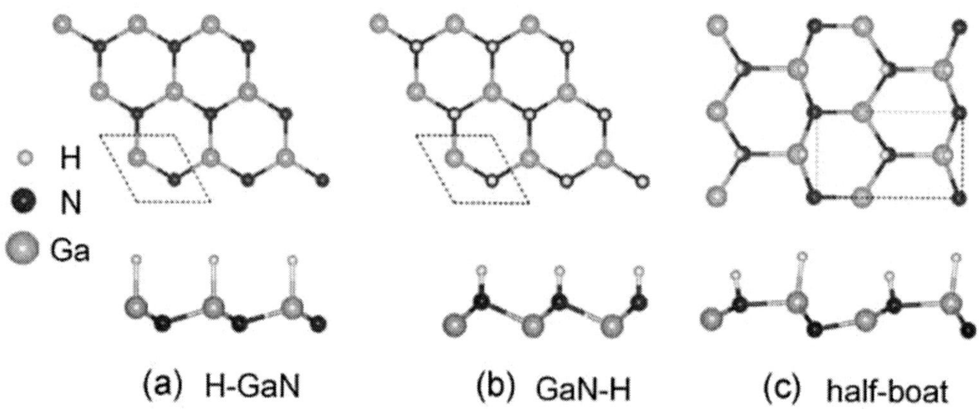

Figure 2. Top and side view of the optimized structures of semihydrogenated GaN sheets: (a) the H–GaN conformation with hydrogen atoms on only Ga atoms, (b) the GaN–H conformation with hydrogen atoms on only N atoms, and (c) the half-boat conformation with hydrogen atoms alternating in pairs on the same side of the GaN sheet. Reprinted with permission from [31]. Copyright 2011, WILEY-VCH Verlag GmbH & Co. KGaA, Weinheim.

For the partially passivated GaN monolayer, the decorated hydrogen (fluorine) atoms are adsorbed at Ga (H-GaN) or N (GaN-H) or with hydrogen atoms alternating in pairs on the same side of the GaN sheet (half boat). For the semifluorinated GaN sheet, the configuration in which the F atoms occupy the atop positions of all the Ga sites is most stable [30], while for the semihydrogenated GaN sheet, the H atoms energetically prefer to adsorb on N atoms rather than Ga atoms [31]. This can be explained by the electronegativity of the atoms. H and Ga tend to donate electrons, while F and N tend to accept electrons. Therefore, Ga favors to bond to F and N, while N favors to bond to H and Ga. This fourth coordination brings the GaN sheet to a buckled configuration similar to the one seen in the bulk. The half-boat conformation is predicted to be the most stable structure for the half-hydrogenated GaN monolayer followed by the conformation where H atoms are adsorbed at N (GaN-H), and the conformation where H atoms are adsorbed at Ga (H-GaN)is least favorable [31].

For the case of fully decorated hydrogenated honeycomb GaN monolayers [32], both Ga and N atoms prefer to have a tetrahedral coordination leading to the distorted buckled monolayer. Two stable structures were found from first-principles calculations, the chair conformation with the hydrogen atoms alternating on both sides of the plane and the boat conformation with the hydrogen atoms alternating in pairs (Figure 3). The boat conformation is energetically more favored [32].

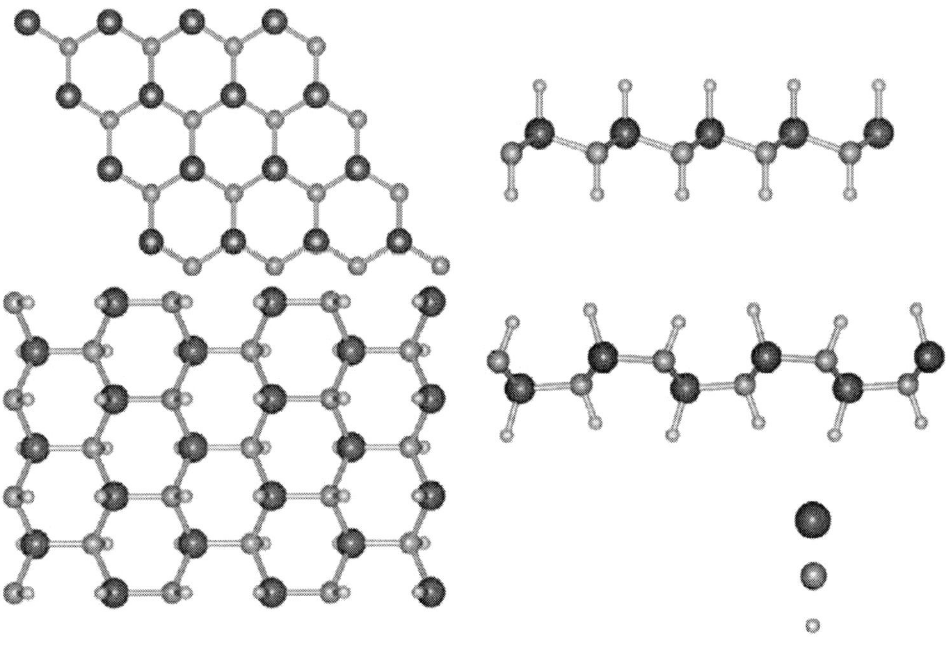

Figure 3. Structures of the (a) chair conformation and (b) boat conformation for the hydrogenated GaN sheets. The large green (light gray), medium blue (gray) and small pink (white) balls represent Ga, N and H (F), respectively. Reprinted with permission from [32]. Copyright 2010, Elsevier Ltd.

Therefore, surface decoration is feasible to modulate the geometrical structures and electronic properties of GaN. Even though the GaN monolayer prefers the flat graphene-like structure, after the under-coordinated surface Ga (N) atoms are saturated with F (H) atoms, the pristine wurtzite structure is restored [16,30-32].

The chemical modification also significantly improves the stability of GaN sheet. The pristine GaN sheet is more chemically active and less stable than graphene. As similarly shown in Figure 3, the N atoms and Ga atoms are saturated by hydrogen and fluorine atoms separately. The hydrogen prefers the top site of N atoms with N–H bonds arranged normal to GaN sheet, similar to the semihydrogenated graphene sheet. GaN sheet modified with H and F atoms on both sides is wrinkled and the F atoms prefer to stay on the hollow sites of the hexagonal GaN sheet and bond with the three nearest Ga atoms, which is analogous with the fluorinated ZnO layers [33]. The presence of H and F atoms can significantly enhance the thermal stability of the GaN sheet [17].

2.2. Nanoribbons

When the 2D sheet is cut into nanoribbons (NRs), edges with large number of unsaturated atoms are created. Moreover, the finite size of the ribbon along one dimension can be a tuning factor for structures and properties as well. These highly active edges also bring new properties to the materials. The percentage of such atoms in a low-dimensional nanoribbon can be a dominant factor in controlling the physical and chemical properties of the system. There are two kinds of edges: armchair and zigzag. The width of a zigzag GaN nanoribbon is defined by the number of zigzag chains across the ribbon width. Likewise, the width of an armchair GaN nanoribbon is classified by the number of dimer lines across the ribbon width. For these bare GaN nanoribbons, it is found that the edge N atoms tend to relax closer to the ribbons at a Ga-N bond length of 1.76 Å. However, the edge Ga atoms relaxes significantly less than N atoms such that the Ga-N bond length changes to about 1.83 Å. Atoms in the inner part relax less as well, and the Ga-N bond length is retained at ~1.82 Å similar to the GaN sheet at the Local-density approximation (LDA) level of theory [11]. It should be noted here that the multilayer GaN nanoribbons, where the edge should play an important role, has different properties compared with the GaN nanosheet. The GaN nanoribbons cleaved from the bulk GaN directly (bulk model) is considered more stable than the stacked GaN nanoribbons layer-by-layer with particular distance (sheet model) [11].

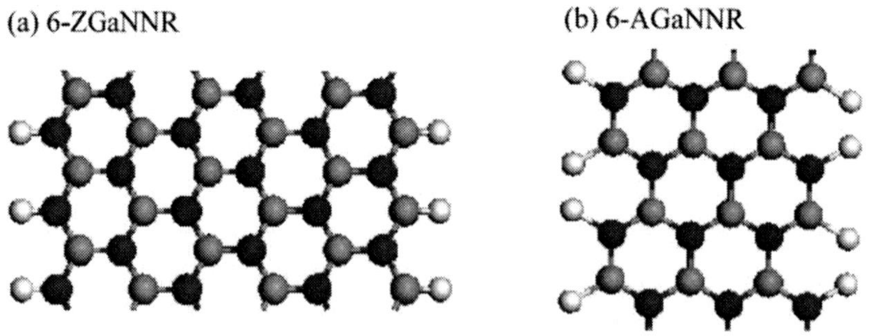

Figure 4. Geometry structures for (a) 6-zigzag GaN nanoribbons and (b) 6-armchair GaN nanoribbons with hydrogen terminations. The large blue (black), large brown (gray) and small white (white) balls represent N, Ga, and H atoms, respectively. Reprinted with permission from [35]. Copyright 2011, American Institute of Physics.

For the bare GaN nanoribbons, there are dangling bonds on N (Ga) atom at the two edges leading to higher energies of the system. To stabilize the nanoribbons, dangling bonds at the ribbon edges can be saturated by hydrogen atoms [11,16,34-36], as shown in Figure 4. By hydrogen termination, the effects of the dangling bond on the electronic states near the Fermi level are avoided. For H-passivated nanoribbons, the modified lengths of Ga-H and N-H bonds were 1.54 Å and 1.02 Å, respectively. The Ga-N bonds close to the edge change at a lesser rate than in bare geometry due to the charge balance at the edge.

2.3. Nanotubes

Following the idea that rolling up a graphite sheet can create a carbon nanotube, the hypothetic graphite-like GaN nanotube can also be formed by rolling graphite-like GaN sheet. Figure 5 shows the typical single-wall armchair and zigzag nanotubes [20]. These single-walled tubular configurations were first explored by DFT methods and the authors suggested the possibility for their formation by comparing the strain energies of GaN nanotubes to those of carbon nanotubes [20]. However, molecular dynamics simulations indicated that this idea is not applicable for creating single-crystal GaN nanotubes [37]. With the MD simulations at room temperature, it was predicted that the so-generated GaN nanotubes breaking into two or more fragments, no matter what the wall thicknesses and tube radii were [37]. Instead, by carving a hexagonal cylinder with [001] axial direction from a wurtzite bulk, a single-crystal nanotubes with the hexagonal shaped cross section can be generated (see Figure 6). It was shown that these GaN nanotubes are thermally stable, and match the experimental observation [19] as well. Furthermore, the MD simulations revealed that a single-crystal GaN nanotubes with lateral facets oriented along [100] direction is more stable than that with lateral facets oriented along [110] direction, under the condition that the wall thickness is the same. However, it was observed in experiments that the lateral facets of the GaN nanotubes are oriented along the [110] direction [19]. Note that the nanotube walls of experimentally synthesized nanotubular GaN are many nanometers thick which means essentially bulk-like atomic layout [19]. More interestingly, for a specific orientation of the lateral facets on the single-crystal GaN nanotubes, the energetic stabilities of the nanotubes obey a P rule, where P is the ratio of the number of four-coordinated atoms to the number of three-coordinated atoms [37].

Another scheme to generate the tubular configuration from bulk GaN is to have a circular cross section [38] instead of a hexagonal one. It is shown at the DFT-GGA level of theory that a multiwall nanotube constructed from the GaN wurtzite crystal along (001) direction retains its original wurtzite form, while a single wall nanotube transforms into a perfect cylindrical tube when relaxed without any symmetry constraint [38,39]. The GaN single wall nanotube was suggested to remain stable at 300 K, which is, however, contradictory to the MD predictions by Xu et al. [37].

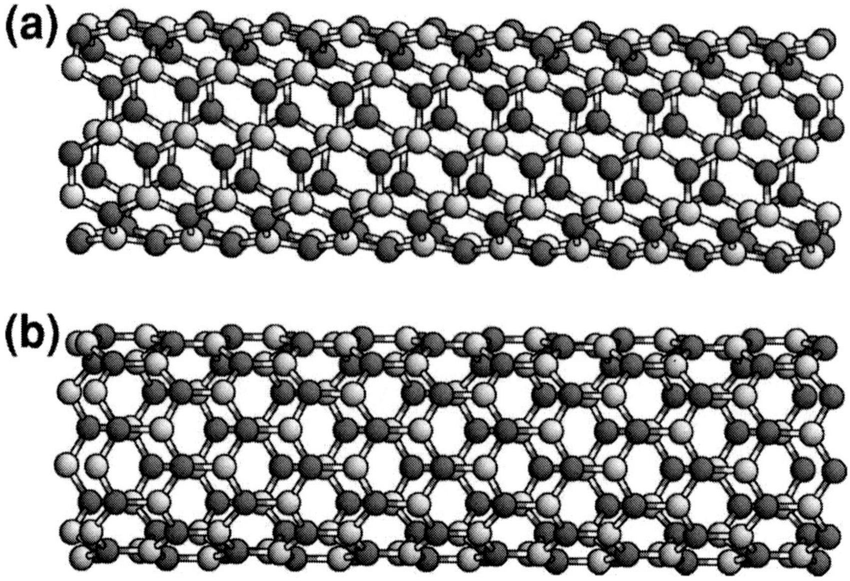

Figure 5. Typical types of GaN nanotubes (a) (5,5) armchair nanotube, and (b) (9,0) zigzag nanotube. Dark-gray and light-gray balls indicate Ga and N, respectively. Reprinted with permission from [20]. Copyright 1999, the American Physical Society.

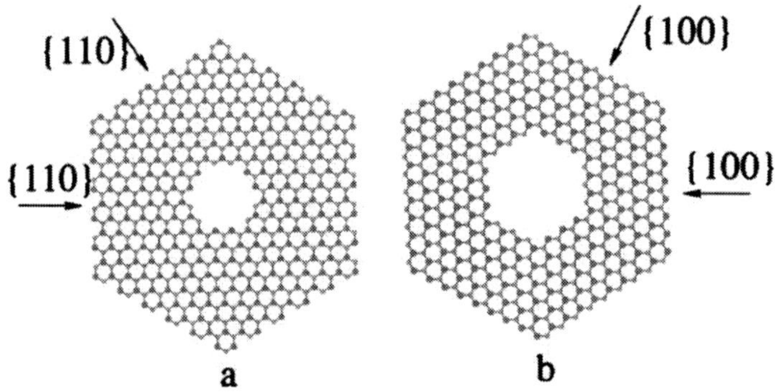

Figure 6. Top view of GaN nanotubes with certain thickness of the wall. (a) The GaN nanotubes with [110]-oriented lateral facets. (b) The GaN nanotubes with [100]-oriented lateral facets. The axial directions of these GaN nanotubes are all along [001] direction. Reprinted with permission from [37]. Copyright 2005, the American Physical Society.

2.4. Nanowires

If the inner part in creating a GaN nanotube is not removed, GaN nanowire is formed. Figure 7 shows the cross section view of GaN nanowire along [0001] direction. The synthesized nanowires show predominantly wurtzite phase [40,41]. Among the various experiment synthesis methods, the resulting nanowire were up to several hundred micrometers in length, typically with diameters of several to a hundred nanometers [24,41-43].

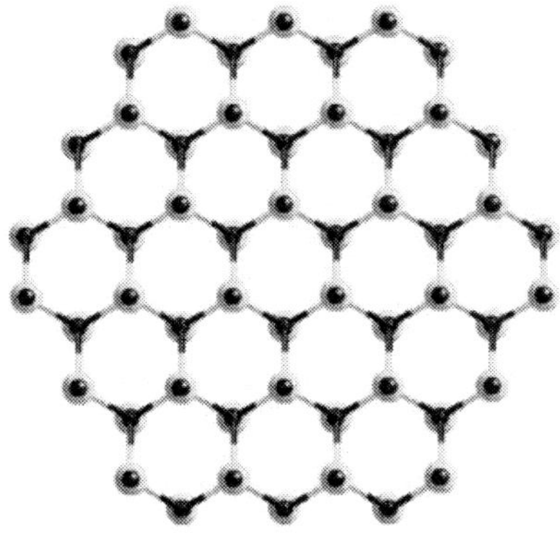

Figure 7. Illustration of the 216-atom nanowires viewed along the [0001] direction (i.e., showing the cross section). Nitrogen and gallium atoms are indicated by the small dark (blue) and large light (aqua) balls, respectively. Reprinted with permission from [44]. Copyright 2009, the American Physical Society.

Studies show that due to the competition between the bond energy, the Coulomb energy, and the energy originating from the dipole field of the wurtzite structure, the structure of nanowire transforms from the graphite-like phase to wurtzite phase as the diameter and length of the nanowire increases [45].

2.5. Clusters

Now if we reduce the size of the system along all three dimensions, it becomes an agglomeration of a few or a dozen of atoms – clusters at nanoscale. Due to the large number of unsaturated atoms on surfaces, these clusters may take a different structural evolution as the size reduces. The smallest GaN clusters are typically planar, whereas in the medium-sized range, the dominant structural motif takes the form of hollow structures with fullerene-like cage and tube-like configurations, where all atoms are three-coordinated. Then the bulk-like structures become more and more favorable with increasing cluster size [46].

A series of DFT studies were conducted on small neutral and anionic structural isomers of Ga_mN_n ($1<m,n<6$) [47-53] clusters. For Neutral GaN clusters, it is found that small polyatomic nitride clusters have a strong tendency to form N-N multiple bonds leading to the weakening of any existent Ga-N or Ga-Ga bonds. A strong preference for N-N multiple bonds is observed in the most stable isomers of triatomic and dimer configurations [52]. In neutral Ga_nN_n ($4<n<6$) clusters, although the competition between N-N dominated structures and Ga-N dominated structures is increasing with the cluster size, there are still not enough Ga-N bonds to break all the N-N bonds (N_3 subunit) in the lowest energy configurations [48]. So the smaller clusters, with a low coordination, display structures dominated by the N-N bond [49] and these clusters present planar-like structures. When an electron is added, the extra electron is located over the metallic atoms instead of being shared by all of the atoms in the cluster. The N_2 and N_3 subunits are still the most important features in the anionic clusters, as they were in the neutral ones [47]. The addition of an electron to the neutral group III nitride clusters induces significant structural changes in GaN. The $Ga_6N_6^-$ cluster is no longer a planar structure [48].

The possible structures of GaN nanocages (a hollow system) and their stabilities are studied using *ab initio* DFT calculation [54]. In contrast to fullerene cages of carbon that have pentagons and hexagons, polar materials such as GaN favor cages that are formed with hexagons and four-membered rings (tetragon) because five-membered rings will have like atoms as nearest neighbors which is energetically unfavorable [55]. Zerenturk and Berber [54] followed a systematic approach and consider all possible nanocages of up to 40 atoms. They found that the stabilities of the GaN nanocages depend mainly on the number of shared edges on the tetragons, and less on the size of the cage. The structures with isolated and ideally buckled tetragons are more stable, while this priority of the nanocages with isolated tetragons may be hampered if non-ideally buckled tetragons occur in the structure. The size effect due to quantum confinement on the electronic structure is dominated by the electronic states localized on the tetragons. Kumar and Zavada studied cage structures as well as bulk fragments to find the low-energy-lying atomic structures of GaN clusters [55]. They found that for Ga_nN_n (n = 12, 16, 22, and 24), cage structures are lower in energy than bulk fragments [55].

The structures of bulk-like stoichiometric GaN clusters shown in Figure 8 are obtained from a first principles study [56]. Note that these initial structures were cut from a bulk. As one can see, most of them maintain the bulk-like structures, which conflicts with the previous studies considering all possible configurations of small clusters [47,48]. The effect of surface states has clearly demonstrated in the optimized structures of Ga_nN_n clusters as shown in Figure 8, where four-membered rings are formed at the surfaces of clusters, and surface atoms have largely adjusted along the polar axis.

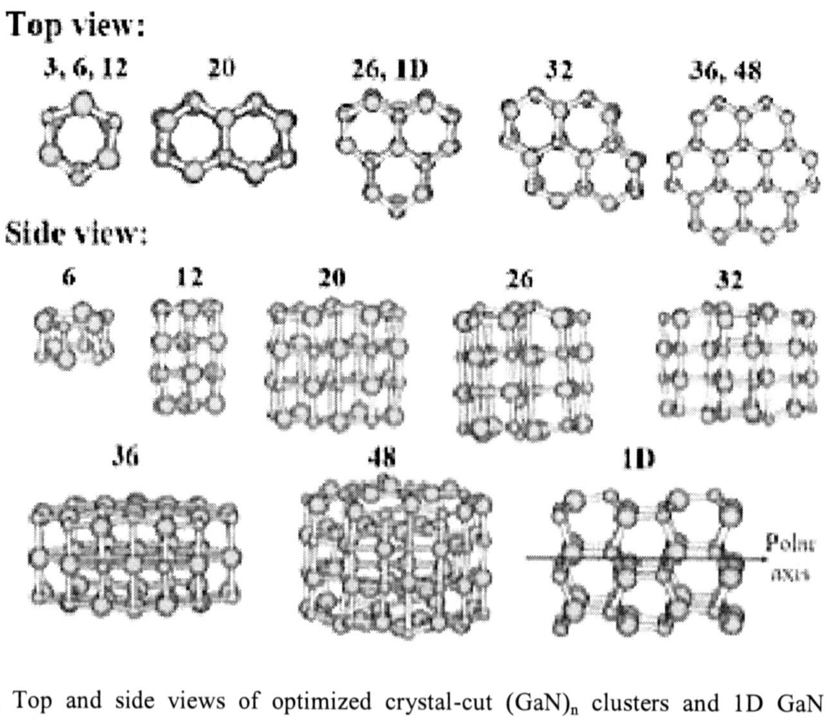

Figure 8. Top and side views of optimized crystal-cut $(GaN)_n$ clusters and 1D GaN nanowire ($(GaN)_{26}$), where the number of GaN units is labeled by *n*. The small blue (gray) and large yellow (light gray) balls refer to N and Ga atoms, respectively. Reprinted with permission from [56]. Copyright 2011, Elsevier B.V.

3. Synthesis

Over the past decade, dramatic efforts have been given in synthesizing GaN-based nanostructures. In this section, we summarize the variety of methods used in synthesizing pure and doped GaN nanostructures ranging from nanoribbons and nanomembranes to nanowires and nanotubes. Both physical and chemical methods used in synthesizing those nano materials are tabulated in Table 1.

Table 1. Physical and chemical synthesis methods of GaN nanostrucutres

	Thin films	Nanoribbons	Nanotubes	Nanowires
Physical methods	Molecular beam epitaxy (MBE) [57]	Radio frequency magnetron sputtering [22]		(MBE) [58-61]
Chemical methods		Catalytic reaction [23]	Chemical vapour deposition (CVD) [62]	Vapor-liquid-solid (VLS) [41,63]
	Metal-organic CVD [64,65]	CVD [21]	Direct reaction of Ga with NH_3 [66]	CVD [24,67-71]

3.1. Nanoribbons and Nanomembranes

The GaN nanoribbons (originally referred to as nanobelts [21]) have been synthesized on a large area of the silicon substrate by a reaction of gallium/gallium oxide and ammonia using nickel and boron oxide as catalysts [21]. The widths of the wurtzite-structured nanobelts are in the range from 100 nm to 1 μm and the thickness is about 1/10 of the width. They usually have an uneven thickness and jagged side edges that are approximately symmetric to the growth direction. It is interesting to note that the growth directions of the nanobelts are not the same, but the [001] direction is always perpendicular to the belt plane.

The substrate is an important issue that needs to be considered in the synthesis of thin films/membranes of GaN. Sapphire and SiC substrates have a large lattice and thermal mismatch with GaN. Silicon is an alternative to these substrates especially because of its low price and electrically and thermally well conducting properties. Cracks can be avoided by introducing ~10-15 nanometer thick low-temperature AlN interlayers, which can also reduce the dislocation [72]. Ultrathin AlN/GaN heterostructure crystalline porous freestanding nanomembranes are fabricated on Si(111) by selective silicon etching, and self-assembled into various geometries such as tubes, spirals, and curved sheets where the driving strain is mainly created by a so-called zipping effect due to island formation during growth [73]. Nanopores with sizes from several to tens of nanometers are produced in nanomembranes of 20-35 nm nominal thickness. The growth of AlGaN/GaN heterostructures on Si(001) by molecular-beam epitaxy (MBE) using ammonia as nitrogen precursor is also reported [74].

To get the freestanding GaN, there are alternative methods of liftoff of GaN. The reported methodologies can be classified into two categories [65]. The first is lateral chemical etching of sacrificial interlayers (CrN, Ga_2O_3, and AlInN) embedded epitaxially in the structure. The second category involves mechanical separation through the creation of a structurally "weakened" layer underneath [65].

3.2. Nanotubes

Hexagonal ZnO nanowires were used as templates in a chemical vapour deposition (CVD) system to successfully grow single-crystal GaN nanotubes with inner diameters of 30–200 nm and wall thicknesses of 5–50 nm in 2003 [62]. The ZnO nanowire templates were subsequently removed by thermal reduction and evaporation, resulting in ordered arrays of GaN nanotubes on the substrates.

As reported, it is also possible to grow large-scale GaN nanotubes (or nanowire) by direct reaction of metal gallium vapor with flowing ammonia in an 850–900 °C horizontal oven [66].

Table 2. Summary of the diameter and length of GaN nanowires synthesized by different methods

Methods	Diameter (nm)	Length (μm)
Catalyst-free metalorganic chemical vapor deposition (CVD) [67]	200	1
Metal–organic CVD process [71]	15-40	1-5
CVD [24]	4-12	
CVD [68]	50-60	Hundreds
CVD [69]	30-70	Hundreds
Chloride-based chemical vapor transport [77]	10-100	Tens
Vapor-liquid-solid (VLS) [76]	5-50	60
VLS [43]	30-150	5-30
VLS [41,63]	6-50	Several
VLS [70]	50-75	Few
Laser-assisted catalytic growth (LCG) [78]	10	10
LCG [40]	10	>1
Direct reaction of Ga with NH3 [66]	26-100	500
Plasma-assisted molecular beam epitaxy [MBE] [59]	50-100	1.9
MBE [60]	10-80	3
MBE [61]	30-350	3

3.3. Nanowires

A significant amount of work is done on the synthesis of GaN nanowires. The diameter of the nanowires ranges from 4 to 350 nm and the length of the wire reaches 500 μm as can be seen from the summary in Table 2. The first GaN nanowire was synthesized by Han and co-workers using carbon nanotubes as reaction template [75]. These nanowires have a diameter of 4 to 50 nanometers and a length of up to 25 micrometers. Later, the preparation of nanowires was reported by a catalytic vapor-liquid-solid (VLS) growth [41, 63]. In 2003, high-quality ultra-fine GaN nanowires were synthesized via the CVD method [24]. The reported nanowires have a diameter ranging from 4 to 12 nm.

It therefore appears that the most common method is based on CVD [24,67-71]. The molecular beam epitaxy (MBE) (more commonly the plasma-assisted molecular beam epitaxy [PAMBE]) method is also widely applied to the growth of GaN nanowires [59-61]. In addition, the vapor-liquid-solid (VLS) method has also found its application in the synthesis of GaN nanowires, including substrate-guided [76] and catalyst-assisted [41] growth of GaN nanowires.

4. Electronic Properties

4.1. Sheets

Unlike the direct band gap feature of the 3D wurtzite bulk GaN, all graphitic GaN sheets (with no more than six layers) are semiconductors with indirect band gaps, and the band gap smoothly decreases as the thickness increases [15,16].

The properties of the GaN nanosheets can be modified by the passivation or electric field. *Ab initio* calculations show that GaN sheet in (0001) face is a planar semiconductor with an indirect band gap of 2.17 eV [15]. Energy gap converts into a direct one and is enlarged when the sheet is modified by H and F atoms [17]. When a vertical electric field is applied to the pristine sheet, no noticeable change in the gap is observed [15]. However, the gap can be efficiently manipulated in a range from 1.8 to 3.5 eV by applying an external electric field on the H and F saturated sheets [17]. Moreover, because of the spontaneous polarization, the gap is remarkably broadened by a positive electric field while it is rapidly decreased under a negative field. This asymmetric response in the electronic structure of the hydrogenated/fluorinated sheet under an electric field may have profound applications in electronics.

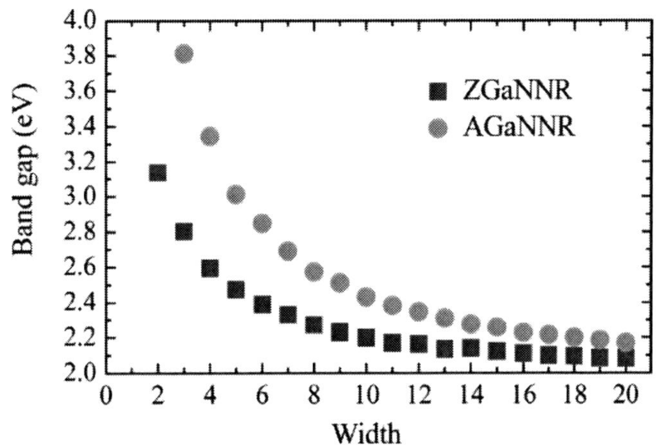

Figure 9. Variation of band gap as a function of the width for zigzag (squares) and armchair GaN nanoribbons (circles). Reprinted with permission from [35]. Copyright 2011, American Institute of Physics.

4.2. Nanoribbons

With the increasing ribbon width, both the indirect band gap of zigzag nanoribbon and the direct band gap of armchair nanoribbon decrease monotonically and become closer to each other approaching their asymptotic limit of a single layer of GaN sheet [35].

DFT calculations show that the H-terminated zigzag nanoribbons are semiconductors with an indirect band gap, which decreases monotonically [11,79] with the increased ribbon width, as can be seen in Figure 9. Dai [79] and Zheng [35] showed that band gap approaches to the gaps of their infinite two dimensional graphitic-like monolayer structures, while Li *et al.* [11] found the band gap of H-passivated zigzag nanoribbon routed through the band gap of the GaN sheet. They attributed this disagreement to the edge states, which contribute primarily to both the top of valence band and the bottom of conduction band sensitively determining the band gap value (see Figure 10). The band gap of the armchair nanoribbon is larger than that of the zigzag nanoribbon for the same ribbon width, because the armchair nanoribbon has stronger interactions between the Ga and N atoms of dimers on each side as well as between the two edges with a shorter separation distance. More interestingly, Li [11] found that the band gap of the bare armchair nanoribbon increased with the width of the Nanoribbon.

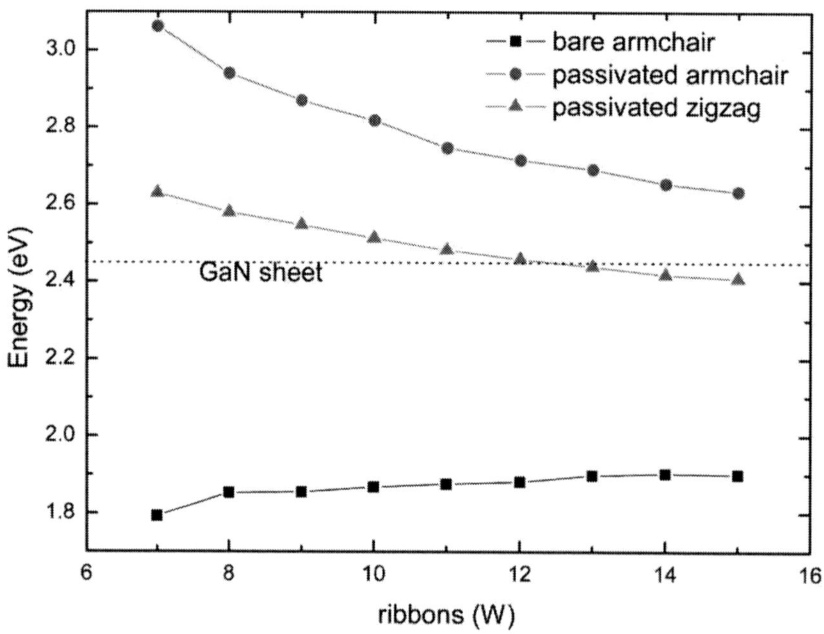

Figure 10. The band gap behavior of bare armchair, H-passivated armchair, and bare zigzag GaN nanoribbon as a function of the width. Reprinted with permission from [11]. Copyright 2010, the American Chemical Society.

Another way of tuning the electronic properties is via defects. The properties of zigzag nanoribbon with vacancies periodically located at different sites across the ribbon are studied by DFT calculations [80]. The energetic data show that the formation of the N-vacancy is easier than that of the Ga-vacancy at each equivalent geometrical site and both of them are endothermic. The vacancy leads to a more or less flat band (localized state) near the Fermi level. The N-vacancy acts as a donor, like the H that triggers various magnetic behaviors, while Ga-vacancy acts as an acceptor. The effects of a vacancy on the electronic and magnetic properties are found to depend on the vacancy positions relative to the edges. More interestingly, due to the low dimensionality and the quantum confinement effect, N-vacancy at different sites activates the local magnetic moment in GaN nanoribbons, but not bulk GaN.

4.3. Nanotubes

In 1999, Lee and co-workers performed first-principles calculations on GaN nanotubes [20]. Different nature of band structures was predicted for GaN nanotubes with different chirality. They found that zigzag nanotubes have a direct band gap, whereas the armchair nanotubes have an indirect band gap. In addition, both band gaps increase with increasing the tube diameter [20], different from the case of carbon nanotubes. The estimated band gap of the GaN nanotubes rises from a value of ~2.2 eV for a nanotube with a diameter of ~ 9 Å ((5,5) and (9,0)) to ~4.0 eV for a nanotube at infinite diameter. The large span of the band gap strongly suggests that ultimate nanoscale optoelectronic devices for full color displays may be realized with varying the diameter of these nanotubes. It is also found that more *p* states are developed by the rehybridization upon wrapping, compared with the graphitic sheet. Guo *et al.* concluded that the indirect band gap for the armchair structure could stem from band sequence change induced by curvature effect [81].

4.4. Nanowires

The DFT calculations on GaN nanowires showed that band gap decreases with the increasing diameter for saturated wires, where the dangling bonds of which at the edge of the nanowires are saturated with hydrogen. While for unsaturated wires, it remains rather constant [44]. Edge states were observed in the band gap for the unsaturated nanowires. These are removed when the dangling bonds at the edge of the nanowires are saturated with hydrogen. When N- or Ga- vacancies are involved, single vacancy prefers to form at the edge of nanowire and multi-vacancies prefer to cluster together rather than remain well separated. By comparing the formation energy of gallium and nitrogen vacancies, it is found that the nitrogen vacancies are significantly more stable than gallium vacancies, suggesting the major defect in GaN nanowire is N vacancy [44]. For unsaturated wires, the nature of the single Ga and N vacancy states is similar to those in the bulk, acting as a triple acceptor and single donor, respectively. The position of the vacancy states relative to the two regions of edge-induced states is similar to their location relative to the conduction band minimum and valence band maximum in the bulk, suggesting that any conductivity arising from the vacancies will be confined to the outer region of the nanowires. The two- and three-nitrogen vacancy complexes induce additional states in the band gap, acting as double and triple donors respectively, while the two-gallium vacancy complex reconstructs to an N_3-like structure and induces several fully occupied and unoccupied singlet states in the band gap. For saturated wires, Ga-vacancy and N-vacancy show similar features as in the GaN bulk, except that the energy levels of the defect states are slightly different from the bulk [44].

5. Magnetic Properties

Magnetic semiconductor materials are considered as very interesting topics in the materials science community due to their high potential applications in spintronics.

Nanoscale GaN offers unique physical properties due to quantum confinement and surface effects. Even pure GaN nanomaterial can have magnetic properties. Li [11] showed that bare zigzag GaN nanoribbons presented magnetic properties, and Ga and N atoms at the edges were observed as the main contributors to the magnetic moment. Multilayers of GaN nanoribbons could possess ferromagnetism even up to six layers.

The magnetic properties can also be obtained by substitution. The conventional dilute magnetic semiconductor is usually created by substituting a cation of a wide bandgap semiconductor compounds with a magnetic transition metal. Some DFT calculations on transition metals absorbing on the surface of the pristine (8,0) zigzag GaN nanotubes showed that the magnetic moment is mainly located at the transition metal [82] and Cr-, Mn-, Ni- and Cu-doped (8,0) GaN nanotubes are all half-metals [83]. Wang [38] *et al.* found that ferromagnetic coupling is energetically favorable in low density Cr-doped GaN nanotubes irrespective of the nanotubes's thickness, but changes to ferrimagnetic as the doping density of Cr increases. The ground states are ferromagnetic for the Mn-doped GaN nanotubes [84] and GaN nanowires [77]. Moreover, the ferromagnetism is isotropic and independent of the chirality and diameter of the nanotubes.

Transition metal free magnetics is, on the other hand, interesting to explore [85]. When a single N atom is substituted by one C atom in either 6-zigzag or 6-armchair GaN nanowires, there exists a magnetic moment which arises mainly from the C atom [35].

Inducing vacancy on the system is another way to obtain magnetics properties. For example, a perfect GaN sheet is nonmagnetic. However, the Ga-defective GaN sheet may induce intrinsic ferromagnetism [11]. The properties of zigzag GaN nanoribbons with periodic vacancy located at different sites across the ribbon are studied by DFT calculations [36]. Except for a typical non-edge N-vacancy, the N-, or Ga-vacancies at other sites induce magnetic moment and spin polarization. The magnetic moment of the N-vacancy is dependent on defect sites, while for the Ga-vacancy, it is less dependent on the defect sites.

Semi-decorating will lead to magnetic properties. Ferromagnetic and metallic properties of the semi-hydrogenated GaN sheet [31] are explored by DFT calculations. Among the three configurations they studied, i.e., H-GaN, GaN-H and half boat (Figure 2), only the H–GaN conformation exhibits ferromagnetic ground state with a magnetic moment of about 1.0 μB per unit cell [30,31]. In the H–GaN configurations, the magnetism is predicted to originate from the unhydrogenated N atoms because of the breaking of the orbital network due to the hydrogenation of Ga atoms [30]. However, as the hydrogen atoms are preferable to adsorb on N atoms rather than Ga atoms, it indicates that the magnetism is not easy to realize experimentally through surface hydrogenation of GaN sheet. The group of Xiao [29] also investigated semi-fluorinated GaN sheets based on spin-polarized DFT calculations. They found that the most stable conformation, fluorine atoms on top of the center of a hexagon of GaN, exhibits half-metallic behavior and ferromagnetic ground state with a magnetic moment of about 1.0 μB per fluorine. The F–GaN conformation in which the F atoms occupy the on-top positions of all the Ga sites also displays ferromagnetic ground states, whereas the GaN–F conformation is a direct band-gap antiferromagnetic semiconductor. One other group, Ma *et al.* [30] predicted that F-GaN configuration is the most stable conformation. They predicted that the ferromagnetic state is more stable than the antiferromagnetic state in the F-GaN sheet. In conclusion, surface decoration can be an effective way of tuning the magnetic properties of GaN sheets. The magnetism of semi-decorated 2D sheet arises from the combined effects of both through-bond and *p–p* direct interactions. [30]

Heterostructure of GaN with other materials may have magnetic properties too. GaN-ZnO nanoribbons are investigated by Chai [86] using DFT methods. It is indicated that the armchair nanoribbons are nonmagnetic semiconductors with band gaps similar to those of armchair GaN–ZnO nanotubes. However, the zigzag nanoribbons are magnetic metals in spite that their corresponding nanotubes are semiconductors. The magnetic moments of zigzag GaN–ZnO nanoribbons decrease as the ribbon width increases.

In conclusion, the magnetic properties of GaN nanomaterial can be obtained in some specific pure nanostructure or by chemical doping (either transition metals or light elements), defect, and passivation or in heterostructures. These discoveries on nanoscale GaN semiconductor materials with magnetism will lead to new developments in the exploration of magnetic semiconductors.

6. Optical Properties

Another interesting application of GaN is use of its optical properties. The change of structure at the nanoscale brings change in the electronic structure, thereby the optical properties. We have discussed the new features of electronic structures of GaN nanostructures in Sec. 4. Here we will discuss more about the optical spectra.

6.1. Nanotubes

Recent experiments found that the optical spectra of single-crystal GaN nanotubes exhibit a slight blue shift relative to that of the bulk GaN [19]. Such a blue shift basically results from the surface effect and/or quantum-size effect related to the thickness of the single-crystal GaN nanotubes. DFT calculations showed that two significant peaks in the ε_2 curves; these peaks originate from the surface atoms and bulk atoms, respectively. Moreover, the positions as well as the relative intensities of the ε_2 peaks depend on the thickness of the GaN nanotube, showing the size effect of single-crystal GaN nanotubes [39].

The indirect band-gap of GaN sheet makes the lowest-energy direct transitions in GaN nanotube optically forbidden (dark) [87]. The lowest-energy excitons in the nanotubes are also found to be dark. The excitation energies decrease with decreasing diameter due to a combination of nanoscopic curvature and the chemistry of Ga. The optical absorption of GaN nanotubes begins at about 3.5–4.0 eV, which is beyond the visible spectrum but can be useful for UV applications. To reduce this energy into the visible, InN or alloyed Ga/InN nanotubes may be considered [87].

6.2. Nanowires

It has been reported that 1D GaN nanowires have higher photoluminescence intensity than 2D GaN sheets due to the larger surface-to-volume ratio [67]. Both are better than the bulk material. It is also found that the luminescence spectra of the GaN nanowire array and the single GaN nanowire are sensitive to the change in temperature and structure of GaN. For the GaN nanowire array and the single GaN nanowire, the near-band-edge line position will have a blue shift with decreasing temperature, and the intensity of the luminescence spectra for the surface state of 3.4 eV will increase with decreasing temperature. However, the Y7 (the 3.21-eV line, denoted as Y7 [88]) defect line did not appear in the single GaN nanowire. Therefore, it was deduced that the source of the Y7 line came from the structural defect existing between the GaN nanowires and the Si substrate. Furthermore, the position-dependent luminescence spectra of the single GaN nanowire revealed that the surface state of the single GaN nanowire is strongly influenced by the diameter of the GaN nanowire [59]. The results reveal that the 3.2-eV peak comes from the structural defect at the interface between the GaN nanowire and Si substrate. The surface state emission of the single GaN nanowire is stronger as the diameter of the GaN nanowire becomes smaller due to an increased surface-to-volume ratio [59]. In addition, it was observed that the optical properties of the nanowires would quickly degrade at a high growth rate in the lateral direction (nanowire diameter), for which defects are easily induced. The critical diameter of defect-free GaN nanowires is estimated to be below ~140 nm under the N-rich conditions [61].

Doping was an efficient way of tuning the optical transitions. In the Mg-doped GaN nanowires grown on Si(111) substrates, a shallow Mg acceptor level was identified at ~220 meV above the valence band edge. Two emission lines related to Mg at about 3.26 and 3.18 eV are assigned to a conduction band to shallow acceptor (or e–A) transitions, and structural defect-related Mg complexes, respectively [58].

The optical properties of the anisotropic GaN nanowires depend not only on their dimensions, but on their crystallographic orientations as well. By metal–organic CVD and appropriate substrate selection, the growth direction of the nanowires can be controlled [71]. Temperature-dependent photoluminescence studies show that the emission for the $[1\bar{1}0]$ wires is blue-shifted by about 100 meV from that of [001] wires. The emission difference is considered from the different wurtzite growth directions. The polar [001] and non-polar $[1\bar{1}0]$ directions represent two orthogonal crystallographic orientations within the wurtzite GaN crystal structure [71]. In addition, it was suggested that the difference in tensile stress, defects introduced into wires when grown along different directions and the unique isosceles triangle cross section of $[1\bar{1}0]$ wire could also contribute to this emission difference.

The photoluminescence properties of GaN nanowires grown on Si(111) substrates by radio-frequency plasma-assisted MBE is studied by the Taylor group [89]. They measured the photoluminescence at 10 K and found that the main emission lines can be assigned as donor bound exciton transitions and are blue shifted, causing spectral overlap with the free exciton emission as the diameter is decreased.

7. Applications and Challenges

In the applications of GaN materials, the defects of GaN will affect the electrical and optical properties of the GaN-based devices greatly and thus affect the performance and reliability of the devices [88]. GaN-based nanostructures have potential applications in light-emitting diode (LED) devices [90]. A recent research showed that metal–oxide–semiconductor-type porous GaN diode on silicon could be formed by CVD in a single growth regime [91]. GaN nanowires are very promising in novel device applications because the geometry is optically active and highly luminescent due to high surface area to volume ratios as well as good crystal quality [58]. The achievement of p-type doping of III-nitride semiconductors was an important step in the development of optoelectronic devices such as p–n junction diodes.

Measurement on the transport property of the individual GaN nanowire field effect transistors was carried out [42,69]. Mobilities calculated from these reliable nanowire field effect transistors indicated that surface scattering plays a dominant role in smaller diameter nanowires, whereas for intermediate diameter devices transport is dominated by grain boundary scattering. Research on single crystal GaN nanowires showed that they could serve as building blocks for the assembly of devices at nanocale. The FETs made from individual n-type GaN nanowires exhibit good switching behavior with gate-dependent conductance variations exceeding three orders of magnitude [78]. The Lieber group [78] also showed that the p-n diodes can be readily assembled from n-type GaN and p-type Si nanowires and these junctions exhibit highly reproducible rectifying behavior.

With all the new development, however, there are great challenges lying ahead. The major issues that need to be tackled in the future are as follows. First of all, we need to gain a better understanding of the structure-property correlation of these low-dimensional materials, especially the size-dependence of their properties. Our current understanding is primarily from theoretical simulation under ideal conditions. And the size range being considered is often case-dependent and far from complete. In addition, it appears that there are contradictory results due to the different theoretical approaches being used. A careful validation from experiments is needed. The precise control of sizes is, however, out of the capability of current experimental techniques. Secondly, we need to have better knowledge of the effect of defects on the structure and properties of low-dimensional GaN materials. Those defects may either be unintentionally generated during the synthesis or purposely introduced to tune the electronic structure and physical properties. In either case, a clear sense of the probability of having these defects and the consequences of having them in the system is highly desired. Studies have shown that defects play extremely important roles in determining the electronic, magnetic and optical properties through newly-induced electronic states. Current theoretical studies in this respect are not comprehensive, while experimental studies are often lack of precise characterization of those defects at the atomic level. Therefore, a direct correlation between the defect and the effect on properties is still lacking. Lastly, we need to develop synthesis techniques that can highly control the structure, size and defects of the GaN low-dimensional materials. The studies of GaN nanostructures are pioneered by computational simulations and experimental synthesis are still in its infancy. It is critical to develop sustainable, low-cost, and highly controllable synthesis routes in order to put GaN low-dimensional materials in real-world applications.

References

[1] Morkoç, H. *Handbook of Nitride Semiconductors and Devices*, GaN-based Optical and Electronic Devices; John Wiley & Sons, 2009; Vol. 3.
[2] Shur, M. S.; Gaska, R.; Bykhovski, A. *Solid-State Electronics* 1999, 43, 1451.
[3] Ponce, F. A.; Bour, D. P. *Nature* 1997, 386, 351.
[4] Bonini, N.; Lazzeri, M.; Marzari, N.; Mauri, F. *Physical Review Letters* 2007, 99, 176802.
[5] Ferrari, A. C. *Solid State Communications* 2007, 143, 47.
[6] Partoens, B.; Peeters, F. *Physical Review B* 2006, 74, 075404.
[7] Karpan, V.; Giovannetti, G.; Khomyakov, P.; Talanana, M.; Starikov, A.; Zwierzycki, M.; Van Den Brink, J.; Brocks, G.; Kelly, P. *Physical Review Letters* 2007, 99, 176602.
[8] Terrones, M.; Botello-Méndez, A. R.; Campos-Delgado, J.; López-Urías, F.; Vega-Cantú, Y. I.; Rodríguez-Macías, F. J.; Elías, A. L.; Muñoz-Sandoval, E.; Cano-Márquez, A. G.; Charlier, J.-C. *Nano Today* 2010, 5, 351.
[9] Son, Y.-W.; Cohen, M. L.; Louie, S. G. *Nature* 2006, 444, 347.
[10] Kan, E.-j.; Li, Z.; Yang, J.; Hou, J. *Journal of the American Chemical Society* 2008, 130, 4224.
[11] Li, H.; Dai, J.; Li, J.; Zhang, S.; Zhou, J.; Zhang, L.; Chu, W.; Chen, D.; Zhao, H.; Yang, J. *The Journal of Physical Chemistry* C 2010, 114, 11390.
[12] Li, Y.; Zhou, Z.; Zhang, S.; Chen, Z. *Journal of the American Chemical Society* 2008, 130, 16739.
[13] Sun, L.; Li, Y.; Li, Z.; Li, Q.; Zhou, Z.; Chen, Z.; Yang, J.; Hou, J. *The Journal of Chemical Physics* 2008, 129, 174114.
[14] Botello-Méndez, A. R.; López-Urías, F.; Terrones, M.; Terrones, H. *Nano letters* 2008, 8, 1562.
[15] Xu, D.; He, H.; Pandey, R.; Karna, S. P. *Journal of Physics: Condensed Matter* 2013, 25, 345302.
[16] Tang, Q.; Cui, Y.; Li, Y.; Zhou, Z.; Chen, Z. *The Journal of Physical Chemistry C* 2011, 115, 1724.
[17] Chen, Q.; Hu, H.; Chen, X.; Wang, J. *Applied Physics Letters* 2011, 98, 053102.
[18] Liliental-Weber, Z.; Chen, Y.; Ruvimov, S.; Washburn, J. *Physical Review Letters* 1997, 79, 2835.
[19] Goldberger, J.; He, R.; Zhang, Y.; Lee, S.; Yan, H.; Choi, H.-J.; Yang, P. *Nature* 2003, 422, 599.
[20] Lee, S. M.; Lee, Y. H.; Hwang, Y. G.; Elsner, J.; Porezag, D.; Frauenheim, T. *Physical Review B* 1999, 60, 7788.
[21] Bae, S. Y.; Seo, H. W.; Park, J.; Yang, H.; Song, S. A. *Chemical Physics Letters* 2002, 365, 525.
[22] Yang, L.; Zhang, X.; Huang, R.; Zhang, G.; An, X. *Solid State Communications* 2004, 130, 769.
[23] Xiang, X.; Cao, C.; Huang, F.; Lv, R.; Zhu, H. Journal of Crystal Growth 2004, 263, 25.
[24] Chen, X.; Xu, J.; Wang, R.; Yu, D. *Advanced Materials* 2003, 15, 419.
[25] Mahmood, A.; Sansores, L. E.; Muhl, S. *Superficies y vacío* 1999, 9, 259.

[26] Freeman, C. L.; Claeyssens, F.; Allan, N. L.; Harding, J. H. *Physical Review Letters* 2006, 96, 066102.
[27] Şahin, H.; Cahangirov, S.; Topsakal, M.; Bekaroglu, E.; Akturk, E.; Senger, R. T.; Ciraci, S. *Physical Review B* 2009, 80, 155453.
[28] Li, H.; Li, J.; Wang, Z.; Zou, G. *Chemical Physics Letters* 2012, 550, 130.
[29] Xiao, W.-Z.; Wang, L.-L.; Luo, H.-J.; Yang, J.-Y. *physica status solidi* (b) 2012, 249, 1465.
[30] Ma, Y.; Dai, Y.; Guo, M.; Niu, C.; Yu, L.; Huang, B. *Applied Surface Science* 2011, 257, 7845.
[31] Xiao, W.-Z.; Wang, L.-L.; Xu, L.; Wan, Q.; Pan, A.-L.; Deng, H.-Q. *physica status solidi* (b) 2011, 248, 1442.
[32] Wang, Y.; Shi, S. *Solid State Communications* 2010, 150, 1473.
[33] Chen, Q.; Wang, J.; Zhu, L.; Wang, S.; Ding, F. *The Journal of Chemical Physics* 2010, 132.
[34] Zhang, J.; Sun, C.; Xu, K. *Sci. China Phys. Mech. Astron.* 2012, 55, 631.
[35] Zheng, F.-L.; Zhang, Y.; Zhang, J.-M.; Xu, K.-W. *Journal of Applied Physics* 2011, 109, 104313.
[36] Chen, G.-X.; Wang, D.-D.; Zhang, J.-M.; Xu, K.-W. *physica status solidi* (b) 2013, 250, 1510.
[37] Xu, B.; Lu, A.; Pan, B.; Yu, Q. *Physical Review B* 2005, 71, 125434.
[38] Wang, Q.; Sun, Q.; Jena, P.; Kawazoe, Y. *Physical Review B* 2006, 73, 205320.
[39] Xu, B.; Pan, B. Physical Review B 2006, 74, 245402.
[40] Duan, X.; Lieber, C. M. *Journal of the American Chemical Society* 2000, 122, 188.
[41] Chen, C.-C.; Yeh, C.-C.; Chen, C.-H.; Yu, M.-Y.; Liu, H.-L.; Wu, J.-J.; Chen, K.-H.; Chen, L.-C.; Peng, J.-Y.; Chen, Y.-F. *Journal of the American Chemical Society* 2001, 123, 2791.
[42] Motayed, A.; Vaudin, M.; Davydov, A. V.; Melngailis, J.; He, M.; Mohammad, S. N. *Applied Physics Letters* 2007, 90, 043104.
[43] Nam, C.-Y.; Jaroenapibal, P.; Tham, D.; Luzzi, D. E.; Evoy, S.; Fischer, J. E. *Nano letters* 2006, 6, 153.
[44] Carter, D.; Stampfl, C. Physical Review B 2009, 79, 195302.
[45] Wu, Y.; Chen, G.; Ye, H.; Zhu, Y.; Wei, S.-H. *Applied Physics Letters* 2009, 94, 253101.
[46] Yong, Y.; Song, B.; He, P. *Physical Chemistry Chemical Physics* 2011, 13, 16182.
[47] Costales, A.; Pandey, R. *The Journal of Physical Chemistry A* 2003, 107, 191.
[48] Costales, A.; Kandalam, A. K.; Pandey, R. *The Journal of Physical Chemistry B* 2003, 107, 4508.
[49] Costales, A.; Blanco, M. A.; Martín Pendás, Á.; Kandalam, A. K.; Pandey, R. *Journal of the American Chemical Society* 2002, 124, 4116.
[50] Kandalam, A. K.; Blanco, M.; Pandey, R. *The Journal of Physical Chemistry B* 2001, 105, 6080.
[51] Kandalam, A. K.; Blanco, M. A.; Pandey, R. *The Journal of Physical Chemistry B* 2002, 106, 1945.
[52] Kandalam, A. K.; Pandey, R.; Blanco, M.; Costales, A.; Recio, J.; Newsam, J. M. *The Journal of Physical Chemistry B* 2000, 104, 4361.

[53] Costales, A.; Kandalam, A. K.; Martín Pendás, A.; Blanco, M.; Recio, J.; Pandey, R. *The Journal of Physical Chemistry B* 2000, 104, 4368.
[54] Zerenturk, A.; Berber, S. *EPL (Europhysics Letters)* 2013, 103, 16003.
[55] Kumar, V.; Zavada, J. M. *Proc. of SPIE* 2010, 7602, 76020E.
[56] Yu, H.-G. *Chemical Physics Letters* 2011, 512, 231.
[57] Takagaki, Y.; Sun, Y. J.; Brandt, O.; Ploog, K. H. *Applied Physics Letters* 2004, 84, 4756.
[58] Park, Y. S.; Na, J. H.; Taylor, R. A.; Park, C. M.; Lee, K. H.; Kang, T. W. *Nanotechnology* 2006, 17, 913.
[59] Tsai, C.-C.; Li, G.-H.; Lin, Y.-T.; Chang, C.-W.; Wadekar, P.; Chen, Q.; Rigutti, L.; Tchernycheva, M.; Julien, F.; Tu, L.-W. *Nanoscale Research Letters* 2011, 6, 631.
[60] Tu, L. W.; Hsiao, C. L.; Chi, T. W.; Lo, I.; Hsieh, K. Y. *Applied Physics Letters* 2003, 82, 1601.
[61] Park, C. M.; Park, Y. S.; Im, H.; Kang, T. W. *Nanotechnology* 2006, 17, 952.
[62] Kominis, I. K.; Kornack, T. W.; Allred, J. C.; Romalis, M. V. *Nature* 2003, 422, 596.
[63] Chen, C. C.; Yeh, C. C. *Advanced Materials* 2000, 12, 738.
[64] Jang, L.-W.; Jeon, D.-W.; Polyakov, A. Y.; Cho, H.-S.; Yun, J.-H.; Jo, D.-S.; Ju, J.-W.; Baek, J.-H.; Lee, I.-H. *Applied Physics Express* 2013, 6, 061001.
[65] Zhang, Y.; Leung, B.; Han, J. *Applied Physics Letters* 2012, 100, 181908.
[66] He, M.; Minus, I.; Zhou, P.; Mohammed, S. N.; Halpern, J. B.; Jacobs, R.; Sarney, W. L.; Salamanca-Riba, L.; Vispute, R. *Applied Physics Letters* 2000, 77, 3731.
[67] Zang, K.; Chua, S. J. *physica status solidi* (c) 2010, 7, 2236.
[68] Lyu, S.; Cha, O.; Suh, E.-K.; Ruh, H.; Lee, H.; Lee, C. *Chemical Physics Letters* 2003, 367, 136.
[69] Kim, J.-R.; So, H. M.; Park, J. W.; Kim, J.-J.; Kim, J.; Lee, C. J.; Lyu, S. C. *Applied Physics Letters* 2002, 80, 3548.
[70] Madapu, K. K.; Dhara, S.; Amirthapandian, S.; Polaki, S.; Tyagi, A. K. *The Journal of Physical Chemistry C* 2013, 117, 21930.
[71] Kuykendall, T.; Pauzauskie, P. J.; Zhang, Y.; Goldberger, J.; Sirbuly, D.; Denlinger, J.; Yang, P. *Nature materials* 2004, 3, 524.
[72] Dadgar, A.; Poschenrieder, M.; Biasing, J.; Contreras, O.; Bertram, F.; Riemann, T.; Kunze, A. R. M.; Daumiller, A.; Diez, A.; Kaluza, A. In State-of-the-Art Program on Compound Semiconductors XXXVII (SOTAPOCS XXXVII), and Narrow Bandgap Optoelectronic Materials and Devices: *Proceedings of the International Symposia*; The Electrochemical Society: 2002; Vol. 2002, p 300.
[73] Mei, Y.; Thurmer, D. J.; Deneke, C.; Kiravittaya, S.; Chen, Y.-F.; Dadgar, A.; Bertram, F.; Bastek, B.; Krost, A.; Christen, J. r.; Reindl, T.; Stoffel, M.; Coric, E.; Schmidt, O. G. *ACS Nano* 2009, 3, 1663.
[74] Joblot, S.; Semond, F.; Cordier, Y.; Lorenzini, P.; Massies, J. *Applied Physics Letters* 2005, 87, 133505.
[75] Han, W. Science 1997, 277, 1287.
[76] Tsivion, D.; Joselevich, E. *Nano letters* 2013, 13, 5491.
[77] Choi, H. J.; Seong, H. K.; Chang, J.; Lee, K. I.; Park, Y. J.; Kim, J. J.; Lee, S. K.; He, R.; Kuykendall, T.; Yang, P. *Advanced Materials* 2005, 17, 1351.
[78] Huang, Y.; Duan, X.; Cui, Y.; Lieber, C. M. *Nano letters* 2002, 2, 101.
[79] Dai, Y.; Chen, X.; Jiang, C. *Physica B: Condensed Matter* 2012, 407, 515.

[80] Chen, G. X.; Wang, D. D.; Zhang, J. M.; Xu, K. W. *Physica status solidi* (b) 2013, 250, 1510.
[81] Guo, Y.; Yan, X.; Yang, Y. *Physics Letters A* 2009, 373, 367.
[82] Chen, G.-X.; Zhang, Y.; Wang, D.-D.; Zhang, J.-M. *Physica E: Low-dimensional Systems and Nanostructures* 2010, 43, 22.
[83] Chen, G.-X.; Zhang, Y.; Wang, D.-D.; Zhang, J.-M.; Xu, K.-W. *Solid State Communications* 2011, 151, 139.
[84] Guo, Y.; Chen, M.; Guo, Z.; Yan, X. *Physics Letters A* 2008, 372, 2688.
[85] Yang, K.; Wu, R.; Shen, L.; Feng, Y. P.; Dai, Y.; Huang, B. *Physical Review B* 2010, 81, 125211.
[86] Chai, G.-L.; Lin, C.-S.; Cheng, W.-D. *Journal of Materials Chemistry* 2012, 22, 7708.
[87] Ismail-Beigi, S. *Physical Review B* 2008, 77, 035306.
[88] Reshchikov, M. A.; Morkoç, H. *Journal of Applied Physics* 2005, 97, 061301.
[89] Park, Y. S.; Kang, T. W.; Taylor, R. A. *Nanotechnology* 2008, 19, 475402.
[90] Kang, M. S.; Lee, C.-H.; Park, J. B.; Yoo, H.; Yi, G.-C. *Nano Energy* 2012, 1, 391.
[91] Bilousov, O. V.; Carvajal, J. J.; Vilalta-Clemente, A.; Ruterana, P.; Díaz, F.; Aguiló, M.; O'Dwyer, C. *Chemistry of Materials* 2014, 26, 1243.

In: Gallium Nitride
Editor: Kiera Olivia Peak

ISBN: 978-1-63321-387-6

Chapter 4

WURTZITE GALLIUM NITRIDE NANOSTRUCTURES: SYNTHESES AND GROWTH MECHANISMS

***Yucheng Lan*[1,*], *Jianye Li*[2] *and Zhifeng Ren*[3]**
[1]Department of Physics and Engineering Physics,
Morgan State University, Baltimore, Maryland, US
[2]Department of Materials Science and Engineering,
University of Wisconsin - Madison, Madison, Wisconsin, US
[3]Texas Center for Superconductivity and Department of Physics,
University of Houston, Houston, Texas, US

Abstract

Gallium nitride is a kind of well-known semiconductor with a direct and wide band-gap of 3.4 eV. It has wide applications in white light-emitting diodes, blue lasers, and field effect transistors because of excellent optoelectronic properties, high mobility, and super thermal stability. The nitride nanomaterials inherit the excellent bulk properties while possess novel properties of nanomaterials, such as quantum effects, being the promising nanostructures for future applications on renewable energy. In the chapter we first introduce various GaN nanostructures, including syntheses and growth mechanisms. Then several popular characterization techniques are introduced. The future outlook of the GaN nanostructures is talked about at the end of the chapter.

PACS: 81.05.Ea, 81.07.-b, 81.07.Bc, 81.16.-c, 77.84.Bw

Keywords: III-V semiconductors, nanoscale materials and structures: fabrication and characterization, nanocrystalline materials, methods of micro- and nanofabrication and processing, gallium nitride

1. Introduction

Gallium nitride (GaN) based semiconductors have attracted great attentions for application on light sources of short wavelength and high-speed/high-power electron devices. [1–4]

*E-mail address: yucheng.lan@morgan.edu

GaN is a binary III/V direct band-gap semiconductor. Wurtize GaN has a space group of $P6_3mc$ with lattice constants $a = 3.1891(1)$ Å and $c = 5.1853(3)$ Å. [5] Its wide band gap of 3.4 eV [6,7] affords the nitride special properties for applications in optoelectronic, high-power and high-frequency devices, high-temperature microelectronic devices, and in bright lighting sources since the 1990s. For example, hexagonal GaN films have been fabricated as blue light emitting diodes and UV laser diodes [3, 4, 8, 9] because of its special emission [10, 11]. The alloys of GaN with In (InGaN) or Al (AlGaN) allows the manufacture of light-emitting diodes with colors that can go from red to ultra-violet. GaN is also the material which makes violet (about 405 nm) laser diodes possible, without use of nonlinear optical frequency-doubling. The first gallium nitride metal semiconductor field-effect transistors were experimentally demonstrated in 1993 [12] and the first enhancement mode gallium nitride transistors became generally available in 2010. High-speed field-effect transistors and high-temperature microelectronic devices were also developed using the material. [2, 4, 13] GaN high-electron-mobility transistors have been offered commercially since 2006, due to their high efficiency and high voltage operation. GaN materials have also been employed as thermoelectric devices [14–17] to harvest waste heat and as betavoltaic microbatteries [18–20] to collect energy from radioactive sources emitting beta particles. More applications of GaN materials can be found in some review papers. [1–4]

To date, GaN nanoparticles, [21] nanorods, [22, 23] nanowires, [24–26] nanotubes, [27–30] as well as nanocomposites [31, 32] have been synthesized using a variety of techniques. These nanomaterials have been gained considerable efforts due to their potential applications in both mesoscopic research and future development of nanodevices.

Compared with GaN single crystal bulks and films, GaN nanomaterials usually show tuning optical properties [33] and other more interesting behaviors, having more wider optoelectronics applications. [34] Here we discuss the syntheses, structures, and growth mechanisms of various GaN nanomaterials.

The quantum confinement of these nanomaterials yields unique physical properties over bulk crystals, making the GaN nanomaterials wider applications. Now GaN nanomaterials have been applied on photocatalytic water splitting, [35–44] photovoltaic devices, [45–47] and piezoelectric nanogenerators [48–52] besides light emission devices and laser devices.

2. GaN Nanowires

One-dimensional semiconductor nanostructures such as nanowires and nanotubes have stimulated their functional applications and are emerging as building blocks for next generation of electronic and optoelectronic nanodevices due to their unique physical and chemical properties. As a particularly important III-V semiconductor with a direct band gap of 3.4 eV at room temperature, GaN nanowires have been widely studied in the past decade. The successful syntheses of GaN nanowires have been demonstrated via various methods including carbon nanotube-confined reaction, [53] arc discharge, [54] sublimation, [22] pyrolysis, [55] chemical vapor deposition, [24, 53, 56–64], laser ablation, [65, 66] metal-organic chemical vapor deposition, [25, 67–70] molecular beam epitaxy, [71–73] and hydride vapor phase epitaxy [61, 74]. Researchers have proposed various possible growth mechanisms of GaN nanowires. The applications of GaN nanowires were focused on the research of photovoltaic devices, field effect transistors, light emitting diodes, lasers, photocatalysts and nanogenerators. In this sub-chapter, we will introduce the growth methods, growth mechanisms, characterizations and applications of GaN nanowires.

2.1. Chemical Vapor Deposition

Chemical vapor deposition (CVD) is a chemical process often used to produce high-purity, high-performance solid materials. The CVD method has been widely used by researchers to synthesize nanowires because of great achievement, simple design and low cost. In the past decade, GaN nanowires have been prepared by many groups using the technique. Recently GaN nanowires with a mean diameter of $6-8$ nm were synthesized by the CVD method. [58,63]

Synthesizing GaN nanowires using CVD method was first demonstrated by Han et al. [53] In their work, carbon nanotubes (CNTs) were used as templates to confine the reaction in the CNTs, leading to the GaN nanowire morphology complementary to that of the carbon nanotubes. A 4:1 molar mixture of Ga-Ga_2O_3 powders were used as precursor of Ga_2O. The reaction in the GaN nanowires growth process was carried out at high temperatures in a conventional furnace with a horizontal tube and can be expressed as

$$2Ga_2O(g) + C(CNT) + 4NH_3(g) \rightarrow 4GaN(NW) + H_2O(g) + CO(g) + 5H_2(g) \quad (1)$$

High-resolution transmission electron microscopy (HRTEM) images showed that the grown GaN nanowires had a diameter similar to that of the original carbon nanotubes. Besides CNT templates, other templates such as GaAs nano-columns, [56] and nanosized-pore Al_2O_3 templates [57] as well were used to prepare GaN nanowires.

Chen et al. reported an approach of synthesizing straight and smooth GaN nanowires via NiO catalyzed CVD technique. [24] The nanowires were deposited onto single crystal $LaAlO_3$ substrates without using any templates. Before the deposition, the substrates were quickly dipped into $Ni(NO_3)_2$ ethanol solutions and then heated at 900 °C in a long quartz tube in a stream of Ar for 2 hours to decompose $Ni(NO_3)_2$ into NiO nanoparticles. Then GaN nanowires were grown in a long quartz tube at $920-940$ °C in a flow of NH_3 for $5-20$ min. The synthesized GaN nanowires had diameters of $10-40$ nm and a maximum length of about 500 μm.

Johnson et al. [62] also demonstrated that long GaN nanowires can be synthesized on sapphire substrates via the CVD method. Using gallium metal and ammonia as sources of Ga and N, GaN nanowires were redeposited on sapphire substrates at 900 °C using nickel catalysts. The produced wires had lengths up to several hundred micrometers and diameters between 30 and 150 nm.

As mentioned above, the growth of GaN nanowires using template-based CVD or synthesizing on expensive single crystal substrates can only get little amounts of GaN nanowires. Chen et al. [75] described a simple method for large-scale production of GaN nanowires on silicon wafers by reacting gallium and ammonia using polycrystalline indium powders as catalysts. In the report, molten gallium and powdered indium in toluene were placed on the Si substrates. After evaporating toluene, the substrates were transferred into a quartz tube in a furnace and kept temperatures at 910 °C for 12 hours under a flow of ammonia. After the reaction, large quantities of GaN nanowires with diameters ranged from 20 to 50 nm and lengths up to several micrometers were synthesized.

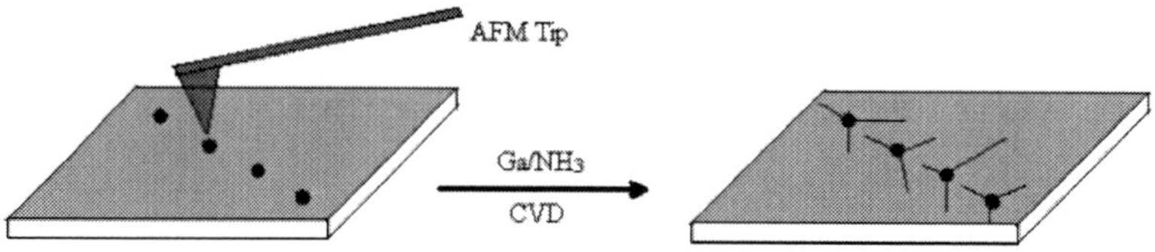

Figure 1. Schematic of the dip-pen nanolithographic process for the patterning of catalysts and growth of GaN nanowires. Reprinted with permission from Ref. [64]. Copyright 2004 American Chemical Society.

In the preparation of nanoelectronic devices, it is very important using nanowires with low defect density and long length. However, the processing, dispersal, and patterning techniques, such as ultrasonication and lithography, could introduce defects in nanowires. To resolve this problem, Li et al. [64] described a method to grow nanowires suitable for *in situ* fabrication of nanowire devices. In the report, a dip-pen nanolithographic technique was developed. A contact-mode atomic force microscopy (AFM) tip was dip-coated with $Ni(NO_3)_2$ solution and was brought into contacts with a substrate to precisely deliver a small amount of $Ni(NO_3)_2$ as demonstrated in a sketch shown in Figure 1. The position of catalyst islands was controlled by moving the tip from one location to another location using a piezoelectric scanner. Then, GaN nanowires were synthesized in a horizontal quartz tube inside a tube furnace with a flow of ammonia using elemental gallium and catalyst-patterned substrates.

2.2. Laser Ablation

Laser ablation is a laser-assisted method and the key feature of this technique is that the catalyst used to define nanowire vapor-liquid-solid (VLS) growth can be selected from phase diagram data and/or knowledge of chemical reactivity. [65] In the method, a pulsed laser is used to vaporize a solid target containing desired materials and catalysts at high temperatures. The resulting liquid nanoclusters directed the growth and defined the diameter of crystalline nanowires through a VLS growth mechanism. [65, 76–78] Single-crystalline semiconductors were successfully synthesized using the laser ablation method. [65, 76, 77]

Duan et al. described a laser ablation process in preparing single crystalline GaN nanowires. [65] A GaN/Fe (GaN: Fe = $95{:}5$ atom%) composite target was positioned within a quartz tube at the center of a furnace. A pulsed Nd:YAG laser (1064 nm, 8 ns pulse width, 10 Hz repetition, 2.5 W average power) was used to ablate the GaN/Fe target and the resulting Fe nanoclusters formed at elevated temperatures. The Fe nanoclusters directed the growth direction and defined the diameter of GaN nanowires as shown in Figure 2. The grown GaN nanowires deposited at the end of the inner quartz tube wall had diameters on the orders of 10 nm and lengths greatly exceeding 1 μm. [65]

Using the laser ablation method, Ng et al. synthesized GaN nanowires [66] with lengths from 300 to 500 nm and diameters around 50 nm. In the report, a KrF (wavelength of 248 nm and pulse width of 23 nm) excimer laser was used at laser intensity of 2.2 J/cm^2 to ablate a GaN target and the produced plasmatic ions were delivered directly to sapphire

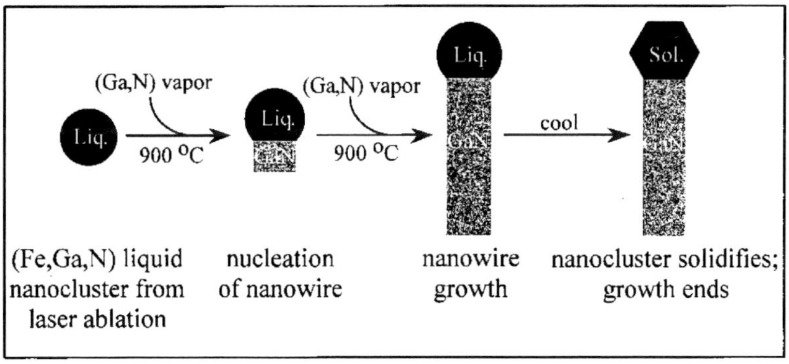

Figure 2. Overall evolution of nanowire growth following the generation of the catalytic nanocluster. Reprinted with permission from Ref. [65]. Copyright 2000 American Chemical Society.

substrates coated with 10 nm of Au. Then GaN nanowires grew on the substrates. The grown GaN nanowires had tips made up of the Au/Ga eutectic, confirming the VLS growth mechanism.

2.3. Metal-Organic Chemical Vapor Deposition

Metal-organic chemical vapor deposition (MOCVD) is a chemical vapour deposition method of epitaxial growth of materials from surface reaction of organic compounds, metalorganics and hydrides containing the required chemical elements. Alternative names for this method include metal-organic vapor phase epitaxy (MOVPE), organometallic vapor phase epitaxy (OMVPE) and organometallic chemical vapor deposition (OMCVD). In the past decade, MOCVD technique has been successfully used to synthesize GaN nanowires both by metal-catalyzed [25, 67, 68] and non-catalyzed procedures. [69, 70]

Yang et al. synthesized high-quality GaN nanowires in 2003 via metal-initiated MOCVD method. [25] Trimethylgallium (TMG) and ammonia source materials were used as Ga and N precursors in the nanowire growth. Silicon substrates, c-plane sapphire substrates, or a-plane sapphire substrates were thermally coated with Ni, Fe, or Au thin films and used as substrates for growth of GaN nanowires. The nanowires growth occurred at a substrate temperature of $800 - 1000$ °C in the form of vapor-liquid-solid growth mechanism. The reaction was carried out in an oxygen-free environment at atmospheric pressure. Nitrogen was used as a carrier gas and percolated through the TMG precursor and coupled with a second nitrogen line. The grown GaN nanowires on gold-coated c-plane sapphire substrates had diameters of $15 - 100$ nm and lengths of $1 - 5$ μm. It was observed that there were preferred nanowire orientations on the substrates, indicating the epitaxial growth of nanowires on sapphire substrates. [25]

Su et al. [67] reported flexible synthesis of GaN nanowires and nanostructures by MOCVD via a catalytic VLS growth mechanism. In this report, sapphire, alumina, and SiO_2 were used as substrates during the growth. The substrates were dipped in 0.01 M $Ni(NO_3)_2$ aqueous solutions and dried in air on filter papers before loading into an MOCVD reactor. It was noted that the anisotropic one-dimensional growth depends sensitively on the growth stoichiometry, or V/III ratio in the MOCVD environment. [67]

Hersee et al. demonstrated [70] a scalable process for the growth of high-quality GaN nanowires and uniform nanowire arrays in which the position and diameter of each nanowire was precisely controlled. The approach was based on a conventional MOCVD using regular precursors without additional metal catalyst. In their report, a 30 nm silicon nitride growth-mask layer was deposited by low-pressure chemical vapor deposition (LPCVD) onto 600 nm GaN films that had been previously grown by MOCVD on sapphire, silicon carbide, and silicon (111) substrates. The GaN growth carried out at 1050 °C and a pressure of 100 torr with a hydrogen/nitrogen carrier gas mixture in an MOCVD reactor. GaN nanowires array emerged from the growth mask and the diameter of each nanowire remained constant during the growth process. Figure 3 shows SEM images of a GaN nanowire array. [70] Lower magnification SEM images revealed a long-range order of the GaN nanowire arrays.

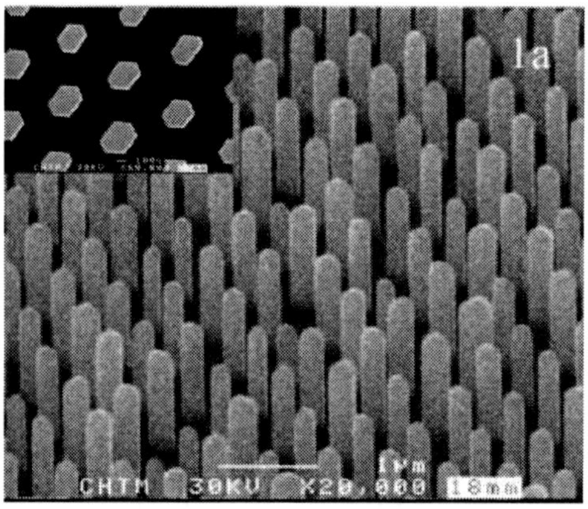

Figure 3. Scanning electron micrograph of a GaN nanowire array consisting of 1 μm GaN nanowires (inset shows plan view and reveals the hexagonal symmetry of the nanowires). Reprinted with permission from Ref. [70]. Copyright 2006 American Chemical Society.

2.4. Molecular Beam Epitaxy

Molecular beam epitaxy (MBE) was invented in the late 1960s to grow single crystals. [79] Now the MBE method is a useful technique to synthesize GaN nanowires. GaN nanowires with extremely good crystal quality have already been synthesized by the MBE method on different substrates.

Using plasma-assisted molecular beam epitaxy method, Calarco et al. prepared GaN nanowires on Si (111) substrates. [80] Their results indicated that the GaN nanowire site-density increased rapidly with growth time and then saturated. Bertness et al. analyzed the spontaneous growth of c-axis GaN nanowires in MBE procedures under a number of different growth parameters, including relative flux of N species to Ga, type of N species, AlN buffer layer thickness, and substrate orientation. They concluded that the nucleation mechanism for GaN nanowires included 1) the formation of nanocolumns in AlN buffer layers, and 2) the propagation of the GaN nanowires. The GaN growth appeared to be driven by different growth rates of different crystallographic planes under N-rich conditions. [71]

Tchernycheva et al. [72] studied the growth of free-stranding GaN nanowires synthesized in catalyst-free mode on Si (111) substrates by plasma-assisted molecular beam epitaxy. Cylindrical nanowires with a hexagonal cross-section defined by $\{10\bar{1}0\}$ planes and diameters down to 20 nm were observed. The growth rate in the lateral direction was studied and the result showed that the lateral over axial growth rate ratio can be tuned from $\sim 1\ \%$ to $\sim 10\ \%$ by changing the III/V flux ratio. Mata et al. also investigated [73] the growth of GaN nanowires on Si (111) substrates during a plasma assisted molecular beam epitaxy procedure. It was concluded that the site-density, diameter, length and length dispersion of GaN nanowires were strongly dependent on growth temperatures, [73] through statistically analyzing SEM images of GaN nanowires grown at different temperatures.

2.5. Hydride Vapor Phase Epitaxy

Hydride vapour phase epitaxy (HVPE) is an epitaxial growth technique often employed to synthesize semiconductors such as GaN, GaAs, InP and their related compounds. Ammonia, hydrogen and various chlorides are commonly used as carrier gases in HVPE processes. HVPE provides a better process control than the commonly used CVD techniques and is of particular interest due to its high growth rates and simplicity of basic reactor design.

Kim et al. [61] proposed a simple approach for a controllable growth of GaN nanorods by the hydride vapor phase epitaxy technique. In their growth process, GaCl precursor was formed by a reaction of Ga metal and HCl gas. Then GaCl was transported to substrate surfaces and mixed with NH_3 to form GaN. GaN nanorods were directly grown on bare sapphire or silicon substrates without a catalyst or template layer. After the growth, a high density of straight and well-aligned GaN nanorods with diameters in the range of $80 - 120$ nm was formed uniformly over the entire substrates.

Seryogin et al. [74] demonstrated a catalytic growth of GaN nanowires by the HVPE method. In the work, nickel-gold was used as catalysts and the nanowire growth was limited to catalyst-patterned areas. The growth of GaN nanowires was carried out in a custom designed HVPE rector at atmospheric pressure. The reactor consisted of a 75 mm quartz tube and placed in a three-zone Mellen horizontal furnace. Two heating zones were used for hydride reactions of metals with HCl gas. Only gallium was used as the source metal to form GaCl in a direct reaction between molten Ga and HCl gas diluted with nitrogen. The chemical reaction can be expressed as

$$\mathrm{Ga(l) + HCl(g) \rightarrow GaCl(g) + 1/2H_2(g)} \tag{2}$$

The third zone was the growth zone and ammonia was delivered to this zone to react with GaCl gas. GaN nanowires were formed via the HVPE process on catalyst-patterned silicon (111) or c-plane sapphire substrates:

$$\mathrm{GaCl(g) + NH_3(g) \rightarrow GaN(NW) + HCl(g) + H_2(g)} \tag{3}$$

2.6. Growth Mechanism of GaN Nanowires

The growth of GaN nanowires is a process involving the anisotropic growth. To obtain anisotropic growth of GaN materials many methods have been adopted, which can be roughly cataloged into two kinds, vapor methods and solution methods, leading to vapor based mechanisms and solution based mechanisms respectively. While due to the nonbeing of solution methods for GaN nanowires by now, only vapor mechanisms have been suggested. The vapor based mechanisms include vapor-liquid-solid (VLS), vapor-solid (VS), and vapor-solid-solid (VSS) procedures. The solution based mechanisms include solution-liquid-solid (SLS) and solution-solid (SS) procedures and will not be discussed here.

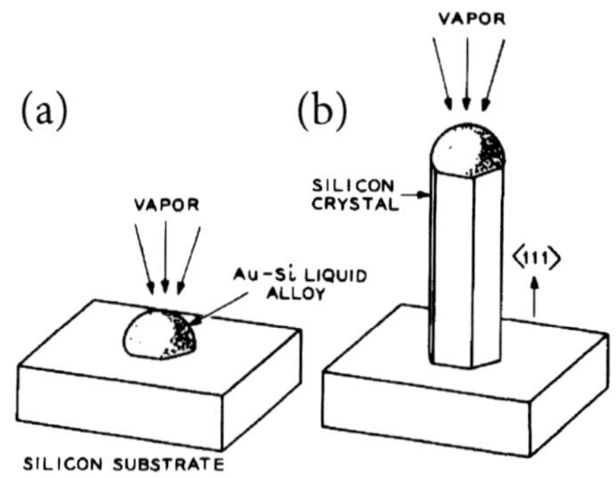

Figure 4. Growth of a silicon nanowire by VLS mechanism. (a) Initial condition with catalytic liquid droplet on substrates. (b) Growing nanowire with liquid droplet at the tip. Reprinted with permission from Ref. [81]. Copyright 1964 American Institute of Physics.

VLS mechanism

The VLS mechanism, shown in Figure 4, is the most widely accepted mechanism first described by Wagner and Ellis in 1964 [81] based on the appearance of Au nanoparticles on the top of crystalline silicon nanowires. Here vapor (V), liquid (L), solid (S) stand for reactant vapor, liquid metal catalyst, and solid nanowire product respectively.

A few key factors should be seriously considered in this mechanism, such as vapor supersaturation, selection of liquid, liquid-substrate interfaces, liquid-solid interfaces, and solid-substrate interfaces. First the vapor supersaturation should be high enough to dissolve the vaporous reactant into the liquid metal catalyst to form a supersaturated alloy and then a nanowire seed can crystallize out of the liquid metal alloy. The control of the vapor supersaturation can be realized by adjusting pressures, temperatures, and raw materials, etc. Secondly the metal catalyst should be seriously considered, generally based on the alloy phase diagram and chemical reactivity. [65] Thirdly the interfaces between the liquid metal catalyst and substrates should also be seriously considered. Because the wettability between them dominates the spatial distribution of the liquid metal catalyst over the substrates, [82] which can be a continuous film, a porous film, a connected network, or isolated islands. The morphology of the grown nanowires can be seriously affected by these key factors.

The nanowire seed maybe crystallize from the surface of the liquid metal, interfaces between the liquid metal and substrates, or from somewhere else. The most common region is the interface between the liquid metal and substrates or the interface between the liquid metal catalyst and the solid nanowires. All these liquid-solid interfaces will lead the epitaxial growth of GaN nanowires. The interface between the solid nanowires and substrates is also an important interface. The crystal nucleus, lattice match, lattice mismatch/dismatch usually happen at the solid-solid interfaces. Therefore the lattice matching and surface atom condition of substrate should be adjusted to satisfy the epitaxial growth of nanowires.

VLS growth occurs in many GaN nanowire growth procedures, such as in laser ablation growth and MOCVD growth. Duan et al. synthesized single crystalline GaN nanowires [65] using laser-assisted catalytic method. A GaN/Fe composite target was ablated by a pulsed Nd:YAG laser and the resulting vapor containing Ga, N, Fe formed a miscible liquid phase. With the saturation of Ga and N in the liquid phase, GaN precipitated and crystallized from the saturated liquid phase. GaN nanowires grew in the crystallization directed by the Ga, N, Fe liquid nanoclusters through a VLS growth mechanism. Yang et al. reported [25] a VLS growth of GaN nanowires by MOCVD technique. Seryogin et al. demonstrated [74] a VLS growth of GaN nanowires catalyzed by nickel-gold using HVPE growth method.

The VLS mechanism was confirmed by high temperature transmission electron microscopy. Stach et al. directly observed [83] the growth of GaN nanowires at high temperatures. Ga liquid droplets and gallium/nitrogen vapor were produced by a thermal decomposition of GaN in a vacuum. Then a liquid-Ga/solid-GaN interface was formed and GaN supersaturated in the Ga liquid, initiating a VLS growth of GaN nanowires.

VSS mechanism

VSS mechanism is an analogous mechanism to VLS with the only difference of catalyst states. In VSS procedures the catalyst is in solid state while in VLS procedures the catalyst is in liquid-state. Here V (vapor), S (solid), S (solid) stand for reactant vapor, solid metal catalyst, and solid nanowire product respectively. The VSS mechanism was first proposed in 2000 by Kamins et al. for Ti-catalyzed silicon nanowires. [84]

VSS growth is common in GaN nanowire syntheses. Hou et al. first reported a Au-assisted GaN nanowire growth [85] by the VSS mechanism. The control of the migration of Au seeds was controlled and detached during the growth of GaN nanowires in a plasma-enhanced chemical vapor deposition system. It was observed that at a low gallium partial pressure, the GaN nanowire growth was driven by a VSS process, with the Au migration rate dramatically reduced due to the slow diffusion of Au in the solid Au-Ga alloy. Weng et al. reported a growth of GaN nanowires using Ni-assisted meta-organic chemical vapor deposition method. [86] As described in the report, *c*-plane sapphires with sputtered Ni thin films were used as substrates and trimethylgallium (TMG) and ammonia as the group III and V precursors respectively. During the reacting process, the nickel thin film transformed into Ni_3Ga nanoparticles and the nanoparticles kept capturing Ga species and became supersaturated. Subsequently, GaN nanowires were formed by a reaction of extra Ga and the decomposed ammonia at the surface of the Ni_3Ga through the VSS mechanism.

VS mechanism

VS is a totally different mechanism from VLS and VSS due to the absence of catalyst. Here V (vapor) and S (solid) stand for the reactant vapor and the solid nanowire product

Figure 5. SEM images of (a) GaN nanorods at low-magnification, displaying the uniform rodlike morphology and (b) an individual rod with well-faceted outer surfaces and a pyramidal end. Reprinted with permission from Ref. [89]. Copyright 2007 Springer.

respectively. The VS mechanism was observed in the growth of Ga-based one-dimensional nanomaterials. [87, 88]

In general the growth temperature of VS procedures is higher than VLS and VSS procedures due to the absence of energetically favored catalyst sites. However the VS growth has the advantage of the VLS and VSS growth: no catalyst addition. Therefore the catalyst pollution is avoided in VS procedures. In principle one dimensional nanomaterial of any material can be grown by the VS mechanism as long as suitable supersaturation and nucleation sites are provided. [85]

The VS procedure is much simpler than VLS and VSS procedures because many interfaces are avoided and no catalyst is employed. Only three kinds of interfaces are considered in VS procedures, vapor-substrate interfaces, vapor-solid interfaces, and solid-substrate interfaces. Vapor-substrate interfaces should be first considered because GaN nanowires nucleate here. The vapor condenses into solid to form nucleus in the vapor-substrate interfaces. So any parameters provided for suitable nucleus sites should be carefully thought over, such as deposition temperatures, pressures, substrate surface conditions, etc. After the key nucleation step finished, two more interfaces are followed, vapor-solid interfaces and solid-substrate interfaces. The vapor-solid interfaces are new and different interfaces lying between reactant vapor and solid nanowire nucleus coming from the vapor-substrate interfaces. The nucleation does not guarantee any epitaxy nanowire growth. If the reactant vapor can not condense into nucleus or condensation does not lead to one directional epitaxy, only nanoparticles would be obtained. The solid-substate interfaces are also important. Crystal nucleus, lattice match, mismatch/dismatch always occur here.

3. GaN Nanorods

Very similar to GaN nanowires, GaN nanorods with a low length/diameter ratio have also been synthesized using the CVD method and other methods. Figure 5 shows typical SEM images of GaN nanorods.

GaN nanorods can be synthesized from Ga_2O_3 and ammonia from a CVD reaction.

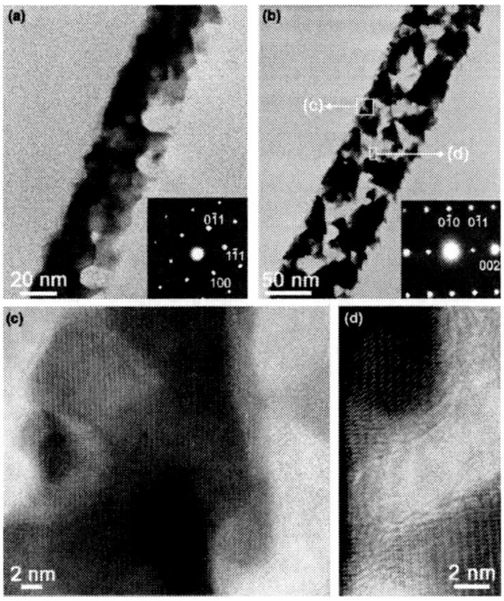

Figure 6. (a) TEM image of a porous GaN nanowire. The corresponding SAED pattern (inset) shows the single-crystalline wurtzite structure with the [011] growth direction. (b) TEM image and SAED pattern of another GaN nanowire. The growth direction is also [011]. (c) HRTEM image showing crystalline (001) planes of GaN nanocrystals. (d) HRTEM image of BCN layers randomly surrounding GaN nanocrystals. Reprinted with permission from Ref. [26]. Copyright 2003 Elsevier.

[53, 89] It is believed that Ga_2O_3 source powder should be firstly reduced at elevating temperatures into metastable gaseous Ga_2O species. Then Ga_2O and NH_3 molecules reacted to produce GaN nuclei in catalytic clusters. When graphite crucibles were employed in such procedures, the whole reaction can be expressed as [89]

$$Ga_2O_3(s) + 2C(s) \rightarrow Ga_2O(g) + 2CO(g) \tag{4}$$

and

$$Ga_2O(g) + 2NH_3(g) \rightarrow 2GaN(s) + H_2O(g) + 2H_2(g) \tag{5}$$

4. Porous GaN Nanowires

In the field of optoelectronics, the porous semiconductors spark a lot of interest and show a great promise, since they have shown to be capable of shifting the wavelength of emission and improving luminescence efficiency. [90, 91] Therefore porous GaN nanowires have attracted lots of interesting recently.

Porous single-crystalline GaN nanowires were synthesized using a thermal CVD method [26] through a reaction of Ga/C-containing ball-milled Ga_2O_3 and B_2O_3 powder mixture with NH_3. White wool-like GaN porous nanowires were homogeneously synthesized on large area of substrates. TEM studies indicated that the diameter of the porous nanowires was in the range $10 - 80$ nm with an average value of 40 nm and all of the nanowires contained many pores inside. Two typical nanowires with the diameter of 30 nm

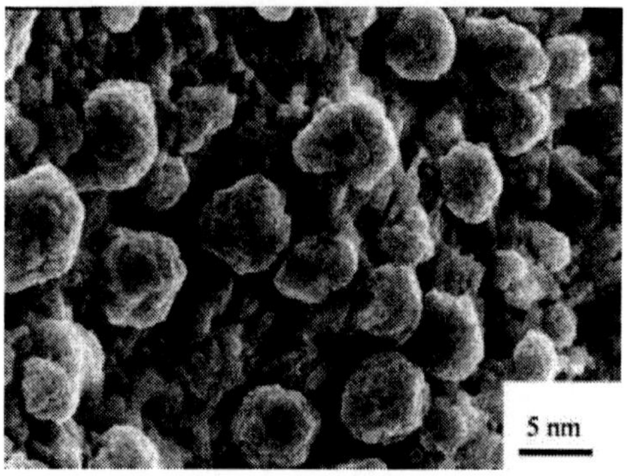

Figure 7. GaN nanoparticles synthesized by gas reactions on rough $LaAlO_3$ substrates. Reprinted with permission from Ref. [94]. Copyright 2000 Springer.

and 70 nm are shown in Figure 6a and Figure 6b, respectively. HRTEM images of individual nanowires revealed the morphology of the nanosize pores. The pore size was in the range $5 - 20$ nm. The corresponding selected-area electron diffraction (SAED) patterns are displayed as insets of the figure, which are consistent with that of the wurtzite structured GaN. The SAED pattern is identical over the entire nanowire, confirming the single GaN crystal. The HRTEM image for the boxed area shown in Figure 6b reveals almost no defects in the lattice fringes of the GaN nanowire (Figure 6c). The wire axis (growth direction) is parallel to the $< 0\bar{1}1 >$ direction of the hexagonal unit cell, for both nanowires. The rotation of the nanowires around the long axis showed a round cross-section. All of the synthesized nanowires had uniformly $< 011 >$ growth direction, irrespective of the morphology of pores and the diameter of nanowires. In contrast to the GaN crystalline nanowires, the GaN nanocrystals of the porous nanowires were coated with layers of BCN compound (Figure 6d).

The growth mechanism of the porous GaN nanowires was proposed as this: (1) Ga_2O vapor was generated as follows: $4Ga(l) + Ga_2O_3(s) \rightarrow 3Ga_2O(g)$ over starting materials [92] when Ga and Ga_2O_3 mixture was used. (2) Then the vapor Ga_2O reacted with ammonia, $Ga_2O(g) + 2NH_3(g) \rightarrow 2GaN(s) + H_2O(g) + 2H_2(g)$ [93], to grow GaN nanowires. The growth should follow the typical VLS growth mechanism. (3) During the growth of the nanowires, a reaction of amorphous C and B_2O_3 powder with NH_3 would yield the deposition of BCN outerlayers on the GaN nanoparticles, causing nanopores (Figure 6d). The reaction between C/B_2O_3 with NH_3 should be $4C(s) + B_2O_3(s) + 2NH_3(g) \rightarrow 2BC_2N(s) + 3H_2O(g)$.

5. GaN Nanoparticles

GaN nanoparticles can be easily prepared from a reaction of Ga and ammonia by CVD methods at $900 - 1150$ °C, [5, 93–96] from GaN nanoparticles by sublimation method, from organic gallium compounds by detonations or pyrolysis [97–99], from GaN powders by dry milling, [100] etc. Figure 7 shows spherical GaN particles synthesized from gallium by a CVD method. The diameters of the nanopaticles were about $5 - 8$ nm, smaller than the exciton Bohr radius of about 10 nm. The crystalline size of the traditioanlly grown nanoparticles are usually smaller than the Bohr radius. However it is challenging to produce massive GaN nanoparticles using these traditional methods because of high cost of starting materials and low product yields.

Ammonothermal technique is an effective method to produce massive GaN nanoparticles from gallium at low temperatures. [21, 101] Gallium metal can be converted to gallium nitride in stainless steel autoclaves at 350 – 500 °C in the presence of lithium mineralizators (pressure about 2000 atm at reaction temperatures), [21] using potassium and $LiNH_2$ as mineralizators, [102] employing NH_4X (X = F, Cl, Br, I) compounds as mineralizators [32, 103] to synthesize micro- and nanocrystalline GaN powders under base or acidic conditions. The average grain size was about 32 nm or less.

GaN nanocrystaline particles can also be produced by a benzene-thermal synthetic route. [104] A benzene-thermal reaction of Li_3N and $GaCl_3$ was carried out in which benzene was used as the solvent under high pressures at 280 °C. The size of the prepared GaN nanoparticles was 30 nm. The reaction temperature was much lower than that of traditional methods, and the yield of GaN reached 80 %. The X-ray powder diffraction pattern indicated that the produced nanomaterials was mainly hexagonal-phase GaN with a small fraction of rocksalt-phase GaN.

GaN nanoparticles can also be produced from Ga_2O_3 over 1100 °C in flowing NH_3. [93, 105] Ga_2O_3 is much cheaper than Ga and the total cost of GaN nanoparticles would be reduced. Pure GaN nanoparticles were synthesized through [93]

$$Ga_2O_3(s) + 2NH_3(g) \rightarrow 2GaN(s) + 3H_2O(g) \qquad (6)$$

6. GaN Nanotubes

Single-crystal GaN hollow nanotubes are advantageous in potential nanoscale electronics, optoelectronics and biochemical-sensing applications. Therefore some techniques have been developed recently to grow GaN nanotubes.

The GaN nanotubes was first synthesized from templates. An epitaxial casting approach [28] was achieved for the synthesis of single-crystal GaN nanotubes with inner diameters of 30 – 200 nm and wall thicknesses of 5 – 50 nm. Figure 8 a-b show SEM and TEM images of the crystalline GaN nanotubes respectively. Hexagonal ZnO nanowires were used as templates for the epitaxial overgrowth of thin GaN layers in a chemical vapor deposition system. The ZnO nanowire templates were subsequently removed by thermal reduction and evaporation, resulting in ordered arrays of GaN nanotubes on the substrates (Figure 8c).

Porous GaN nanotubes were also synthesized using templates. [106] Zinc oxide arrays were used as templates. The precursor of gallium (Ga_2O) was generated from a mixture of Ga/Ga_2O_3. When the temperature of the furnace reached 950 °C, Ga_2O reacted with ammonia to release GaN and hydrogen. The epitaxial growth occurred at the surface of ZnO nanowires. Figure 9a-b show SEM and TEM images of porous GaN nanotubes respectively. There were lots of nanopores in the tubes. The inner diameters of these GaN tubes were accurately controlled by varying the diameters of zinc oxide wires.

The deoxidization of zinc oxide at the beginning of the epitaxial growth was the key step to obtain these porous single crystal micro/nanotubes. [106] At the beginning of the

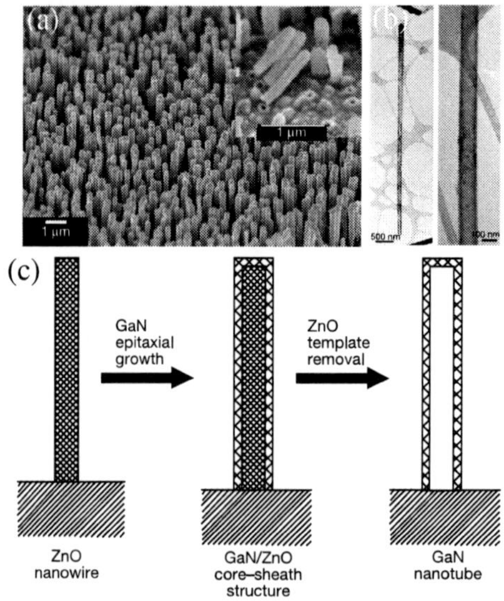

Figure 8. (a) SEM images of GaN nanotube arrays using ZnO nanowires as templates. Inset: fractured interface between the GaN nanotubes and the substrate. (b) TEM images of the GaN nanotubes. (c) Schematic illustration of the growth process for making single-crystal GaN nanotubes. Reprinted with permission from Ref. [28]. Copyright 2003 Macmillan Publishers.

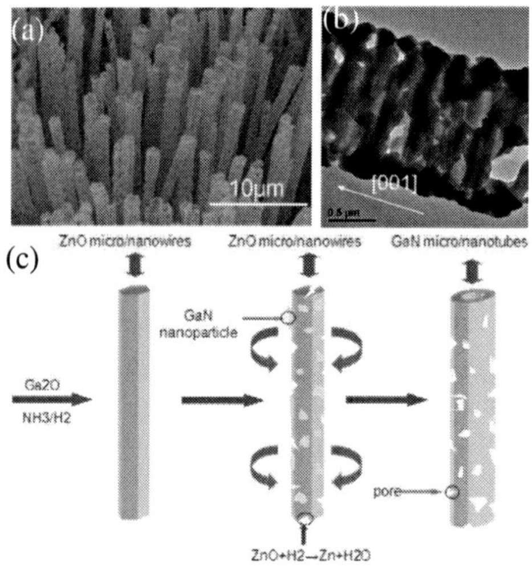

Figure 9. (a) SEM image of GaN porous nanotube arrays using ZnO nanowires as templates. (b) TEM image of a porous GaN nanotube. (c) Schematic illustration of the growth process. Reprinted with permission from Ref. [106]. Copyright 2013 Royal Society of Chemistry.

reaction, large amounts of synthetic GaN molecules were adsorbed onto the surface of ZnO nanowires to self-assemble into nanoparticles. At the same time, partial ZnO nanowire surfaces without GaN were reduced by hydrogen to produce gaseous Zn (boiling point of zinc is 906 °C) and H_2O slowly. These deoxidization areas were the sites of the GaN pores.

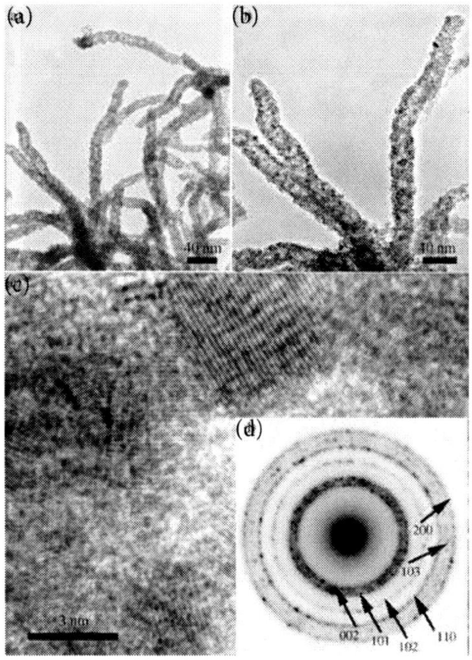

Figure 10. (a - b) TEM images of hollow GaN nanotubes synthesized at a temperature above 1100 °C by a gas interface reaction route. (c) An HRTEM image taken from the wall of a single nanotube. The inset is a SAED pattern of a single nanotube, suggesting that the synthesized nanotube is polycrystalline. Reprinted with permission from Ref. [107]. Copyright 2005 WILEY-VCH Verlag GmbH & Co.

As the process proceeds, the synthetic GaN nanoparticles grew and connected to each other epitaxially to form single crystal tubes. It was believed that the introduced hydrogen was the key to obtain the porous structures. If no hydrogen was introduced, ZnO/GaN core-shell structures were obtained.

GaN nanotubes can also be assembled from GaN nanoparticles with smaller diameter. [107] Figure 10 shows GaN nanotubes with wall thicknesses of $3.5 - 5.0$ nm. The GaN nanotubes had a length of several hundreds of nanometers, were polycrystalline and composed of very fine GaN nanocrystalline particles with a diameter of $3.0 - 3.5$ nm. It believed that gallium droplet did not act as a catalyst but rather as a reactant and a template for the formation of hollow GaN structures. There were a lot of vacancies and dangling bonds on the surface of the Ga liquid droplets because the surface area of nanocrystals is much larger than that of bulk materials. When the NH_3 gas was introduced, the NH_3 molecules were easily adsorbed onto the surface of the starting nanosized Ga droplets. The surface of the metallic gallium droplet was the location where GaN nucleates. NH_3 reacted with gallium to form GaN on the surface of the Ga droplets. With increasing reaction time, more and more NH_3 passed though the previously formed GaN-Ga interfaces to react with Ga and form GaN to finally give hollow spheres composed of small, polycrystalline particles. The formed hollow GaN spheres aggregated, and the hollow GaN nanotubes were formed by the coalescence of the nanosized hollow GaN spheres.

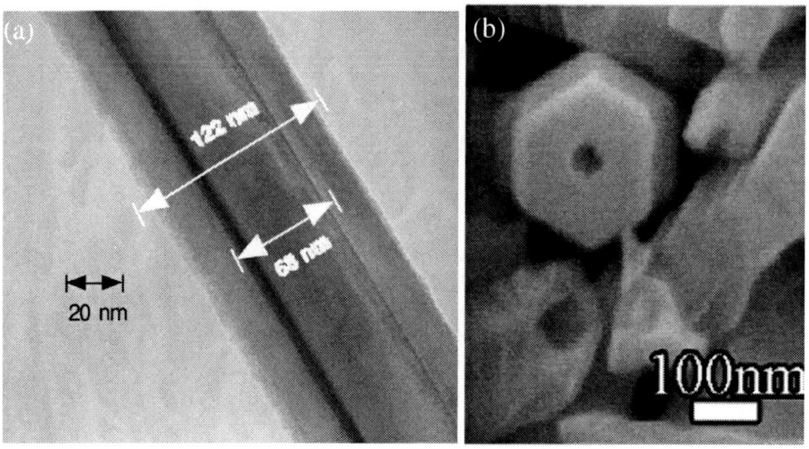

Figure 11. (a) TEM image of GaN nanotubes synthesized by direct reaction of Ga with NH_3. Reprinted with permission from Ref. [27]. Copyright 2000 American Institute of Physics. (b) SEM image of GaN nanotubes produced fromGa_2O_3 and GaN under NH_3. Reprinted with permission from Ref. [108]. Copyright 2009 IOP Publishing.

GaN nanotubes have also been synthesized without any templates. Unfortunately such produced GaN nanotubes always have thick walls. Figure 11a shows GaN nanotubes directly synthesized from Ga and NH_3. The wall thickness was about 30 nm. [27] Sometimes the walls were even thicker. Figure 11b shows GaN nanotubes produced from a mixed powder of Ga_2O_3 and GaN under NH_3 at 1150 °C. The typical outer diameter of the GaN nanotubes was larger than 200 nm, whereas the inner diameter was in the range of 20 − 100 nm, with a wall thickness of 100 nm. [108] More researches are carried out to synthesize GaN nanotubes with a thinner wall.

Beside wurtzite GaN nanotubes, cubic GaN nanotubes have been produced from Ga_2O_3. [29] The cubic GaN nanotubes had rectangular cross-sections with wall thicknesses of 20 − 50 nm and cross-section dimensions of 50 − 150 nm. The microtubes were several micrometers long.

7. GaN Microtubes

GaN microtubes can be produced on templates, similar to the GaN nanotubes grown on templates. At the same time, GaN microtubes were also produced without templates. [109, 110]

Figure 12 shows wurtzite GaN microtubes grown by a CVD method without templates. The synthesis procedures of GaN microtubules were carried out in a hot-walled CVD system. Gallium was placed on quartz substrates. Then Ga reacted with ammonia to form GaN microtubes at 820 − 840 °C. After reactions yellow deposits were observed on the substrates. SEM observation indicated that the yellow deposits consisted of many microtubes. The synthesized structures were hollow microtubes with cross sections about 8 μm and lengths up to 100 μm. The ends of most of these microtubes were open, but some were closed (Figure 12b-d).

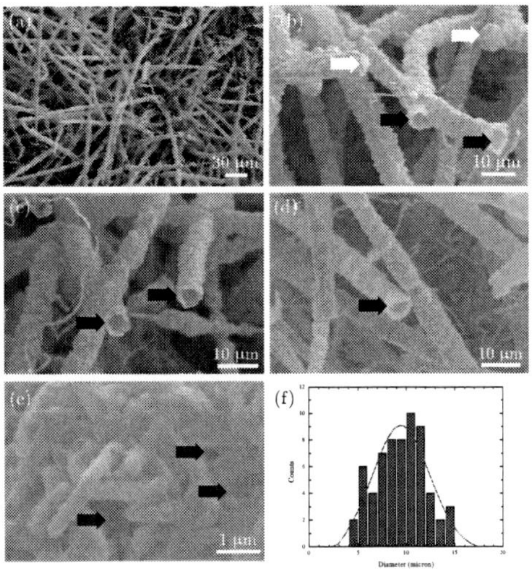

Figure 12. SEM images of (a-d) GN microtubes grown at different regions and (e) wall surface of a microtube. White arrows mark the closed ends, black arrows the open ends, and the blue arrows the nanopores of the microtubes. (f) Histogram of microtubule diameter. Reprinted with permission from Ref. [110]. Copyright 2014 Elsevier.

From open ends of the microtubes, the thicknesses of the microtube walls were found to be about 100 nm. More detailed examination (Figure 12d) indicated the walls consisted of single layer of randomly oriented sub-micron irregular structures. The diameters of the sub-micron irregular structures were about 100 nm and their lengths were about 1 μm.

High magnification SEM images (Figure 12e) indicated that, as a result of the random packing of the sub-micron irregular structures, the thin walls of the microtubes were porous. Some nanopores of these walls are notated by blue arrows in Figure 12e. Such porous structures should have large surface areas. Unfortunately it is very difficult to transport the synthesized microtubes from the substrates because of brittleness.

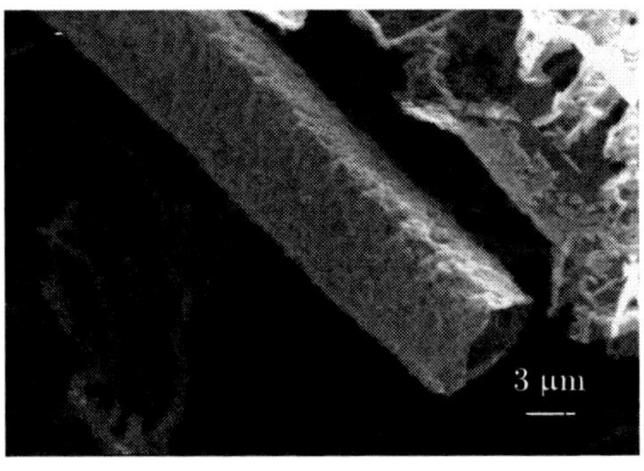

Figure 13. SEM image a cubic GaN microtubes. Reprinted with permission from Ref. [109]. Copyright 2013 Springer.

Cubic GaN microtubes (Figure 13) were also prepared using the CVD method. [109] The walls consisted of single layer of randomly oriented submicron irregular structures with a cubic phase. The thicknesses of the microtube walls were about 100 nm.

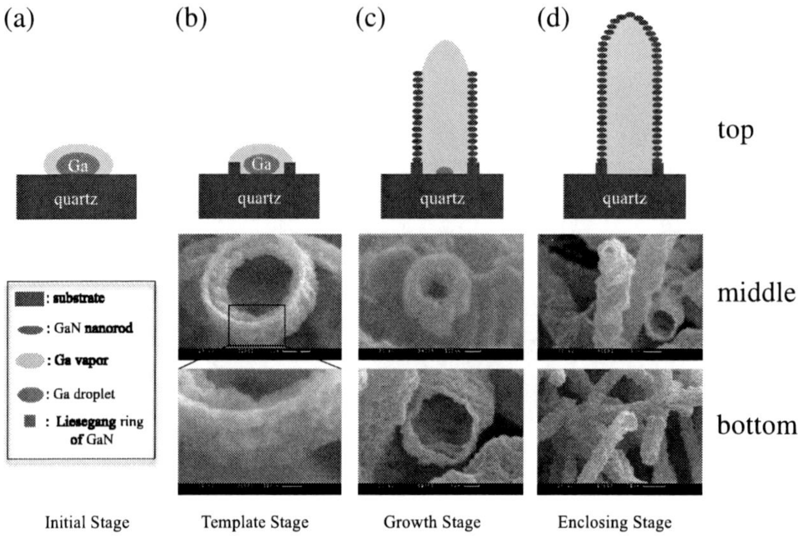

Figure 14. Possible growth mechanism of the self-assembled GaN microtubules. (a) Initial stage. No GaN nanomaterials are synthesized. (b) Template stage. Liesegang rings of GaN nanomaterials are formed. (c) Growth stage. GaN nanorods self-assemble into microtubules. (d) Enclosing stage. Ga sources are exhausted and the ends of GaN microtubules are closed. Top: model of each stage. Middle and bottom: SEM images at each stage. Reprinted with permission from Ref. [110]. Copyright 2014 Elsevier.

The formation of the porous microtubules can be explained by the Liesegang phenominon and diffusion limited growth. Gallium has a low melting point of 303 K and a boiling point of 2520 K. Therefore large amounts of gallium liquid droplets can condense on quartz substrates from gaseous gallium (Figure 14a), as observed on silicon substrates [111]. The diameter of the condensed droplets is usually $5-50$ μm [112], depending on the reaction temperature and substrates. The behavior of gallium is very similar to that of other low melting point metals, such as selenium [113]. Gallium droplets evaporate at high temperatures to produce gallium vapor. Once ammonia passes the gallium droplets, gallium vapor would react with ammonia to produce GaN nanoparticles (Figure 14b). The number of the GaN nanoparticles per unit volume is so dense around the gallium droplets that the GaN nanoparticles around the gallium droplets can be considered as a supersaturated aerosol, similar to charged nanoparticels in solutions [114] and nanoparticles in vapor phases [115,116]. These supersaturated GaN nanoparticles would spontaneously aggregate into Liesegang rings because of thermodynamatic stability, as observed in other kinds of ordered Liesegang nanoparticle-rings. [114,117]

GaN nanoparticle Liesegang rings were observed on SEM. The middle panel in Figure 14b shows an SEM image of a typical GaN nanoparticle ring with diameter of several micrometers. Higher magnification SEM image (bottom panel in Figure 14b) indicated that the microring was consisted of many GaN nanoparticles with size below several ten nanometers. Other GaN rings with different diameters were also observed. These survived

GaN rings indicated that the Liesegang rings can self-assemble from GaN nanomaterials during the CVD procedures when GaN nanoparticles were supersaturated enough.

Then porous GaN microtubules grow on the Liesegang ring templates, as shown in Figure 14c. Based on the diffusion-limited aggregation mechanism, the protruding parts of the GaN microrings would easily attract the GaN nanorods and grow quickly to build a new nanorod-wall from the Liesegang ring edges, leaving the hollow interior. SEM images in Figure 14c show tips of two GaN microtubules, indicating the growth on the Liesegang rings. With the consuming of gallium, supersaturated GaN nanorods would be continuously produced from the reaction of ammonia and gallium vapor as long as the gallium vapor pressure are high enough, resulting in the continuous deposition of nanorods on the GaN microrings to form GaN microtubules. During the whole growth stage, the gallium droplets would evaporate to generate the gallium vapor to keep the microtubules open.

The gallium droplets would be smaller and smaller with the reaction time because of the vaporation. The vapor pressure of the gallium droplets would decrease near the open ends of GaN tubules once the gallium droplets vanish. Then the gallium vapor pressure front near the microtubule open tips would shrink back, resulting in the aggregation of GaN nanorods along the gallium vapor pressure front to enclose the microtubule ends, as shown in Figure 14d. SEM images in Figure 14d show two microtubules with half-closed end.

Additionally, the GaN microtubules were synthesized in a short time, only two minutes. It was an indirect evidence for the supersaturation growth, indicating rapidity of the Liesegang ring formation and the microtubule growth. These two phenomena (ring formation and rapid growth) are the specific characteristics of supersaturation growth caused by the Liesegang effect.

8. GaN Nanocomposites

The diameter of most GaN nanowires is usually larger than Bohr radius of GaN (11 nm). So the quantum confinement effects were not much observed in GaN nanowires. The produced GaN nanoparticles were usually smaller than the Bohr radius and show quantum confinement. If GaN nanoparticles can be self-assembled into millimeter scale bulk nanocomposites, such GaN nanocomposites should have the potential for incorporating a number of unique properties of nanocrystal while taking advantage of properties of bulk condensed matters. [31,32]

GaN nanocomposites have been directly produced through reactions of Ga metal and liquid ammonia with addition of NH_4Cl in autoclaves. [32] Figure 15a shows a typical photograph of a fragment obtained by mechanical smashing assemblies. The size of the fragment was about 0.7×0.6 cm with an average thickness of about 0.2 cm. It can be seen clearly that the fragment was transparent to visible light. SEM observations revealed the existence of terrace steps which resemble characteristics of fractures of brittle metal materials with coarse-sized grains. The assemblies were bulk materials like ceramic materials, not simple aggregation of grains.

Figure 15b shows an HRTEM image of a nanopore resulting from the packing of three grains with [001] orientation. The size of these pores was estimated to be around 5 nm, roughly half of the average diameter of grains. The measured density for GaN nanocomposites was 72 % of that for the GaN single crystals. This value approached the theoretical

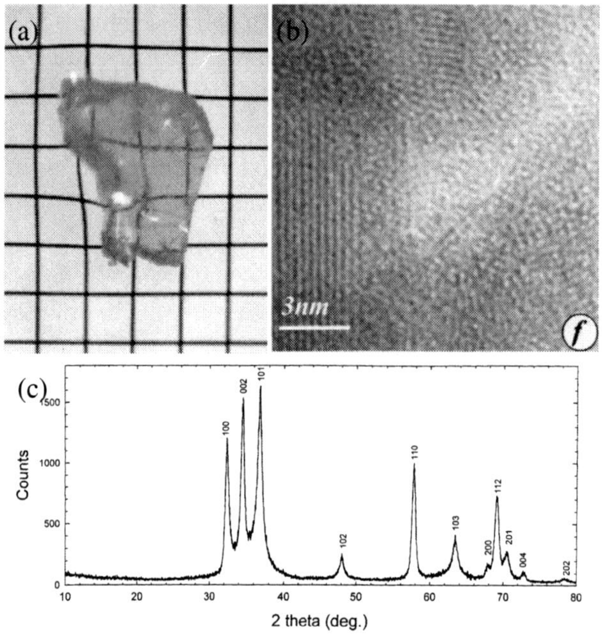

Figure 15. (a) Optical photograph of a fragment of GaN nanocomposites. The grid spacings are 0.2×0.2 cm. (b) HRTEM image showing a nanopore resulting from the packing of three grains of [001] orientation. (b) An XRD pattern of GaN nanocomposites. Reprinted with permission from Ref. [32]. Copyright 2000 Elsevier.

limit for close packing of identical spheres. Since the size of pores was well below the wavelengths of visible light, the GaN nanocomposites was transparent to visible light.

XRD patterns of the assemblies further indicated that the assemblies were hexagonal GaN. No impurity phase was detected. Figure 15c shows a typical XRD pattern for the powder obtained by milling the fragments of the GaN nanocomposites. All reflections were broadened due to the size effect. The average grain size according to the Scherrer formula was 11.8 nm.

The formation mechanism of nanocomposites and nanocrystalline GaN is not clear. A possible mechanism was proposed as follows. [32] 1) NH_4Cl is known to be very soluble in liquid ammonia (124 g / 100 g NH_3) and the resulting solution possesses a strong acidity. Metal Ga can be easily oxidized to produce Ga^{3+} ions and form soluble intermediate compounds such as $Ga(NH_3)_nCl_3$ ($n = 1 - 14$) [118] in such a solution. 2) Then nano-sized GaN particles would precipitate from either the decomposition or further reactions of these intermediate compounds with liquid ammonia. 3) Since there was a strong aggregating tendency for nanosized clusters in the solution like the case with small clusters in a colloid, these nanocrystals were consolidated into GaN nanocomposites when the pressure in an autoclave was high enough.

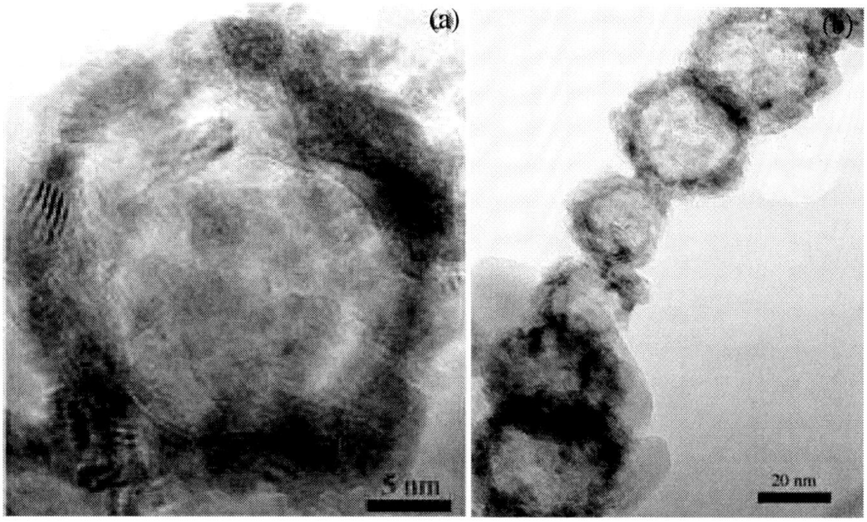

Figure 16. TEM images of (a) a nanosphere and (b) aligned hollow GaN spheres. Reprinted with permission from Ref. [107]. Copyrigh 2005 WILEY-VCH Verlag GmbH & Co.

9. Others Nanostructures

Other interesting GaN nanomaterials have also been reported, such as nanopheres. Figure 16 shows GaN nanospheres [107] with diameters of 20 – 25 nm, which was self-assembled from GaN nanoparticles with diameter of several nanometers. It was assumed that GaN nanocrystals nucleated and grew on the surface of Ga liquid droplets to form the hollow GaN spheres.

10. Characterizations of GaN Nanomaterials

In the previous sub-chapters, syntheses and structures of GaN nanomaterials have been discussed. Some advanced techniques are employed to characterize these nanomaterials. For example, transmission electron microscopy plays an important role in the microstructures of GaN nanomaterials. Morphology of the GaN nanomaterials have been characterized by scanning electron microscopy that is the most popular method to characterize nanomaterials. In this sub-chapter, we introduce the details of these advanced techniques, including X-ray diffraction, scanning electron microscopy, transmission electron microscopy, Raman spectroscopy, and photoluminescence spectroscopy.

10.1. X-Ray Diffraction

The physical basics of X-ray diffraction were formulated by Max von Laue, William H. Bragg and William L. Bragg at the beginning of the twentieth century, based on the X-ray findings of W. Conrad Röntgen. Characteristic X-rays are emitted when high-energy electrons knock off an inner-shell electron from specimens, causing out-shell electrons to fill inner shells to release energy. When such an X-ray beam incidents on crystalline samples,

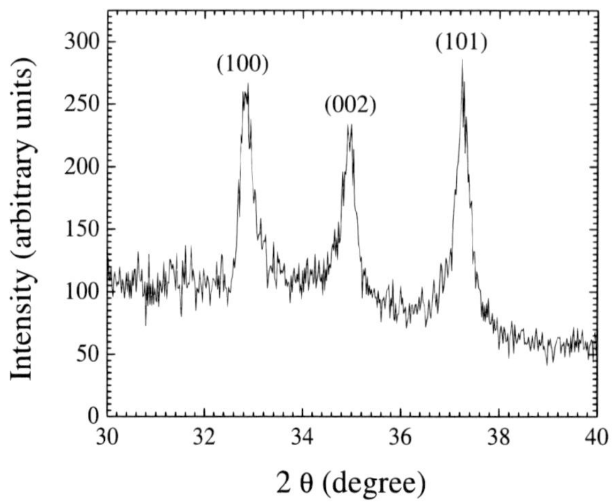

Figure 17. XRD pattern of the as-prepared GaN microtubules. Reprinted with permission from Ref. [110]. Copyright 2014 Elsevier.

X-ray is diffracted by the crystallographic planes of the sample, as described by Bragg's law expressed in

$$2d \sin\theta = n\lambda \tag{7}$$

where λ is the wavelength, d is the distance between crystal planes, θ is the angle of the diffracted wave, and n is an integer known as the order of the diffracted beam. XRD has been widely used to detect the crystallographic structures of GaN nanomaterials as well as to identify crystallinity, measure lattice parameters, calculate crystalline size and strain, etc. Usually, GaN nanomaterials are stable in a hexagonal wurtzite phase with lattice constants of $a = 0.3186$ nm and $c = 0.5178$ nm. [22, 64, 65] The XRD patterns of GaN nanomaterials have these reflection peaks: (100), (002), (101), (102), (110), (112) indicating hexagonal wurtzite phase. [22, 64] (100), (002), and (101) peaks are the three highest intensity peaks in the XRD patterns. The crystallographic orientation of the GaN nanomaterials can also be determined from the intensity of diffraction peaks. [119]

Figure 17 shows a typical X-ray powder diffraction pattern of GaN microtubes with a wurtzite structure. All of the XRD peaks are broadened because of size effect and can be indexed on the basis of the hexagonal wurtzite GaN phase with a space group of $P6_3mc$. [5]

Crystalline size D of GaN nanomaterials can be calculated from Scherrer's equation:

$$\mathrm{D} = \frac{0.9\lambda}{\beta \cos\theta} \tag{8}$$

where $\lambda = 1.5415$ Å when a Cu radiation source is used, θ the diffracted angle in degree, and β is full width at half maximum in rad.

10.2. Scanning Electron Microscopy

The basic concept of scanning electron microscopy was proposed by M. Knoll in 1935 [120] and various experimental SEMs were constructed in the late 1930s and early 1940s.

[121–123] The resolution of SEM was achieved at 50 nm in 1952 [124] and about 1 nm at present. Up to now, SEM has been widely employed to detect the surface of various materials, [125, 126] including GaN nanomaterials.

The high-energy electrons interact with the specimen atoms to produce signals that contain information about the specimens surface topography, composition, and electrical conductivity etc. in the SEM raster scan mode. The produced signals include secondary electrons, back-scattered electrons, characteristic X-rays, cathodoluminescence, specimen current, etc. Secondary electron imaging (SEI) is the most common detection mode in SEM and widely employed to observe the morphology of nanomaterials. The interesting readers on other type of imaging and related principles are referred to some textbooks [127–132] and references therein.

10.3. Transmission Electron Microscopy

Electron was discovered as a kind of particles in 1897. According to de Broglies hypothesis of the wave-particle duality, electrons can act as waves with a wavelength λ of

$$\lambda = \frac{h}{\sqrt{2m_0eU}}\frac{1}{\sqrt{1+\frac{eU}{2m_0c^2}}} \tag{9}$$

where c is the speed of light, h is Planck's constant, U is the electric potential to accelerate the electrons, m_0 is the mass of the electron, and e is the elementary charge, after taking relativistic effects into account. The wavelength of electrons is 0.25 Å at 200 kV. The de Broglie wavelength of electrons is many orders of magnitude smaller than that for light, theoretically allowing for imaging at atomic scales. More recently, a resolution below 0.5 Å is achieved in an aberration corrector design. [133]

Transmission electron microscopy (TEM) is one of the most powerful and versatile techniques for the characterization of nanostructured materials. Its unique features allow researchers to achieve morphology, crystallographic structures, atomic structures, as well as chemical composition of nanomaterials. Crystalline phases, crystallographic planes, growth direction, interplanar spacing and chemical composition of GaN nanomaterials have been determined on TEM. For example, TEM has been utilized to characterize atomic structures and growth directions of GaN nanowires, [65] real-time observation of GaN nanowire growth at high temperatures [83] to confirm the self-catalytic growth of GaN nanowires via the VLS mechanism.

10.4. Raman Spectroscopy

When photons are scattered from an atom or molecule, most photons are elastically scattered, Rayleigh scattering. The frequency and wavelength of the Rayleigh scattered photons are same as that of the incident photons. At the same time, a small fraction (approximately 1 in 10 million photons) of the scattered photons is inelastically scattered by an excitation. The frequency of the inelastically scattered photons is different from the incident photons. The inelastic scattering of light was predicted in 1923, [134] first observed in liquids [135, 136] and later in solids [137] in 1928, named as Raman effect later. Up to now, the Raman spectroscopy has been widely used in chemistry to identify molecules, in solid

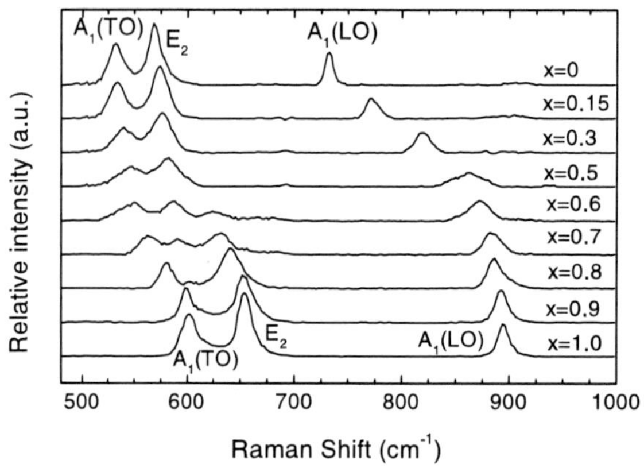

Figure 18. Room-temperature Raman spectra for $Al_xGa_{1-x}N$ alloys with various x values. Reprinted with permission from Ref. [145]. Copyright 2001 Springer.

state physics to characterize inorganic materials, etc. Raman scattering is also a useful tool to characterize nanosized materials and a qualitative probe of the presence of lattice defects in solids. For example the crystalline quality can be judged from the peak shapes and the selection rules. [138]

For GaN nanomaterials, Raman spectroscopy provides information on the vibrational states that are sensitive to the crystalline quality, the stress, carrier concentration and mobility and therefore can be employed for process and growth monitoring. [139]

GaN crystallizes in the hexagonal wurtzite structure space group C_{6v} - $P6_3mc$ [5] with two formula units per primitive cell. Group theory predicts four Raman active modes [140, 141]: A1, E1, $2E_2$ (E_2^L and E_2^H). The A1 and E1 are polar modes: i.e., the interaction of the vibration with the long-range Coulomb field leads to energy differences between phonons polarized longitudinally and transverse to the direction of propagation of the phonon. The A1 and E1 modes are therefore split into LO and TO modes. The peaks at 143, 556, 568, and 724 cm^{-1} are usually assigned to first-order phonons of E_2^L, E1(TO), E_2^H [140] and A1(LO) [58,60,142] respectively. These Raman modes were used to investigate the effects of growth temperatures, [143] crystalline size effect, [142] and phonon modes [144] of GaN nanomaterials.

Figure 18 shows the Raman scattering spectra of $Al_xGa_{1-x}N$ nanocrystalline solids with various alloy compositions. It can be seen that A1(LO), A1(TO), and E2 phonon mode frequencies shift to higher energy as the aluminum content in the alloy increases. A1(LO) phonon mode shifts smoothly from the GaN to AlN energy position with increasing aluminum content, parabolically following an empirical formula $A1(LO) = 729.1 + 349.6x - 186.9x^2(\text{cm}^{-1})$. [145] The linewidth of E2 (GaN) mode increases with increasing aluminum content, indicating the disorder state with increasing aluminum.

10.5. Photoluminescopy

Photoluminescence (PL) technique is a non-contact, nondestructive method of probing the electronic structure of semiconductor materials. When light is directed onto a sample, it is absorbed and photo-excitation process can occur. The photo-excitation causes the material to jump to a higher electronic state, and will then release energy (photons) as it relaxes and returns back to a lower energy level. PL spectrocopy is an effective way to study the electronic structure, optical and photochemical properties of semiconductor materials, by which information such as surface oxygen vacancies and defects, as well as the efficiency of charge carrier trapping, immigration and transfer can be obtained. [146] The PL spectrum of GaN nanowires usually has a near band-gap emission peak blue-shifted compared with the value of 365 nm of bulk GaN, which can be ascribed to the quantum confinement effect. [147]

Photoluminescence of GaN materials is affected by many parameters, such as defects [24] and crystallographic orientations [148].

11. Summary

This chapter introduces the achievements on GaN nanomaterials in recent decades. Syntheses of GaN nanowires have been widely researched and a variety of growth methods have been developed. Several growth mechanisms including VLS, VSS and VS have been proposed to explain the growth of GaN nanomaterials. Various characterization methods are also presented. These GaN nanomaterials have been applied in many fields such as photovoltaic devices, field effect transistor, light emitting diodes, lasers, photocatalyst and nanogenerators.

Acknowledgment

YCL and ZFR acknowledge the financially supporting by Defense Threat Reduction Agency under grant HDTRA122221 and the support of Small Grant Program from University of Houston. JYL thanks the financial supports from the Fundamental Research Funds for the Central Universities of China, the Research Fund for Doctoral Program of Higher Education of China (No. 20110006110002), Ministry of Education of China.

References

[1] H. Morkoç, S. Strite, G. B. Gao, M. E. Lin, B. Sverdlov, and M. Burns, "Large band gap SiC, III-V nitride, and II-VI ZnSe-based semiconductor device technologies," *Journal of Applied Physics*, vol. 76, no. 3, pp. 1363 – 1398, 1994.

[2] S. Mohammad and H. Morkoç, "Progress and prospects of group-III nitride semiconductors," *Progress in Quantum Electronics*, vol. 20, no. 5 - 6, pp. 361 – 525, 1996.

[3] I. Akasaki and H. Amano, "Crystal growth and conductivity control of group III nitride semiconductors and their application to short wavelength light emitters," *Japanese Journal of Applied Physics*, vol. 36, no. 9R, p. 5393, 1997.

[4] S. Nakamura, "The roles of structural imperfections in InGaN-based blue light-emitting diodes and laser diodes," *Science*, vol. 281, no. 5379, pp. 956 – 961, 1998.

[5] X. L. Chen, Y. C. Lan, J. K. Liang, X. R. Cheng, Y. P. Xu, T. Xu, P. Z. Jiang, and K. Q. Lu, "Structure and heat capacity of wurtzite GaN from 113 to 1073 K," *Chin. Phys. Lett.*, vol. 16, no. 2, pp. 107 – 108, 1999.

[6] B. Monemar, "Fundamental energy gap of GaN from photoluminescence excitation spectra," *Phys. Rev. B*, vol. 10, pp. 676 – 681, 1974.

[7] K. P. Korona, A. Wysmołek, K. Pakuła, R. Stępniewski, J. M. Baranowski, I. Grzegory, B. Łucznik, M. Wróblewski, and S. Porowski, "Exciton region reflectance of homoepitaxial GaN layers," *Applied Physics Letters*, vol. 69, no. 6, pp. 788 – 790, 1996.

[8] S. Nakamura, T. Mukai, and M. Senoh, "High-power GaN p-n junction blue-light-emitting diodes," *Japanese Journal of Applied Physics*, vol. 30, no. 12A, p. L1998, 1991.

[9] S. Nakamura, M. Senoh, and T. Mukai, "High power InGaN/GaN double heterostructure violet light emitting diodes," *Applied Physics Letters*, vol. 62, no. 19, pp. 2390 – 2392, 1993.

[10] I. Akasaki, H. Amano, S. Sota, H. Sakai, T. Tanaka, and M. Koike, "Stimulated emission by current injection from an AlGaN/GaN/GaInN quantum well device," *Japanese Journal of Applied Physics*, vol. 34, no. 11B, p. L1517, 1995.

[11] H. Amano, T. Asahi, and I. Akasaki, "Stimulated emission near ultraviolet at room temperature from a GaN film grown on sapphire by MOVPE using an AlN buffer layer," *Japanese Journal of Applied Physics*, vol. 29, no. 2A, p. L205, 1990.

[12] M. Asif Khan, J. N. Kuznia, A. R. Bhattarai, and D. T. Olson, "Metal semiconductor field effect transistor based on single crystal GaN," *Applied Physics Letters*, vol. 62, no. 15, pp. 1786 – 1787, 1993.

[13] F. A. Ponce and D. P. Bour, "Nitride-based semiconductors for blue and green light-emitting devices," *Nature*, vol. 386, no. 6623, pp. 351 – 359, 1997.

[14] M. S. Brandt, P. Herbst, H. Angerer, O. Ambacher, and M. Stutzmann, "Thermopower investigation of n- and p-type GaN," *Phys. Rev. B*, vol. 58, pp. 7786 – 7791, 1998.

[15] W. Liu and A. A. Balandin, "Thermoelectric effects in wurtzite GaN and $Al_xGa_{1-x}N$ alloys," *J. Appl. Phys.*, vol. 97, no. 12, p. 123705, 2005.

[16] S. Adachi, "Lattice thermal conductivity of group-IV and III-V semiconductor alloys," *Journal of Applied Physics*, vol. 102, no. 6, p. 063502, 2007.

[17] E. Hurwitz, M. Asghar, A. Melton, B. Kucukgok, L. Su, M. Orocz, M. Jamil, N. Lu, and I. Ferguson, "Thermopower study of GaN-based materials for next-generation thermoelectric devices and applications," *J. Electron. Mater.*, vol. 40, no. 5, pp. 513 – 517, 2011.

[18] J. Li, F. Li, S. Lin, S. Zhong, Y. Wei, X. Meng, and X. Shen, "Theoretical study on $In_xGa_{1-x}N$/Si hetero-junction solar cells," *Proc. SPIE*, vol. 7409, p. 740910, 2009.

[19] M. Lu, G. guang Zhang, K. Fu, G. hao Yu, D. Su, and J. feng Hu, "Gallium nitride schottky betavoltaic nuclear batteries," *Energ. Convers. Manage.*, vol. 52, no. 4, pp. 1955 – 1958, 2011.

[20] Z. Cheng, X. Chen, H. San, Z. Feng, and B. Liu, "A high open-circuit voltage gallium nitride betavoltaic microbattery," *J. Micromech. Microeng.*, vol. 22, no. 7, p. 074011, 2012.

[21] Y. C. Lan, X. L. Chen, Y. P. Xu, Y. G. Cao, and F. Huang, "Syntheses and structure of nanocrystalline gallium nitride obtained from ammonothermal method using lithium metal as mineralizator," *Mater. Res. Bull.*, vol. 35, no. 14 - 15, pp. 2325 – 2330, 2000.

[22] J. Y. Li, X. L. Chen, Z. Y. Qiao, Y. G. Cao, and Y. C. Lan, "Formation of GaN nanorods by a sublimation method," *J. Cryst. Growth*, vol. 213, no. 3 - 4, pp. 408 – 410, 2000.

[23] S. Li and A. Waag, "GaN based nanorods for solid state lighting," *J. Appl. Phys.*, vol. 111, no. 7, pp. 071101/1 – 071101/23, 2012.

[24] X. Chen, J. Li, Y. Cao, Y. Lan, H. Li, M. He, C. Wang, Z. Zhang, and Z. Qiao, "Straight and smooth GaN nanowires," *Adv. Mater.*, vol. 12, no. 19, pp. 1432 – 1434, 2000.

[25] T. Kuykendall, P. Pauzauskie, S. Lee, Y. Zhang, J. Goldberger, and P. Yang, "Metalorganic chemical vapor deposition route to GaN nanowires with triangular cross sections," *Nano Lett.*, vol. 3, no. 8, pp. 1063 – 1066, 2003.

[26] S. Y. Bae, H. W. Seo, J. Park, H. Yang, and B. Kim, "Porous GaN nanowires synthesized using thermal chemical vapor deposition," *Chem. Phys. Lett.*, vol. 376, no. 3 - 4, pp. 445 – 451, 2003.

[27] M. He, I. Minus, P. Zhou, S. N. Mohammed, J. B. Halpern, R. Jacobs, W. L. Sarney, L. Salamanca-Riba, and R. D. Vispute, "Growth of large-scale GaN nanowires and tubes by direct reaction of Ga with NH_3," *Appl. Phys. Lett.*, vol. 77, no. 23, pp. 3731–3733, 2000.

[28] J. Goldberger, R. He, Y. Zhang, S. Lee, H. Yan, H.-J. Choi, and P. Yang, "Single-crystal gallium nitride nanotubes," *Nature*, vol. 422, no. 6932, pp. 599 – 602, 2003.

[29] J. Q. Hu, Y. Bando, J. H. Zhan, F. F. Xu, T. Sekiguchi, and D. Golberg, "Growth of single-crystalline cubic GaN nanotubes with rectangular cross-sections," *Adv. Mater.*, vol. 16, no. 16, pp. 1465 – 1468, 2004.

[30] B. Liu, Y. Bando, C. Tang, G. Shen, D. Golberg, and F. Xu, "Wurtzite-type faceted single-crystalline GaN nanotubes," *Appl. Phys. Lett.*, vol. 88, no. 9, p. 093120, 2006.

[31] R. W. Siegel, *Physics of new materials*, vol. 27 of *Springer Series in Materials Science*, ch. Synthesis of Nanoparticles and their Agglomerates, pp. 66 – 106. Springer Berlin Heidelberg, 1998.

[32] X. L. Chen, Y. G. Cao, Y. C. Lan, X. P. Xu, J. Q. Li, K. Q. Lu, P. Z. Jiang, T. Xu, Z. G. Bai, Y. D. Yu, and J. K. Liang, "Synthesis and structure of nanocrystal-assembled bulk GaN," *J. Cryst. Growth*, vol. 209, no. 1, pp. 208 – 212, 2000.

[33] A. Di Carlo, "Tuning optical properties of GaN-based nanostructures by charge screening," *Physica Status Solidi (a)*, vol. 183, no. 1, pp. 81 – 85, 2001.

[34] Y. Arakawa, "Progress in GaN-based quantum dots for optoelectronics applications," *Selected Topics in Quantum Electronics, IEEE Journal of*, vol. 8, no. 4, pp. 823 – 832, 2002.

[35] S. S. Kocha, M. W. Peterson, D. J. Arent, J. M. Redwing, M. A. Tischler, and J. A. Turner, "Electrochemical investigation of the gallium nitride - aqueous electrolyte interface," *J. Electrochem. Soc.*, vol. 142, no. 12, pp. L238 – L240, 1995.

[36] I. M. Huygens, K. Strubbe, and W. P. Gomes, "Electrochemistry and photoetching of n-GaN," *J. Electrochem. Soc.*, vol. 147, no. 5, pp. 1797 – 1802, 2000.

[37] K. Maeda, T. Takata, M. Hara, N. Saito, Y. Inoue, H. Kobayashi, and K. Domen, "GaN:ZnO solid solution as a photocatalyst for visible-light-driven overall water splitting," *Journal of the American Chemical Society*, vol. 127, no. 23, pp. 8286 – 8287, 2005.

[38] K. Maeda, K. Teramura, D. Lu, T. Takata, N. Saito, Y. Inoue, and K. Domen, "Photocatalyst releasing hydrogen from water," *Nature*, vol. 440, no. 7082, pp. 295 – 295, 2006.

[39] K. Fujii and K. Ohkawa, "Bias-assisted h_2 gas generation in HCl and KOH solutions using n-type GaN photoelectrode," *J. Electrochem. Soc.*, vol. 153, no. 3, pp. A468 – A471, 2006.

[40] H. S. Jung, Y. J. Hong, Y. Li, J. Cho, Y.-J. Kim, and G.-C. Yi, "Photocatalysis using GaN nanowires," *ACS Nano*, vol. 2, no. 4, pp. 637 – 642, 2008.

[41] D. Wang, A. Pierre, M. G. Kibria, K. Cui, X. Han, K. H. Bevan, H. Guo, S. Paradis, A.-R. Hakima, and Z. Mi, "Wafer-level photocatalytic water splitting on GaN nanowire arrays grown by molecular beam epitaxy," *Nano Lett.*, vol. 11, no. 6, pp. 2353 – 2357, 2011.

[42] S.-W. Ryu, Y. Zhang, B. Leung, C. Yerino, and J. Han, "Improved photoelectrochemical water splitting efficiency of nanoporous gan photoanode," *Semiconductor Science and Technology*, vol. 27, no. 1, p. 015014, 2012.

[43] F. Cavalca, A. B. Laursen, B. E. Kardynal, R. E. Dunin-Borkowski, S. Dahl, J. B. Wagner, and T. W. Hansen, "*In situ* transmission electron microscopy of light-induced photocatalytic reactions," *Nanotechnology*, vol. 23, no. 7, p. 075705, 2012.

[44] B. Adeli and F. Taghipour, "A review of synthesis techniques for gallium-zinc oxynitride solar-activated photocatalyst for water splitting," *ECS Journal of Solid State Science and Technology*, vol. 2, no. 7, pp. Q118 – Q126, 2013.

[45] Y. B. Tang, Z. H. Chen, H. S. Song, C. S. Lee, H. T. Cong, H. M. Cheng, W. J. Zhang, I. Bello, and S. T. Lee, "Vertically aligned p-type single-crystalline GaN nanorod arrays on n-type Si for heterojunction photovoltaic cells," *Nano Lett.*, vol. 8, no. 12, pp. 4191 – 4195, 2008.

[46] F. Li, S. Lee, J. You, T. Kim, K. Lee, J. Lee, Y. Kwon, and T. Kang, "UV photovoltaic cells fabricated utilizing GaN nanorod/Si heterostructures," *J. Cryst. Growth*, vol. 312, no. 16 - 17, pp. 2320 – 2323, 2010.

[47] H. Wang, H. Chen, Y. Chang, C. Lin, H. Han, M. Tsai, H. C. Kuo, P. Yu, and S. Lin, "Conversion efficiency enhancement of GaN/$In_{0.11}Ga_{0.89}N$ solar cells with nano patterned sapphire and biomimetic surface antireflection process," *Photonics Technology Letters, IEEE*, vol. 23, no. 18, pp. 1304 – 1306, 2011.

[48] C.-T. Huang, J. Song, W.-F. Lee, Y. Ding, Z. Gao, Y. Hao, L.-J. Chen, and Z. L. Wang, "GaN nanowire arrays for high-output nanogenerators," *J. Am. Chem. Soc.*, vol. 132, no. 13, pp. 4766 – 4771, 2010.

[49] L. Lin, C.-H. Lai, Y. Hu, Y. Zhang, X. Wang, C. Xu, R. L. Snyder, L.-J. Chen, and Z. L. Wang, "High output nanogenerator based on assembly of GaN nanowires," *Nanotechnology*, vol. 22, no. 47, p. 475401, 2011.

[50] M. Minary-Jolandan, R. A. Bernal, I. Kuljanishvili, V. Parpoil, and H. D. Espinosa, "Individual GaN nanowires exhibit strong piezoelectricity in 3D," *Nano Lett.*, vol. 12, no. 2, pp. 970 – 976, 2012.

[51] C.-Y. Chen, G. Zhu, Y. Hu, J.-W. Yu, J. Song, K.-Y. Cheng, L.-H. Peng, L.-J. Chou, and Z. L. Wang, "Gallium nitride nanowire based nanogenerators and light-emitting diodes," *ACS Nano*, vol. 6, no. 6, pp. 5687 – 5692, 2012.

[52] Z. L. Wang, "Progress in piezotronics and piezo-phototronics," *Advanced Materials*, vol. 24, no. 34, pp. 4632 – 4646, 2012.

[53] W. Han, S. Fan, Q. Li, and Y. Hu, "Synthesis of gallium nitride nanorods through a carbon nanotube-confined reaction," *Science*, vol. 277, no. 5330, pp. 1287 – 1289, 1997.

[54] W. Han, P. Redlich, F. Ernst, and M. Rühle, "Synthesis of GaN-carbon composite nanotubes and GaN nanorods by arc discharge in nitrogen atmosphere," *Applied Physics Letters*, vol. 76, no. 5, pp. 652 – 654, 2000.

[55] W.-Q. Han and A. Zettl, "Pyrolysis approach to the synthesis of gallium nitride nanorods," *Applied Physics Letters*, vol. 80, no. 2, pp. 303 – 305, 2002.

[56] A. Hashimoto, T. M. amd H. Wada, and A. Yamamoto, "Silicon carbide, III-nitrides and related materials," *Mater. Sci. Forum*, vol. 264 - 268, pp. 1129 – 1132, 1998.

[57] G. S. Cheng, L. D. Zhang, Y. Zhu, G. T. Fei, L. Li, C. M. Mo, and Y. Q. Mao, "Large-scale synthesis of single crystalline gallium nitride nanowires," *Appl. Phys. Lett.*, vol. 75, no. 16, pp. 2455 – 2457, 1999.

[58] C.-C. Chen, C.-C. Yeh, C.-H. Chen, M.-Y. Yu, H.-L. Liu, J.-J. Wu, K.-H. Chen, L.-C. Chen, J.-Y. Peng, and Y.-F. Chen, "Catalytic growth and characterization of gallium nitride nanowires," *Journal of the American Chemical Society*, vol. 123, no. 12, pp. 2791 – 2798, 2001.

[59] J. Zhang, L. D. Zhang, X. F. Wang, C. H. Liang, X. S. Peng, and Y. W. Wang, "Fabrication and photoluminescence of ordered GaN nanowire arrays," *The Journal of Chemical Physics*, vol. 115, no. 13, pp. 5714 – 5717, 2001.

[60] H. W. Seo, S. Y. Bae, J. Park, H. Yang, K. S. Park, and S. Kim, "Strained gallium nitride nanowires," *The Journal of Chemical Physics*, vol. 116, no. 21, pp. 9492 – 9499, 2002.

[61] H.-M. Kim, D. Kim, Y. Park, D. Kim, T. Kang, and K. Chung, "Growth of GaN nanorods by a hydride vapor phase epitaxy method," *Adv. Mater.*, vol. 14, no. 13 - 14, pp. 991 – 993, 2002.

[62] J. C. Johnson, H.-J. Choi, K. P. Knutsen, R. D. Schaller, P. Yang, and R. J. Saykally, "Single gallium nitride nanowire lasers," *Nat. Mater.*, vol. 1, no. 2, pp. 106 – 110, 2002.

[63] X. Chen, J. Xu, R. Wang, and D. Yu, "High-quality ultra-fine GaN nanowires synthesized via chemical vapor deposition," *Advanced Materials*, vol. 15, no. 5, pp. 419 – 421, 2003.

[64] J. Li, C. Lu, B. Maynor, S. Huang, and J. Liu, "Controlled growth of long GaN nanowires from catalyst patterns fabricated by dip-pen nanolithographic techniques," *Chem. Mater.*, vol. 16, no. 9, pp. 1633 – 1636, 2004.

[65] X. Duan and C. M. Lieber, "Laser-assisted catalytic growth of single crystal GaN nanowires," *J. Am. Chem. Soc.*, vol. 122, no. 1, pp. 188 – 189, 2000.

[66] D. Ng, L. Tan, and M. Hong, "Synthesis of GaN nanowires on gold-coated substrates by pulsed laser ablation," *Curr. Appl. Phys.*, vol. 6, no. 3, pp. 403 – 406, 2006.

[67] J. Su, G. Cui, M. Gherasimova, H. Tsukamoto, J. Han, D. Ciuparu, S. Lim, L. Pfefferle, Y. He, A. V. Nurmikko, C. Broadbridge, and A. Lehman, "Catalytic growth of group III-nitride nanowires and nanostructures by metalorganic chemical vapor deposition," *Appl. Phys. Lett.*, vol. 86, no. 1, p. 013105, 2005.

[68] G. T. Wang, A. A. Talin, D. J. Werder, J. R. Creighton, E. Lai, R. J. Anderson, and I. Arslan, "Highly aligned, template-free growth and characterization of vertical GaN nanowires on sapphire by metal-organic chemical vapour deposition," *Nanotechnology*, vol. 17, no. 23, p. 5773, 2006.

[69] P. Deb, H. Kim, V. Rawat, M. Oliver, S. Kim, M. Marshall, E. Stach, and T. Sands, "Faceted and vertically aligned GaN nanorod arrays fabricated without catalysts or lithography," *Nano Lett.*, vol. 5, no. 9, pp. 1847 – 1851, 2005.

[70] S. D. Hersee, X. Sun, and X. Wang, "The controlled growth of GaN nanowires," *Nano Lett.*, vol. 6, no. 8, pp. 1808 – 1811, 2006.

[71] K. Bertness, A. Roshko, L. Mansfield, T. Harvey, and N. Sanford, "Nucleation conditions for catalyst-free GaN nanowires," *J. C*, vol. 300, no. 1, pp. 94 – 99, 2007.

[72] M. Tchernycheva, C. Sartel, G. Cirlin, L. Travers, G. Patriarche, J.-C. Harmand, L. S. Dang, J. Renard, B. Gayral, L. Nevou, and F. Julien, "Growth of GaN free-standing nanowires by plasma-assisted molecular beam epitaxy: structural and optical characterization," *Nanotechnology*, vol. 18, no. 38, p. 385306, 2007.

[73] R. Mata, K. Hestroffer, J. Budagosky, A. Cros, C. Bougerol, H. Renevier, and B. Daudin, "Nucleation of GaN nanowires grown by plasma-assisted molecular beam epitaxy: The effect of temperature," *J. Cryst. Growth*, vol. 334, no. 1, pp. 177 – 180, 2011.

[74] G. Seryogin, I. Shalish, W. Moberlychan, and V. Narayanamurti, "Catalytic hydride vapour phase epitaxy growth of GaN nanowires," *Nanotechnology*, vol. 16, no. 10, p. 2342, 2005.

[75] C.-C. Chen and C.-C. Yeh, "Large-scale catalytic synthesis of crystalline gallium nitride nanowires," *Adv. Mater.*, vol. 12, no. 10, pp. 738 – 741, 2000.

[76] A. M. Morales and C. M. Lieber, "A laser ablation method for the synthesis of crystalline semiconductor nanowires," *Science*, vol. 279, no. 5348, pp. 208 – 211, 1998.

[77] M. S. Gudiksen and C. M. Lieber, "Diameter-selective synthesis of semiconductor nanowires," *J. Am. Chem. Soc.*, vol. 122, no. 36, pp. 8801 – 8802, 2000.

[78] P. Yang, "The chemistry and physics of semiconductor nanowires," *MRS Bulletin*, vol. 30, pp. 85 – 91, 2005.

[79] J. Stangl, V. Holý, and G. Bauer, "Structural properties of self-organized semiconductor nanostructures," *Rev. Mod. Phys.*, vol. 76, pp. 725 – 783, 2004.

[80] R. Calarco, R. J. Meijers, R. K. Debnath, T. Stoica, E. Sutter, and H. Lüth, "Nucleation and growth of GaN nanowires on Si (111) performed by molecular beam epitaxy," *Nano Lett.*, vol. 7, no. 8, pp. 2248 – 2251, 2007.

[81] R. S. Wagner and W. C. Ellis, "Vapor-liquid-solid mechanism of single crystal growth," *Appl. Phys. Lett.*, vol. 4, no. 5, pp. 89 – 90, 1964.

[82] H. T. Ng, J. Li, M. K. Smith, P. Nguyen, A. Cassell, J. Han, and M. Meyyappan, "Growth of epitaxial nanowires at the junctions of nanowalls," *Science*, vol. 300, no. 5623, p. 1249, 2003.

[83] E. A. Stach, P. J. Pauzauskie, T. Kuykendall, J. Goldberger, R. He, and P. Yang, "Watching GaN nanowires grow," *Nano Lett.*, vol. 3, no. 6, pp. 867 – 869, 2003.

[84] T. I. Kamins, R. S. Williams, Y. Chen, Y.-L. Chang, and Y. A. Chang, "Chemical vapor deposition of Si nanowires nucleated by $TiSi_2$ islands on Si," *Appl. Phys. Lett.*, vol. 76, no. 5, pp. 562 – 564, 2000.

[85] W.-C. Hou, L.-Y. Chen, W.-C. Tang, and F. C. N. Hong, "Control of seed detachment in Au-assisted GaN nanowire growths," *Cryst. Growth Des.*, vol. 11, no. 4, pp. 990 – 994, 2011.

[86] X. Weng, R. A. Burke, and J. M. Redwing, "The nature of catalyst particles and growth mechanisms of GaN nanowires grown by Ni-assisted metal-organic chemical vapor deposition," *Nanotechnology*, vol. 20, no. 8, p. 085610, 2009.

[87] Y. Lan, M. Crimp, and J. Zhang, "Anisotropic vapor phase growth of Ga_2O_3 crystalline nanobelts," *J. Cryst. Growth*, vol. 290, no. 2, pp. 585 – 591, 2006.

[88] H. Wang, Y. Lan, J. Zhang, M. A. Crimp, and Z. Ren, "Growth mechanism and elemental distribution of β-Ga_2O_3 crystalline nanowires synthesized by cobalt-assisted chemical vapor deposition," *J. Nanosci. Nanotechnol.*, vol. 12, no. 4, pp. 3101 – 3107, 2012.

[89] X. Xiang and H. Zhu, "One-dimensional gallium nitride micro/nanostructures synthesized by a space-confined growth technique," *Applied Physics A*, vol. 87, no. 4, pp. 651 – 659, 2007.

[90] L. T. Canham, "Silicon quantum wire array fabrication by electrochemical and chemical dissolution of wafers," *Applied Physics Letters*, vol. 57, no. 10, pp. 1046 – 1048, 1990.

[91] V. Lehmann and U. Gösele, "Porous silicon formation: A quantum wire effect," *Applied Physics Letters*, vol. 58, no. 8, pp. 856 – 858, 1991.

[92] C. J. Frosch and C. D. Thurmond, "The pressure of Ga_2O over gallium-Ga_2O_3 mixtures," *The Journal of Physical Chemistry*, vol. 66, no. 5, pp. 877 – 878, 1962.

[93] C. M. Balkaş and R. F. Davis, "Synthesis routes and characterization of high-purity, single-phase gallium nitride powders," *Journal of the American Ceramic Society*, vol. 79, no. 9, pp. 2309 – 2312, 1996.

[94] Y. G. Cao, X. L. Chen, J. Y. Li, Y. C. Lan, and J. K. Liang, "Observation of a quantum-confinement effect with GaN nanoparticles synthesized through a new gas reaction route," *Applied Physics A: Materials Science and Processing*, vol. 71, no. 2, pp. 229 – 231, 2000.

[95] W. C. Johnson, J. B. Parson, and M. C. Crew, "Nitrogen compounds of gallium. III," *The Journal of Physical Chemistry*, vol. 36, no. 10, pp. 2651 – 2654, 1931.

[96] Y. Lan, X. Chen, Y. Cao, Y. Xu, T. Xu, J. Li, Z. Tao, and J. Liang, "Morphology of GaN in ammonia," *J. Mater. Sci. Lett.*, vol. 19, no. 24, pp. 2215 – 2217, 2000.

[97] A. C. Frank, F. Stowasser, H. Sussek, H. Pritzkow, C. R. Miskys, O. Ambacher, M. Giersig, and R. A. Fischer, "Detonations of gallium azides: A simple route to hexagonal GaN nanocrystals," *Journal of the American Chemical Society*, vol. 120, no. 14, pp. 3512 – 3513, 1998.

[98] J.-W. Hwang, J. P. Campbell, J. Kozubowski, S. A. Hanson, J. F. Evans, and W. L. Gladfelter, "Topochemical control in the solid-state conversion of cyclotrigallazane into nanocrystalline gallium nitride," *Chemistry of Materials*, vol. 7, no. 3, pp. 517 – 525, 1995.

[99] J. L. Coffer, M. A. Johnson, L. Zhang, R. L. Wells, and J. F. Janik, "Influence of precursor route on the photoluminescence of bulk nanocrystalline gallium nitride," *Chemistry of Materials*, vol. 9, no. 12, pp. 2671 – 2673, 1997.

[100] Y. G. Cao, X. L. Chen, Y. C. Lan, J. Y. Li, Y. P. Xu, T. Xu, and J. K. Liang, "Fabrication of nano-sized AlGaN alloy by dry milling and thermal annealing," *J. Alloys Compd.*, vol. 309, no. 1-2, pp. L13 – L15, 2000.

[101] Y. G. Cao, X. L. Chen, Y. C. Lan, Y. P. Xu, T. Xu, and J. K. Liang, "Synthesis, Raman scattering, and infrared spectra of a new condensed form of GaN nanophase material," *J. Mater. Res.*, vol. 15, no. 2, pp. 267 – 269, 2000.

[102] R. Dwilinski, J. A. M. Baranowski, M. Kaminska, R. Doradzinski, J. Garczynski, and L. Siersputowski, "On GaN crystallization by ammonothermal method," *Acta Phys. Pol. A*, vol. 90, p. 763, 1996.

[103] A. P. Purdy, "Ammonothermal synthesis of cubic gallium nitride," *Chemistry of Materials*, vol. 11, no. 7, pp. 1648 – 1651, 1999.

[104] Y. Xie, Y. Qian, W. Wang, S. Zhang, and Y. Zhang, "A benzene-thermal synthetic route to nanocrystalline GaN," *Science*, vol. 272, no. 5270, pp. 1926 – 1927, 1996.

[105] M. R. Lorenz and B. B. Binkowski, "Preparation, stability, and luminescence of gallium nitride," *J. Electrochem. Soc.*, vol. 109, no. 1, pp. 24 – 26, 1962.

[106] S. Jiang, J. Zhang, X. Qi, M. He, and J. Li, "Large-area synthesis of diameter-controllable porous single crystal gallium nitride micro/nanotube arrays," *CrystEngComm*, vol. 15, pp. 9837 – 9840, 2013.

[107] L.-W. Yin, Y. Bando, M.-S. Li, and D. Golberg, "Growth of semiconducting GaN hollow spheres and nanotubes with very thin shells via a controllable liquid gallium-gas interface chemical reaction," *Small*, vol. 1, no. 11, pp. 1094 – 1099, 2005.

[108] B. Liu, Y. Bando, M. Wang, C. Tang, M. Mitome, and D. Golberg, "Crystallography and elasticity of individual GaN nanotubes," *Nanotechnology*, vol. 20, no. 18, p. 185705, 2009.

[109] Y. Lan, H. Wang, F. Lin, Y. Lu, Y. Li, Y. Liu, J. Bao, Z. Ren, and M. Crimp, "Nanoporous gallium nitride square microtubes," *Journal of Materials Science*, vol. 48, no. 21, pp. 7703 – 7707, 2013.

[110] Y. Lan, H. Wang, H. Zhao, and Z. F. Ren, "Gan microtubes," *J. Cryst. Growth*, no. 0, p. submitted, 2014.

[111] M. Sacilotti, L. Imhoff, S. Bourgeois, C. Dumas, J. Decobert, P. Baldeck, and I. Colombier, "MOVPE growth of Ga 3D structures for fabrication of GaN materials," *J. Cryst. Growth*, vol. 261, no. 2-3, pp. 253 – 258, 2004.

[112] Z. W. Pan, Z. R. Dai, C. Ma, and Z. L. Wang, "Molten gallium as a catalyst for the large-scale growth of highly aligned silica nanowires," *J. Am. Chem. Soc.*, vol. 124, no. 8, pp. 1817 – 1822, 2002.

[113] E. Filippo, D. Manno, and A. Serra, "Characterization and growth mechanism of selenium microtubes synthesized by a vapor phase deposition route," *Cryst. Growth Des.*, vol. 10, no. 11, pp. 4890 – 4897, 2010.

[114] I. Lagzi, B. Kowalczyk, and B. A. Grzybowski, "Liesegang rings engineered from charged nanoparticles," *J. Am. Chem. Soc.*, vol. 132, no. 1, pp. 58 – 60, 2010.

[115] E. L. Spotz and J. O. Hirschfelder, "Liesegang ring formation arising from diffusion of ammonia and hydrogen chlorine gases through air," *J. Chem. Phys.*, vol. 19, no. 9, pp. 1215 - 1215, 1951.

[116] C. F. Goodeve, A. S. Eastman, and A. Dooley, "The reaction between sulphur trioxide and water vapours and a new periodic phenomenon," *Trans. Faraday Soc.*, vol. 30, pp. 1127 - 1133, 1934.

[117] L. V. Govor, G. Reiter, G. H. Bauer, and J. Parisi, "Nanoparticle ring formation in evaporating micron-size droplets," *Appl. Phys. Lett.*, vol. 84, no. 23, pp. 4774 – 4776, 2004.

[118] D. Nicholls, *Inorganic chemistry in liquid ammonia*. Topics in inorganic and general chemistry, Elsevier Scientificc Publishing Company, 1979.

[119] D. Tsivion, M. Schvartzman, R. Popovitz-Biro, P. von Huth, and E. Joselevich, "Guided growth of millimeter-long horizontal nanowires with controlled orientations," *Science*, vol. 333, no. 6045, pp. 1003 – 1007, 2011.

[120] M. Knoll, "Aufladepotentiel und sekundäremission elektronenbestrahlter körper," *Z. Tech. Phys.*, vol. 16, pp. 467 – 475, 1935.

[121] M. von Ardenne, "Das elektronen-rastermikroskop. theoretische grundlagen," *Z. Phys*, vol. 109, pp. 553 – 572, 1938.

[122] M. von Ardenne, "Das elektronen-rastermikroskop. praktische ausführung," *Z. Tech. Phys.*, vol. 19, pp. 407 – 416, 1938.

[123] V. A. Zworykin, J. Hillier, and R. L. Snyder, "A scanning electron microscope," *ASTM Bull.*, vol. 117, pp. 15 – 23, 1942.

[124] D. McMullan, "An improved scanning electron microscope for opaque specimens," *Proc. Inst. Electr. Engrs.*, vol. 100, pp. 245 – 249, 1953.

[125] Z. F. Ren, Y. C. Lan, and Y. Wang, *Aligned Carbon Nanotubes: Physics, Concepts, Fabrication and Devices*. Heidelberg, Berlin: Springer-Verlag, 2013.

[126] Y. Lan, Y. Wang, and Z. F. Ren, "Physics and applications of aligned carbon nanotubes," *Adv. Phys.*, vol. 60, no. 4, pp. 553 – 678, 2011.

[127] G. I. Goldstein, D. E. Newbury, P. Echlin, D. C. Joy, C. Fiori, and E. Lifshin, *Scanning Electron Microscopy and X-ray Microanalysis*. New York: Plenum Press, 1981.

[128] I. M. Watt, *The Principles and Practice of Electron Microscopy*. Cambridge: Cambridge University Press, 1985.

[129] L. Reimer, *Scanning Electron Microscopy: Physics of Image Formation and Microanalysis*. Heidelberg: Springer, 1998.

[130] A. R. Clarke and C. Eberhardt, *Microscopy Techniques for Materials Science*. Boca Raton: CRC Press, 2002.

[131] J. Goldstein, *Scanning Electron Microscopy and X-ray Microanalysis*. New York: Kluwer Adacemic/Plenum Pulbishers, 2003.

[132] R. F. Egerton, *Physical Principles of Electron Microscopy: An Introduction to TEM, SEM, and AEM*. New York: Springer, 2005.

[133] R. Erni, M. D. Rossell, C. Kisielowski, and U. Dahmen, "Atomic-resolution imaging with a sub-50-pm electron probe," *Phys. Rev. Lett.*, vol. 102, no. 9, p. 096101, 2009.

[134] A. Smekal, "Zur quantentheorie der dispersion," *Naturwissenschaften*, vol. 11, no. 43, pp. 873 – 875, 1923.

[135] C. V. Raman and K. S. Krishnan, "A new radiation.," *Indian J. Phys.*, vol. 2, p. 387, 1928.

[136] C. V. Raman and K. S. Krishnan, "A new type of secondary radiation," *Nature*, vol. 121, p. 501, 1928.

[137] G. Landsberg and L. Mandelstam, "Eine neue erscheinung bei der lichtzerstreuung in krystallen," *Naturwissenschaften*, vol. 16, no. 28, p. 557, 1928.

[138] J. Y. Li, X. L. Chen, Y. G. Cao, Z. Y. Qiao, and Y. C. Lan, "Raman-scattering spectrum of GaN straight nanowires," *Appl. Phys. A: Mater.*, vol. 71, no. 3, pp. 345 – 346, 2000.

[139] M. Kuball, "Raman spectroscopy of GaN, AlGaN and AlN for process and growth monitoring/control," *Surf. Interface Anal.*, vol. 31, no. 10, pp. 987 – 999, 2001.

[140] H. Harima, "Properties of GaN and related compounds studied by means of Raman scattering," *Journal of Physics: Condensed Matter*, vol. 14, no. 38, p. R967, 2002.

[141] T. Livneh, J. Zhang, G. Cheng, and M. Moskovits, "Polarized Raman scattering from single GaN nanowires," *Phys. Rev. B*, vol. 74, p. 035320, 2006.

[142] H.-L. Liu, C.-C. Chen, C.-T. Chia, C.-C. Yeh, C.-H. Chen, M.-Y. Yu, S. Keller, and S. P. DenBaars, "Infrared and Raman-scattering studies in single-crystalline GaN nanowires," *Chemical Physics Letters*, vol. 345, no. 3 - 4, pp. 245 – 251, 2001.

[143] L. Tan, F. Yam, L. Low, K. Beh, and Z. Hassan, "The influence of growth temperatures on the characteristics of GaN nanowires: The Raman study," *Physica B*, vol. 434, no. 0, pp. 101 – 105, 2014.

[144] P. J. Pauzauskie, D. Talaga, K. Seo, P. Yang, and F. Lagugn-Labarthet, "Polarized raman confocal microscopy of single gallium nitride nanowires," *J. Am. Chem. Soc.*, vol. 127, no. 49, pp. 17146 – 17147, 2005.

[145] Y. G. Cao, X. L. Chen, Y. C. Lan, J. Y. Li, Y. Zhang, Z. Yang, and J. K. Liang, "Synthesis and Raman characteristics of hexagonal $Al_xGa_{1-x}N$ alloy nanocrystalline solids through ammonothermal routes," *Applied Physics A: Materials Science and Processing*, vol. 72, no. 1, pp. 125 – 127, 2001.

[146] L. Jing, Y. Qu, B. Wang, S. Li, B. Jiang, L. Yang, W. Fu, H. Fu, and J. Sun, "Review of photoluminescence performance of nano-sized semiconductor materials and its relationships with photocatalytic activity," *Sol. Energ. Mat. Sol. C.*, vol. 90, no. 12, pp. 1773 – 1787, 2006.

[147] J. Y. Li, X. L. Chen, Z. Y. Qiao, Y. G. Cao, and Y. C. Lan, "Photoluminescence spectrum of straight GaN nanowires," *J. Mater. Sci. Lett.*, vol. 20, no. 8, pp. 757 – 758, 2001.

[148] A. H. Chin, T. S. Ahn, H. Li, S. Vaddiraju, C. J. Bardeen, C.-Z. Ning, and M. K. Sunkara, "Photoluminescence of GaN nanowires of different crystallographic orientations," *Nano Letters*, vol. 7, no. 3, pp. 626 – 631, 2007.

In: Gallium Nitride
Editor: Kiera Olivia Peak
ISBN: 978-1-63321-387-6

Chapter 5

GaN Schottky Contacts and Their Applications

R. X. Wang[1], S. J. Xu[2] and S. Fung[2]
[1]Suzhou Institute of Nano-Tech and Nano-Bionics,
Chinese Academy of Sciences, Suzhou, China
[2]Department of Physics, The University of Hong Kong,
Hong Kong, China

Abstract

Electrical contacts are vital for the fabrication of modern electronic and optoelectronic devices. A junction forms when a metal is in intimate contact with a semiconductor. This junction is either an ohmic or a Schottky contact. Typically, the latter, being more technologically important, exhibits distinct rectifying characteristics and has thus attracted much more interest due to a wide range of practical applications. Currently, Schottky contacts on GaN are still an active topic because the properties of contacts are found to be strongly dependent on not only the GaN materials, such as quality, back ground carrier concentration, and surface status as well as their growth techniques and conditions, but also the deposition techniques and parameters of contact materials as well as post-deposition treatments. This chapter provides an overall review of GaN based Schottky contacts using metals or transparent conductive oxide (TCO) as contact materials. Particularly, the influence of GaN epilayer qualities and preparation conditions such as surface treatment, deposition parameters, selection of contact materials etc., on the microstructure, surface status, and electrical characteristics of Schottky contacts is addressed. Finally, comparison with different materials and the optimized fabrication conditions for selected device applications are discussed.

1. Introduction

GaN and most related III-nitride compounds of either wurtzite or cubic crystallize structures are direct bandgap semiconductor materials. These group-III nitrides form a continuous range of direct energy bandgaps throughout the entire visible spectral range and even into the deep ultraviolet region. They have attracted a great deal of interest in recent years, owing to their wide optoelectronic and electronic applications, such as blue/ultraviolet light-emitting diodes [1, 2], full color display [3], ultraviolet (UV) detectors [4, 5], photovoltaic devices [6], visible and UV laser diodes [7], data storage, field-effect transistors for high-temperature and high-power/high-frequency [8, 9], wireless communication [1]. All these exciting practical applications demand for high performance electrical contacts on GaN based materials. It is known that the current majority of GaN and related compounds for device applications are grown epitaxially on foreign substrates such as sapphire, SiC, or Si etc. due to the lack of large size bulk GaN. These GaN heteroepitaxies usually show high density of defects including stacking faults and threading dislocations and high *n*-type carrier concentration. [9, 10].

In the past two decades, great progress has been achieved in the growth of GaN based materials and many of technical difficulties have been overcome. For example, the two-step low-temperature growth technique, inserting an optimized buffer layer approach, and the usage of completely matched bulk or free-standing GaN substrates have led to significant epilayer defect density reduction [11]. For *n*-GaN materials, presently, the background carrier concentrations have been reduced to be as low as 10^{16} cm^{-3} and the mobility has been improved to be over 1000 cm^2/Vs at room temperature. For *p*-GaN materials, the hole concentration of 3×10^{18} cm^{-3} has been consistently achieved. At the same time, some reliable and appropriate methods and processes for fabricating GaN devices have been developed. These progresses and developments in material qualities and processes have made some GaN-based devices commercially viable. As mentioned in the forgoing, metal/GaN electrical contacts are of fundamentally importance to the GaN-based electronic and optoelectronic devices. Therefore, the fabrication, characterization and understanding of the metal/GaN contacts, especially Schottky contacts are a very important issue.

Usually, metal/GaN electrical contacts can be categorized into two groups, namely, Schottky and ohmic contacts. Schottky contact exhibits distinct rectifying characteristics whereas ohmic contact shows linear *I-V* dependence. The former has attracted much more interest due to a wide range of practical applications, such as solar-blind UV detectors [12], gas sensors [13, 14], high-power devices [15], field-effect transistors [16-19], and high-electron mobility transistors [19]. Currently, Schottky contacts on GaN are still an active topic. Various challenges still exist in the fabrication of high performance metal contacts with GaN because the properties of the contacts have been found to be dependent strongly on GaN materials and contact materials as well as their preparation methods and conditions. [20-22] Indeed an extensive but incomplete survey of the literature on the most extensively studied metal/*n*-GaN Schottky contacts reveals a large spread of Schottky barrier height (SBH) values and ideality factors as well as abnormal current leakage under reverse bias. [1, 2, 9, 23, 24] Among the parameters characterizing the contact quality, the values of ideality factor n and barrier height $q\phi_b$ are the two key indicators. It has been established that the values of the SBHs and ideality factors depend critically on metals used for the Schottky contacts as well as deposition techniques and conditions in addition to the quality and surface status of GaN materials. Typically, when contact metal is deposited onto the GaN surface, the surface atoms of GaN will interact with the arriving metal atoms, leading to the rearranged structures at the interface region, the creation of some alloys, and the formation of interstitial and/or substitutional atoms. During the interaction, variations of strain energy and chemical compositions may change the energy and band bending status of the contact barrier layer. In addition, the strong interaction may result in the production of the defective sites and the inhomogeneity distribution of excess charges, even causes delamination of the contact films. [25-27] In particular, on GaN materials which have ionic-like chemical bonds and surface polarization, i.e., N- and Ga-polarity layer, these variations may cause more complicated effects, such as changing polarization-induced charges. The resulted defects and/or their complexes near the metal–semiconductor interface may act as surface donors and deep level centers so as to influence the characteristics of devices. [27] In practical preparations of Schottky junctions, contact materials can be deposited in single layer or multilayers. One important requirement is that the work function of the contact materials must be of the correct value. It has to be shallower than the conduction-band edge of GaN in the case of *n*-type GaN. [28] So far, good Schottky contacts on *n*-GaN have been realized with some metals with higher work functions among which Au, Pt, Ni, are the most often used metals. [29, 30] After much effort at developing metal/*n*-type GaN Schottky contacts, a comprehensive summary and review on the issue, which is our motivation and also goal in this chapter.

This review is organized as follows: In section 2, a general description of the electrical transport mechanism of Schottky contacts is given. Section 3 contains a detailed review on the effects of GaN materials quality, surface treatment, contact materials and deposition conditions on the microstructure, surface morphology, and electrical characteristics of Schottky contacts. In section 4, an overview of typical applications of GaN-based Schottky contacts is presented.

2. General Description of the Electrical Transport Mechanism of Schottky Contacts

It is well known that a good ohmic contact shows a linear current-voltage (*I-V*) characteristic. For Schottky contacts, however, a good rectifying behavior is the most important *I-V* characteristic. This behavior is caused by electron transporting over an effective potential energy barrier which is a function of a space charge region on the semiconductor side. This barrier is characterized by a Schottky barrier height being the most important parameter determining the current-voltage and capacitance-voltage (*C-V*) characteristics of metal-semiconductor contact. In the following, we will introduce some basic theoretical considerations to the electron transport in terms of *I-V* and *C-V* curves.

A. Current-Voltage Characteristic

First of all, we would like to point out that a number of models have been established for interpreting various transport mechanisms in metal-semiconductor Schottky contacts. [23, 26, 31-39] According to the Schottky-Mott model, the Schottky barrier height is dependent on the difference between the metal work function $q\phi_M$ and electron affinity of semiconductor χ: [40] $q\phi_b = q\phi_M - \chi$. For *n*-type semiconductors: $q\phi_b = (q\phi_M - \chi)$. For *p*-type semiconductors: $q\phi_b = (E_{gap} + \chi) - q\phi_M$. Under the thermionic emission conditions, the current I_{TE} is given by [41, 42]

$$I_{TE} = I_0 \exp\left(\frac{qV}{nkT}\right)\left[1 - \exp\left(\frac{qV}{kT}\right)\right], \tag{1}$$

where the leakage current of the Schottky diode is saturated as:

$$I_0 = A^* T^2 \exp\left(-\frac{q\phi_b}{kT}\right), \tag{2}$$

where q is the electron charge, A^* is the effective Richardson's constant, T is the absolute temperature, k is the Boltzmann's constant, V is the applied voltage and $q\phi_b$ is the SBH and n is the ideality factor. However, some other effects, including generation and recombination of carriers in the charge depletion layer, the field-dependent tunneling through the barrier height, and the defect-assisted conduction, result in a deviation from the ideal thermionic emission. The total current density I through the Schottky barrier is thus the sum of current components from various mechanisms:

$$I = I_{TE} + I_{TUN} + I_{GR}, \tag{3}$$

where I_{TE}, I_{TUN}, and I_{GR} are the thermionic emission current, tunneling current, and generation-recombination current, respectively. The recombination current I_{GR} arises from the minority carrier injection into the Schottky barrier region. The ratio of the thermionic to recombination current in the wide band gap semiconductor can be estimated by [41, 42]

$$T^2 \tau \exp\left[\frac{q(E_g + V - 2\phi_b)}{2kT}\right], \tag{4}$$

where E_g is the band gap of semiconductor and τ is the life time within the depletion region. The tunneling current density I_{TUN} is proportional to the tunneling probability as [41]

$$I_{TUN} \sim \exp\left(-2W\sqrt{\frac{2m^* q(\phi_b - V)}{\hbar^2}}\right), \tag{5}$$

where m^* is the effective electron mass of semiconductor and W is the depletion layer width,

$$W = \sqrt{\frac{2\varepsilon(\phi_b - V)}{qN_d}}, \tag{6}$$

where ε is the dielectric constant of semiconductor and N_d is the carrier concentration of semiconductor.

For contacts with high doping or defect-assisted conduction, the barrier width becomes narrow and the tunneling current may become dominant. The tunneling current for the forward bias is [40]

$$I_{FTUN} = I_0 \exp\left(\frac{qV}{E_0}\right), \tag{7}$$

$$E_0 = E_{00} \coth\left(\frac{E_{00}}{kT}\right), \tag{8}$$

where E_{00} is the characteristic tunneling energy that is related to the tunneling transmission probability:

$$E_{00} = \left(\frac{h}{4\pi}\right)\sqrt{\frac{N_d}{m^* \varepsilon}}, \tag{9}$$

where h is the Planck constant. The current under reverse bias can be expressed by the following equations: [35, 36]

$$I_{RTUN} = I_{STUN} \exp\left[\frac{qV_R}{kT} - \frac{qV_R}{E_{00}\coth\left(E_{00}/kT\right)}\right], \tag{10}$$

$$I_{STUN} = \frac{AA^*T\sqrt{\pi E_{00}}}{k}\left[\left(V_R - V_0\right) + \frac{\phi_b}{\coth^2\left(E_{00}/kT\right)}\right]^{1/2} \exp\left(-\frac{\phi_b}{E_{00}\coth\left(E_{00}/kT\right)}\right), \tag{11}$$

where V_R is the reverse bias voltage, A is the area of the contact.

As emphasized earlier, the SBH and ideality factor are the two most important parameters that determine the carrier transport characteristics of metal/semiconductor contacts. The SBH and the ideality factor can be derived from the linear fit of the current-voltage measurement plot:

$$\phi_b = \frac{kT}{q}\ln\left(\frac{A^*T^2}{I_0}\right), \tag{12}$$

and

$$n = \frac{q}{kT}\left(\frac{dV}{d\ln(I)}\right), \tag{13}$$

respectively. The theoretical value A^* is determined to be 26.4 A/(cm^{2}K^2) by using $A^* = 4\pi m^* qk/h^3$. For *n*-type GaN, $m^* = 0.22m_0$ is the electron's effective mass, and m_0 is the electron rest mass.

To obtain more detailed information about the conduction process or the nature of barrier formation at the metal–semiconductor (MS) interface, the *I–V* characteristics of the Schottky diode at different temperatures are usually analyzed. [37, 40, 43-48] Since accurate extrapolation to zero voltage may be difficult, an approach in determining SBH and A^* is the modified Norde plot [49] and the activation energy plot [50, 51] from Norde function [52]:

$$F(V) = \frac{V}{2} - \frac{kT}{q}\ln\left(\frac{I}{R_s AA^* T^2}\right), \tag{14}$$

where R_s is a series resistance. The zero-bias SBH can be extracted: [48, 52]

$$\phi_{b0} = F(V_{\min}) + \frac{V_{\min}}{2} - \frac{kT}{q}, \tag{15}$$

where V_{min} is the minimum voltage. Using the measured value of the minimum F_{min} and the corresponding current I_{min}, the value of SBH and A^* can be derived.

In reality, the SBH and the ideality factor may be altered by many factors such as structural defects, interface roughness, interfacial reactions and diffusion/interdiffusion of applied materials on the surfaces, inhomogeneous doping, as well as inhomogeneities of thickness and composition of the layers, and non-uniformity of interfacial charges or the presence of a thin insulating layer between metal and semiconductor. [37, 39, 53] For example, it has been reported that the values of the Richardson constant is much scattered and varies with temperature, ranging from 3.3-80 A/(cm^{2}K^2). [29, 47, 54-61] These scattered values suggest that there is an interfacial barrier through which the electron tunnels [50], the decrease or increase of the effective contact area [51], spatially inhomogeneous barrier heights and potential profile fluctuations at the interface that consists of low and high barrier areas. [29, 37] For GaN Schottky devices, the ionic component of bonds is quite substantial. The surface exhibits active chemisorption, which leads to the formation of surface native oxides and hydroxides after the surface exposes in air environments. That not only produces layer of native oxides and hydroxides, but also induces other effects because of charge and nonuniformity of the GaN surface involving many dislocations to alter the Schottky-Mott barrier height and ideality factor. When an insulating layer exists, n should be modified to: [2, 62]

$$n_{mdify} = 1 + \frac{d_{insul}(\varepsilon_s + qN_{sb})}{1 + d_{insul} qN_{sa}}, \tag{16}$$

$$d_{insul} = \frac{D_{insul}}{\varepsilon_{insul}}, \tag{17}$$

where N_{sa} and N_{sb} are respectively a density of interface states in equilibrium with the metal and semiconductor, D_{insul} is the thickness of the insulating layer, and ε_{insul} is the dielectric constant of the insulating layer. [2, 62] When some charges at the Schottky interface act as additional ionized doping atoms which enhance the concentration of ionized doping atoms at the surface, this effect then makes the SBH thinner at the surface and facilitates the tunneling of the carriers through the barrier. An obvious alternative thin Schottky barrier (TSB) model explains the observed anomalous current transport phenomenon. The TSB regions are distributed spatially nonuniformly rather than being one uniform region. [38] The forward saturation current of Schottky barrier diodes are given as [38]

$$I_{FS} = \eta \frac{AA^{*}qT}{k\cosh(E_{00}/kT)}\left[\pi E_{00}(\phi_b - \phi_0)\tanh\left(\frac{E_{00}}{kT}\right)\right]^{1/2} \exp\left(-\frac{qV_n}{kT}\right)\exp\left[-\frac{q(\phi_b - V_n)}{n_F kT}\right], \tag{18}$$

$$V_n = kT\ln\left(\frac{N_C}{N_D}\right), \tag{19}$$

$$n_F = \left(\frac{E_{00}}{kT}\right)\coth\left(\frac{E_{00}}{kT}\right), \tag{20}$$

where η is the ratio of the total TSB area to total sample area, ϕ_0 is minimum potential, N_C and N_D are the effective density of states in the conduction band of the semiconductor and the donor density, respectively. The reverse current in the thermionic emission dominant case can be expressed by the following equations: [35, 36]

$$I_R = I_s \exp\left[\frac{V_R}{kT} - \frac{V_R}{E_{00}\coth(E_{00}/kT)}\right], \tag{21}$$

$$I_s = \frac{A^{*}T\sqrt{\pi E_{00}}}{k}\left[(V_R - V_0) + \frac{\phi_b}{\coth^2(E_{00}/kT)}\right]^{1/2} \exp\left(-\frac{\phi_b}{E_{00}\coth(E_{00}/kT)}\right), \tag{22}$$

Due to this spatially non-uniformity of SBH, a Gaussian distribution of the bias- and temperature-dependent barrier heights exists according to the following expression: [37, 39, 63-67]

$$\phi_{apb}(V,T)=\overline{\phi_{apb}}(0,T)-\frac{q\sigma^2(0)}{2kT}+\left(\rho_2-\frac{q\rho_3}{2kT}\right)V, \quad (23)$$

where $q\phi_{apb}$ is apparent barrier height measured experimentally. $q\overline{\phi_{apb}}$ and σ_0 are the average value and the standard deviation from the mean Schottky barrier height, respectively. The variation of observed ideality factor with temperature in the model is given by [39]

$$n_{ap}=\frac{1}{1-\rho_2+\dfrac{q\rho_3}{2kT}}, \quad (24)$$

$$\rho_2 V=\Delta\overline{\phi_{apb}}(V,T)=\overline{\phi_{apb}}(V,T)-\overline{\phi_{apb}}(0,T), \quad (25)$$

$$\rho_3 V=\Delta\sigma^2(V)=\sigma^2(V)-\sigma^2(0), \quad (26)$$

where $n_{\rm ap}$ is the apparent ideality factor and ρ_2 the voltage coefficient of the mean barrier height, and ρ_3 the voltage coefficient of the standard deviation. At lower temperatures the carriers can easilier overcome the lower SBH and the dominant current passes through these regions. As the temperature increases, more carriers obtain more energy to surmount higher SBH and the current increases.

B. Capacitance-Voltage Characteristic

For metal-semiconductor Schottky contacts, its depletion layer capacitance as function of the applied reverse bias is given by: [41, 42]

$$C=\sqrt{\frac{\varepsilon q N_d A^2}{2(V_b+V_R-kT/q)}}, \quad (27)$$

$$\phi_b=V_b+V_n. \quad (28)$$

For an ideal Schottky barrier, the space charge will become larger than its equilibrium value when the reverse bias increases. The amount of charge in the depletion region increases due to increasing number of ionized defect centers. Thus, the carrier concentration changes appreciably and a local slope of the C^{-2}-V curve decreases gradually with increasing bias voltage. The Schottky barrier height can be derived from the C^{-2}-V curves. It has been reported that there are different values of SBH obtained from the I-V and C-V curves. This is probably due to the presence of an insulating layer or charges existing at the metal–semiconductor interface, deep impurity levels, image force barrier lowering, and edge leakage currents. For a device with surface states, an additional surface-state capacitance should be considered. Thus, the total capacitance of the device is written as

$$C_s = C_{SS} + C_{SC}, \tag{29}$$

where C_{SC} and C_{SS} are, respectively, the differential space-charge and surface-state capacitances. According to the work of Many et al. C_{SC} and C_{SS} are given by [68,69]

$$C_{SC} \equiv \left|\frac{dQ_{SC}}{dV_S}\right| = \frac{q}{kT}\left|\frac{dQ_{SC}}{dv_S}\right|, \tag{30}$$

and

$$C_{SS} \equiv \left|\frac{dQ_{SS}}{dV_S}\right| = \frac{q}{kT}\left|\frac{dQ_{SS}}{dv_S}\right| \tag{31}$$

where Q_{SC} is the space-charge density, Q_{SS} is the surface state-charge density which is equal to $-qN_{SS}$ for acceptor like surface states and $+qN_{SS}$ for donor like surface states, $v_s \equiv qV_S / kT$ is dimensionless barrier height. For the case of fully ionized impurities and single-charge surface states characterized with both the density N_{SS} and effective energy level E_t, one yields [68,69]

$$C_{SC} = \frac{\varepsilon\varepsilon_0}{L}\left|\frac{df_S}{dv_S}\right|, \tag{32}$$

and

$$C_{SS} = \frac{q^2 N_{SS}}{kT}\left|\frac{df(E_t)}{dv_S}\right|, \tag{33}$$

where ε and ε_0 are, respectively, the relative dielectric constant of GaN and the dielectric constant of vacuum, L is the effective Debye length characterizing the width of the space-charge region, f_S is the value of a function f at the surface, and $f(E_t)$ is the Fermi-Dirac distribution function. Usually, the exact expression of the function f for a metal-semiconductor Schottky contact is not easily found. A study of the characteristics of the functions f_S and $f(E_t)$ reveals that C_{SC} achieves a minimum value in the depletion layers and increases rapidly in both accumulation and inversion layers. On the other hand, C_{SS} reaches its maximum value when the Fermi level at the surface crosses the effective energy level E_t which is induced by the surface states. Based on the work of Many et al. [68], C_{SS} can be derived as

$$C_{SS} = \frac{q^2 N_{SS}}{kT} \frac{\exp\left[(E_t - E_i)/kT - u_S\right]}{\{1 + [\exp(E_t - E_i)/kT - u_S]\}^2}, \tag{34}$$

where E_i is the Fermi level of intrinsic semiconductor at 0 K and u_S is the dimensionless surface potential which is related to v_S in a relationship of $v_S = u_S - u_b$. It is worth noting that the interface between the metal and semiconductor is quite complicated as a result of spatial fluctuations of the built-in local effective barrier and distribution of charge at the depletion region boundary.

3. Preparation Effects

3.1. GaN Materials Effects

As mentioned earlier, most of GaN are usually grown on foreign substrates, e.g. sapphire, SiC, Si, and so on, by different techniques. GaN materials grown on these foreign substrates are found to have high densities of dislocations and other structural defects due to large mismatch in lattice constant and large difference in coefficient of thermal expansion between GaN layer and substrates. The final densities of dislocations and defects are dependent largely on the growth methods, such as metal organic chemical vapor deposition (MOCVD), molecular beam epitaxy (MBE), and hydride vapor phase epitaxy (HVPE), and growth conditions, as well as on the substrate used. [70-75] It is known that high density of dislocations plays a crucial role in the carrier transport at metal/GaN interfaces. [9, 27, 48, 54, 59, 72, 73, 75-81] Sawada et al. suggested that surface crystal defects such as dislocations and micropipes are likely the origin of the surface patches and SBH was lowered 0.41 eV when the fraction of the total patchy area was 10^{-5}. [27] Zhang et al. pointed out that carrier transport via conductive dislocations is the dominant source of leakage current at higher temperatures [72].

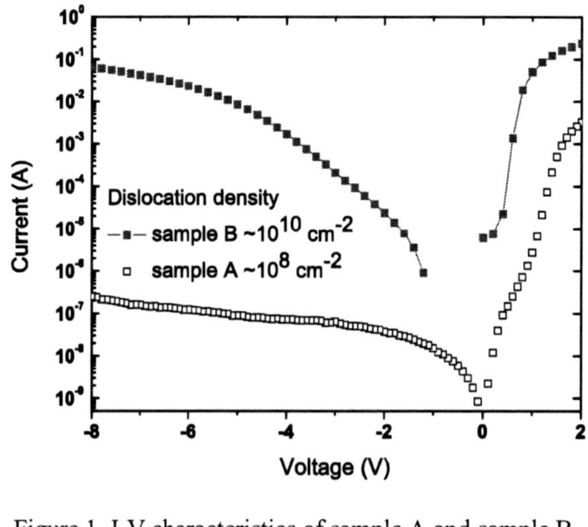

Figure 1. I-V characteristics of sample A and sample B.

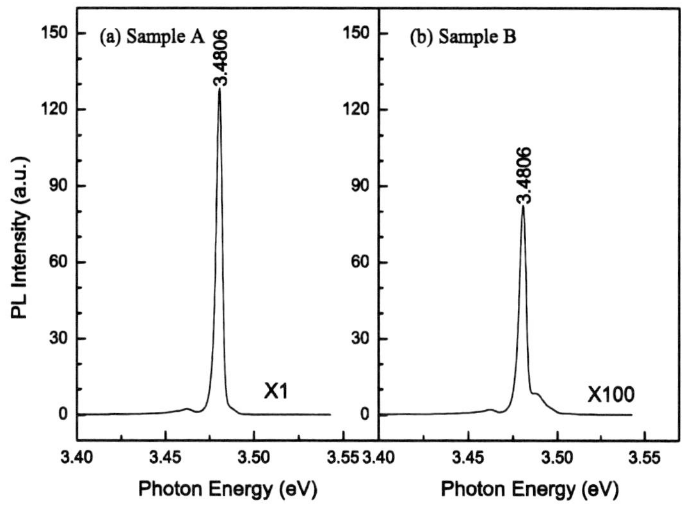

Figure 2. 4K photoluminescence spectra of sample A and sample B. The PL intensity of sample A is 155 times stronger than that of sample B.

In order to investigate the influence of GaN materials quality on the Schottky contact, we prepared Schottky contacts on GaN epilayers growth by different techniques. Figure 1 shows the characteristics of two kinds of Au/GaN Schottky diodes (samples A and B) fabricated on two different *n*-GaN epilayers which were respectively grown by HVPE and MBE. Their carrier concentrations and mobility measured by Hall measurements are respectively 3.3×10^{17} cm^{-3} and 258 cm^2/Vs as well as 1.3×10^{14} cm^{-3} and 41 cm^2/Vs. It can be seen that both of the diodes show Schottky-type diode characteristics. For sample A, the reverse current is less than 1.5 x 10^{-7}A even at 20 V. The barrier height of the Schottky junction extracted from the measured *C-V* curve is 1.09 eV and the ideality factor *n* of the junction derived from the forward characteristic using the standard thermionic emission theory is 1.3. By comparing the *I-V* curves of the two kinds of Schottky diodes, one can conclude that the sample B (MBE grown GaN) shows considerably poorer characteristics. Figure 2 (a) and (b) show the 4 K photoluminescence (PL) spectra of samples A and B, respectively. The sharp peak at 3.4806 eV, which is identified to the neutral donor bound exciton (D^0X) transition, [72, 82] is dominant in the spectra of the two samples. The D^0X peaks of the two samples located at the almost same energetic position and their full width at half maximum height (FWHM) is very close to each other (~3.4 meV). However, the peak intensity of sample A is 155 times stronger than that of sample B, indicating that it has much better optical quality. Generally, non-radiative centers induced by defects near the GaN surface reduce the PL intensity because the penetration depth of the 325 nm UV laser light is quite short, i.e., around 100 nm. [82, 83] It may be concluded that the defect density presented in the near surface region of sample B is much higher than that of sample A. The optical observation concurs with the electrical measurements on the samples. For example, the Hall mobility of the carriers in sample B is much lower than that of sample A. Atomic force microscopic (AFM) images of the two samples also show that their morphologies are very different, as shown in Figure 3. The dislocation sheet densities of sample A and sample B are respectively estimated to be about 4×10^8 cm^{-2} and $\sim10^{10}$ cm^{-2}. These consistent observations thus tend to confirm that the material quality of sample A is significantly better than that of sample B, as demonstrated by other works. [84, 85] Zhang et al. and Hsu et al. independently showed the dominate effect of dislocations on the leakage current mechanisms in the GaN films grown by MBE [72, 73].

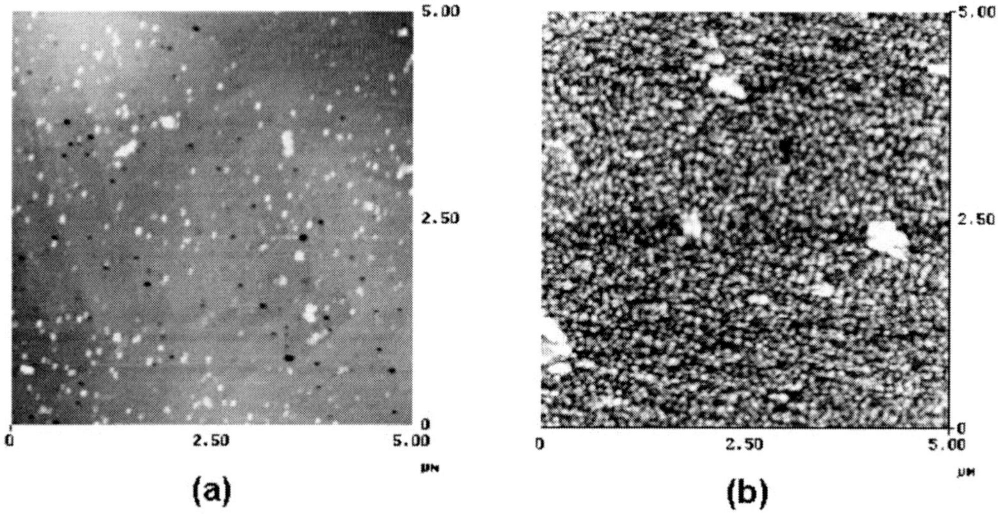

Figure 3. AFM images (5 μm × 5 μm) of the GaN samples etched by H_3PO_4 for 50 minutes at 160 °C. (a) Surface morphology of sample A. Etched pits at the defect sites are formed with a density of 4×10^8 cm^{-2}. (b) Surface morphology of sample B. where approximate density is $\sim10^{10}cm^{-2}$.

In the case of GaN growth by MOCVD with different process conditions, Suresh et al. found that the V/III ratio affects the Schottky barrier height, ideality factor and leakage current as well as the carrier concentration and Hall mobility of the GaN epilayers. [70] Saarinen et al. reported that intrinsic defects increase with increaseing the V/III ratio of GaN. [71] Figure 4 shows the typical room-temperature *I-V* characteristics of samples C and D. [69] Their carrier concentrations and mobility measured by Hall measurements are respectively 3.3×10^{16} $cm^{-3}$and 836 cm^2/Vs as well as 1.9×10^{17} cm^{-3} and 331 cm^2/Vs. The *I-V* curves clearly exhibit rectifying characteristic of metal-semiconductor junctions. However, the reverse-bias characteristics of the two kinds of Au/GaN Schottky contacts are very different. For sample C, the reverse-bias characteristics are not only uniform but also reproducible and stable. The reverse-bias leakage current is around 10^{-11} A at −20 V. In contrast, sample D shows a much larger reverse leakage current and lower breakdown voltage. The reproducibility and stability of the reverse-bias *I-V* characteristics of sample D are also poor. Figure 5 shows the typical *C-V* curves of sample C and sample D at 10 M (left figure) and 1K (right figure) frequency. It can be observed that the frequency-dependent *C-V* curves of both the samples are quite different from each other. The most noteworthy difference is that a peak structure at around 1.3 V appears in the *C-V* curves of sample D. Moreover, the peak structure becomes more prominent with decreasing the applied frequency as shown in Fig 6. This is a typical *C-V* characteristic of metal-semiconductor Schottky contact due to the existence of defect centers on semiconductor surface. [69, 83, 86] In sharp contrast to sample D, the capacitance of the sample C simply and monotonically decreases when the applied voltage is lower than 1.3 V, which is the typical behavior of a nearly ideal metal-semiconductor Schottky contact without surface states.

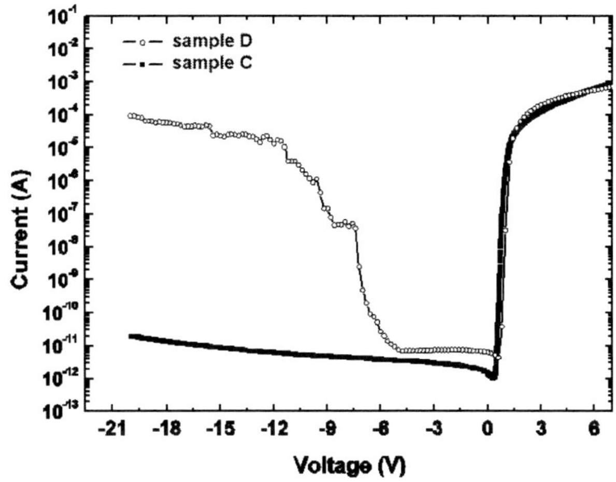

Figure 4. Typical *I-V* characteristics of samples C and D.

Using Eqs. (27)-(34), a fitting to the *C-V* curve measured at 1 kHz for the device fabricated on sample D was shown in Figure 6. N_{SS} of the device fabricated on sample D is determined to be around 1.2×10^{14} cm^{-2} and the effective energy level E_t is 1.3 eV. Room-temperature PL spectra of the two samples were also measured and illustrated in Figure 7. In addition to the great difference in intensity at around 3.4 eV (i.e., the near-band-edge emission I_{BE}), the yellow luminescence band I_{YL} of sample D is much stronger than that of sample C. The ratio I_{YL}/ I_{BE} of sample D is ten folds higher than that of sample C, which indicates that the density of deep level centers of sample D is much higher than that of sample C. Some recent investigations on the microstructural origin of the yellow emission of GaN epilayers show that the emission is closely related to the edge dislocations [69, 86] and other structure defects.

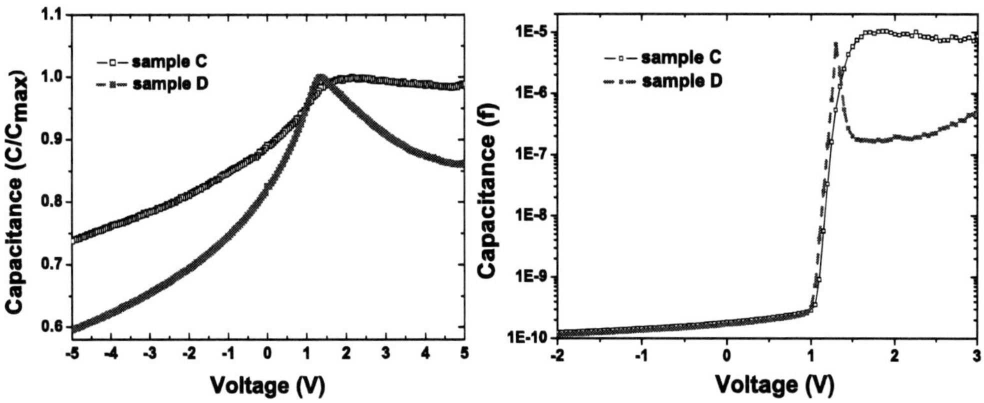

Figure 5. Typical *C-V* curves of sample C and D measured at 10M (left) and 1K (right) frequency.

As shown and argued above, qualities of GaN materials indeed have a big impact on the electrical characteristics of their metal-semiconductor contacts. If a metal layer is deposited on epitaxial GaN with many defects near the surface region, the excess of charge at the defective sites could reduce the Schottky barrier height and enhance the thermionic-tunneling and/or defect-assisted tunneling. [23, 24, 27, 53, 72, 86-89] Ozbek et al. found that the unintentional surface-defect donors cause barrier thinning and thus enhance electrons' tunneling through the Schottky interface, leading to a large leakage current. [81].

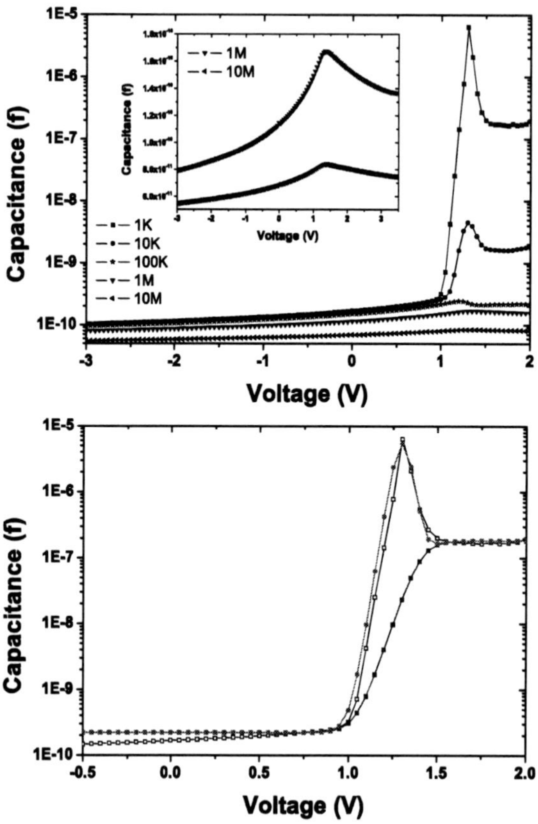

Figure 6. Frequency-dependent *C-V* curves of samples D. The inset in the left figure shows a close-up view of the *C-V* curves at 1M and 10M frequency. The peak structures in the *C-V* curves at ~1.3 V are due to the surface states induced by the deep level centers. In the right figure, the solid colour lines + symbols are the fitting curves using Eqs. (27)–(34) while the open black symbol line is the experimental space-charge capacitance.

In addition, the barrier height values of a metal/GaN contact may exhibit a Gaussian distribution where the strong Schottky barrier inhomogeneity is a consequence of the poor material quality, as indicated by Roccaforte et al.,. [90] Hsu et al. conveyed that a majority of current leakage paths is located on screw and mixed dislocations so that the reverse current distribution is highly nonuniform. [73,76] When a reverse bias is applied to such a Schottky contact, the uniform Fermi level across the entire structure will be split into quasi-Fermi levels for electrons and holes. The energy levels of the deep centers and/or dislocations (most likely defect complexes) vary with respect to the Fermi level in the depletion region of the Schottky barrier where the bands are bent. Hence, the occupation status of the defect levels will also change. As the reverse bias is increased, more electrons can gain enough energy to pass through the barrier, resulting in an enhancement of the tunneling and recombination currents. It is thus concluded that the defects in GaN epilayers have a critical influence on the electrical characteristics of metal/GaN Schottky contacts.

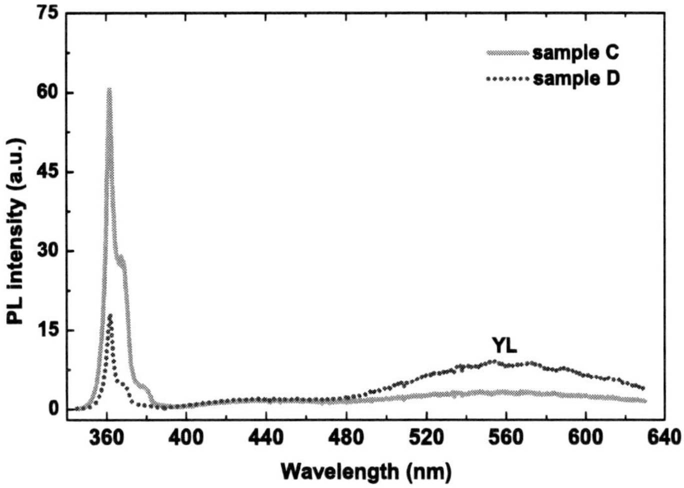

Figure 7. Room-temperature PL spectra of samples C and D.

In order to reduce the densities of dislocations and other defects, several advanced growth techniques, such as two-step grown method, inserting a buffer layer, and lateral epitaxial overgrowth (LEO) etc., have been explored. [70, 89, 91-97] For example, for the LEO growth technique, no threading dislocations are inherited and no lattice mismatch dislocations are observed in the area of lateral epitaxial overgrowth window regions so that the threading dislocation density can be reduced by two to four orders of magnitude. Today, extremely thick layers of GaN can be grown on sapphire substrates, which leads to a significant reduction (i.e., from 10^{10} to 10^{4} cm^{-2}) of naturally and significantly dislocation density in the upper part of the layer. [75] However, the epilayer thickness is much more limited on silicon due to the unacceptable cracks caused by the internal layer constraint. [91] Further studies on advanced GaN heteroepitaxial techniques are certainly needed along the direction of reduction of the dislocations and other defects. Only on high-quality GaN materials with low background electron concentrations and especially with low densities of defects, excellent Schottky diodes can be fabricated on them.

3.2. Surface Treatment Effects

Surface treatments are an important main procedure in the fabrication of electronic and optoelectronic devices. If contamination and growth oxide layer are present at the interface between GaN and contact materials, leakage current will become larger and SBH will be decreased by the induced local nanoscopic patches which locally control device characteristics in different temperature regimes of operation. [40, 55, 90, 98] It is well known that GaN is an ionic-like compound. In other words, the chemical bonds on the GaN surface have an ionic nature. Surfaces of GaN crystals can be terminated as either Ga surface (denoted as Ga-face for simplicity) or N surface (N-face). When GaN is terminated by Ga-face, an electric field due to the spontaneous polarization is built up and directs from surface to the bulk. For the N-face sample, it is opposite. Such a built-in electric field causes band bending near the surface. For example, the Ga-face GaN may have a higher upward surface band bending of 1.4 eV than the N-face GaN. The different surface band bending may have a substantial influence on the properties of Schottky barrier. It has been reported that the Schottky barrier height fabricated on Ga-face sample is higher than that on the N-face sample to influence *I-V* and *C-V* characteristics. [58, 59, 99-101] On the other hand, the electronegativity, where the difference between Ga and N is 1.87, exhibits active chemisorption, easily leading to the formation of surface native oxides and hydroxides layer. [102, 103] Jang et al. reported that the decomposed metallic Ga chemically react with oxygen in air to form a 4 nm Ga oxide layer so that a large number of N vacancies are produced near the surface. [104] Ishikawa et al. also reported that a 2 nm thickness GaO_x layer contains adsorbed-carbons on the GaN substrate grown by MOCVD. [105] The surface modifying layers may affect the electrical characteristics of Schottky contacts via these mechanisms: *(i)* physical or chemical isolation and insulation layer as energy barrier; *(ii)* oxidization-induced surface states; *(iii)* nonstoichiometric surfaces through preferential loss of one of the lattice elements; *(iv)* carbon incorporation via substituting Ga as a shallow donor leading to a high-density doping effect near the GaN surface; *(v)* variation of the band bending. [27, 106, 107] Prior to metal deposition, therefore, some methods have been employed to remove the contamination layer and improve contact qualities on GaN. Among them, wet chemical reaction etching and dry etching are the two major methods.

For the wet chemical reaction cleaning, a number of solutions, such as aqua regia, HCl, HF, BOE, phosphoric acid, KOH, NaOH, NH_3OH and even organic solutions have been tested to clean and treat the surface of GaN. [55, 64, 108-113] Compared with phosphoric acid which did not result in significant improvements, other solutions including aqua regia, HCl, HF, KOH, NaOH and NH_3OH, effectively clean the surface and improve the rectifying characteristics of Schottky barriers. Several groups have investigated the influence of wet clean methods on surface states and surface structures of GaN, which in turn affect the electrical characteristics of Schottky contacts. Mohammad et al. observed that the surface morphology of GaN after KOH cleaning becomes smoother than ones after aqua regia treatment. [26] Kim et al. investigated the variation of surface Ga/N atomic ratio with aqua regia treatment methods. [114] It has been shown that H_2O and methanol effectively remove half of the layer consisting of transparent organic and inorganic contamination and NH_4OH and NaOH remove another half of the 2-5nm presumably oxide layer. HF was demonstrated to be more effective in removing oxide on GaN than HCl, though not completely removing surface oxide. [115-117] Figure 8 shows the *I-V* characteristics of the Au/GaN Schottky diodes using different chemical solution treatments prior to the metal deposition. It can be observed that there are great variations due to the surface chemical-solution treatments in the characteristics of the Schottky contacts. From the available reports, the effects of the wet solution etching are possibly the removal isolation layer of oxide, decreasing defects where high densities of surface states are present as a parallel current path, reducing the number of crystalline orientation facets in the surface, smoothing the surface roughness, changing the surface termination elements and the Ga/N ratio. [26, 114-117] However, it is worth pointing out that the top layers of GaN surfaces which exposure to air after the treatments may not only adsorb oxygen, hydrogen or OH but also reform the oxide layers because of dangling bonds of GaN on the terminating surface, radicals and active chemisorption. [13, 40, 102-104] Sulfides can remove the native oxide at top surface after $(NH_4)_2S_x$ or Na_2S solution treatments [118, 119] while the barrier height decreases as well as the forward and reverse currents become larger in diode structures. [3] These surface re-contaminations restrict the improvement of the electrical properties of the metal/GaN contacts [13, 27, 105].

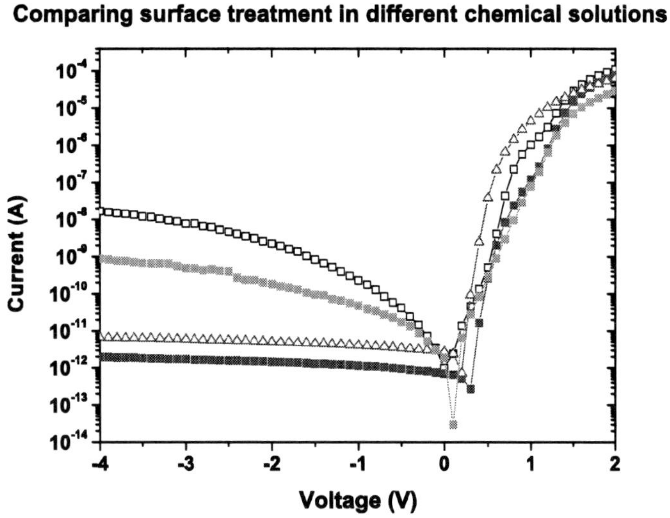

Figure 8. I-V characteristics of the diodes using different chemical solution treatments.

For the dry cleaning treatment, a variety of plasma and ion beam techniques, such as sputtering, plasma etching, ion-beam bombardment, and laser irradiation have been adopted to clean the GaN surface. [99, 119-125] These dry-cleaning methods have been shown to be capable of removing the top layer of atoms and modifying the status of the surface. Hunt et al. reported the usage of N ion-beam sputtering for removing the oxide layer on the GaN surface. [125] High-energy particles, ions, atoms, photons and/or electrons do not only break the strong Ga-N bonds at the top surface and change the ratios of the surface constituents, the surface dipoles, and the surface charge densities, but also affect the work functions and depletion region widths as well as the surface roughness and morphology. In particular in the plasma treatments, at least two mechanisms including the ion bombardment and the chemical reaction between the GaN surface and reactive species, have strong influence on the surface status and microstructure during the treatments. For the effect of ion bombardment, it is a physical sputtering process in which chemical interactions are lacking. As the bias power increases, the electric field in the plasma sheath becomes stronger and the energy of ions becomes higher. These accelerated ions will have higher kinetic energy to efficiently remove the top layer. At the same time, it should be noted that the surface is damaged and usually lattice defects are induced. Some constituent atoms will be forced to move away from their lattice sites, and thus lattice defects including interstitials, antisites, and vacancies, as well as nonstoichiometric surfaces into the GaN subsurface could be produced. [121] In general, the number of defects introduced increases with increasing the energy of the ions. As a side effect, the degradation of electrical and optical characteristics may take place to some extent. Wu et al. reported that both the donor and acceptor states introduced by sputtering cause a ~0.8 eV band bending on both *n*- and *p*-GaN(0001). [122] Pearton et al. and Cao et al. observed that N_2 plasma exposure creates stronger degradation of the diode characteristics than the H_2 plasma does due to the heavier ion mass. [3,123] Usikov et al. showed that Ar and Ar^+/N_2 ion beams produce degradation of both the optical and electrical properties since an intense broad emission band appears at 3.05 eV and the intensity of the yellow band increases in the PL spectrum as well as the electron concentration near the surface increases from 10^{16} to 10^{18} cm^{-3}. [124] Ishikawa et al. compared the variation of N/Ga ratio by Ar ion and N ion sputtering. [105] They found that N atoms on the GaN surface are preferably sputtered by Ar ions to form N vacancies while N atoms are implanted into the GaN by N ion sputtering to suppress formation of N vacancies which act as a shallow donor level in the band gap. [105] Sometimes the contamination layer, such as adsorbed-carbons layer, was not completely removed by the sputtering process. [105, 125] Kimura et al. investigated the effect of carbon incorporation on the electrical properties of *n*-type GaN. [106] It has been shown that carbon can be incorporated to substitute Ga as C_{Ga} which behaves as a shallow donor in GaN. This leads to a high-density doping effect near the GaN surface which reduces the width of the Schottky potential thus enabling the electrons to tunnel through the thinner barrier more efficiently. For the effect of chemical reactions during the dry etching, the primary ions may chemically react with the top atoms of GaN surface. It has been understood that the damage could be reduced when the reactive gases were introduced in the plasma for a given ion energy and plasma density. [3] Two kinds of major reactive plasmas, chloride-based and fluorine-based plasmas including BCl_3, $SiCl_4$, Cl_2+Ar, Cl_2+N_2, BCl_3+Cl_2, SF_6, SF_6+N_2, and so on, are used to remove the several layers at the top of the surface by chemical reactions. [28, 39, 55, 115, 119, 126-142] In the case of the activated chloride-based plasmas, ionized chlorinated atoms and/or excited molecules easily react with both Ga and N atoms at the surface of GaN. The created reactants $GaCl_x$ and NCl_x can be either desorption as vaporization of NCl_3 and $GaCl_3$ or residues of $GaCl_x$ and NCl_x species on the GaN surface. [119] Jang et al. obtained a Ga-rich surface [140] while Kim et al. detected an N-rich surface [141] after the pure Cl_2 inductively coupled plasma (ICP) etching. Effects of mixture plasmas usually become more complicated. If Ar was added in the reactive plasma, the removal rate of N from the GaN surface was enhanced due to the preferential sputter desorption of lighter elements. [105] If N_2 was added into the plasma, compensation effect for the N vacancies may be created during the bombardment, resulting in an N-rich surface. [142] It is also observed that the shallow-donor-like defects including intrinsic defects, i.e., nitrogen vacancies acting as the electron donors, and Cl_2-plasma induced defects, increase at the near-surface region after the Cl-based plasma treatment. [28, 127] However, the effect of chlorine plasma still remains a puzzling issue. For example, several groups claimed that the negatively charged surface decreased the SBH and severely degraded diode characteristics [28, 136, 139] while Rong et al. and Chen et al. reported the enhancement of the barrier height of Schottky diodes after the chlorine base plasma treatment. [137, 138] The band bending on the GaN surface is altered and the intensity of the PL decreases after the reactive ion etching (RIE) treatment. [135] The yellow luminescence (YL) intensity was found to increase after the ICP etching [133, 134]. However, Choi et al. observed a result of decrease in YL intensity. In the case of fluorine-based plasma treatment, nonvolatile GaF_x cannot be purely chemically removed. [120, 136] In order to sufficiently remove this GaF_x layer by strong physical bombardment, other associated atoms would have to be added in SF_6 plasma. [131, 132].

In a short summary, the major effects of plasmas include (*i*) the removal of a native surface oxide layer; (*ii*) varying Ga/N ratio on the surfaces of GaN through preferential loss of one of the lattice elements because of strong differences in the volatility of the respective etching reactant. Some less volatile species or lighter lattice element can be kinked out by strong physical sputtering. The depth of the nonstoichiometric layer is typically 10nm; (*iii*) creation of lattice defects which generally behave as deep level centers. [28,115, 125,126] These defects may bring out the effects of compensation, trapping or recombination in the material as deep as 100 nm from the surface through channeling by striking and rapid diffusion of the defects created, even though the projected range of the ions is less than 1nm. [3] (*iv*) unintentional passivation effect by atomic hydrogen and sulfide; (*v*) deposition of polymeric films from the plasma chemical gas source, or photoresist masks. [3] All these mechanisms may compete for affecting the electrical properties and optical properties of the plasma-treated GaN films.

3.3. Contact Material Effects

Two types of contact materials can be used for fabricating Schottky contacts on GaN. One is the traditionally used metal materials, such as Au. The other is the transparent conducting oxide materials which are well known as a good contacting material in the fabrication of optoelectronic devices.

A. Metal Contacts

Typically, the most used metals for Schottky contacts on *n*-type GaN have high work functions since they should result in a high value of the Schottky barrier. The Schottky barrier heights of metal/GaN contacts with a variety of elemental metals including Au, Pt, Pd, Ni, Cr and so on have been reported. The Schottky barrier heights $q\phi_b$ (on *n*-GaN) reported so far vary significantly with the metal work function and are quite scattered. [30, 143] When a metal is deposited on the GaN surface, it will interact with the surface atoms of GaN. For N atoms at the top layer, there are two possible phenomena. One is that the N atoms are probably kicked away by the depositing metal, which results in the creation of N vacancies V_N. The other is that the metal directly reacts with N to form a compound layer or insulating layer. For Ga atoms at the top layer, deposited metal atoms may react with them to form an alloy layer and to create some new phases. The arriving metal atoms could also directly enter the GaN lattice or the interstitial positions to rearrange the GaN structure in the vicinity of the metal/GaN interface. Dobos et al. observed the formation of the structurally layered ternary compounds $M_{x-1}GaN_x$ which are polycrystalline nanolaminates. [144, 145] If the nanopipes and deep ditches are present near the interface as mentioned in previous section, the arriving metal atoms are able to migrate through the screws and dislocations and go deeper into the GaN epilayers. All these physical and chemical processes may produce dramatic influence on the electrical properties of the contacts. [26] An evident evidence supporting this argument is that the actual barrier heights and properties of the GaN-based Schottky diodes are very different from one another. The reported values rely on the characteristics of the metals, the metal deposition thickness and deposition temperature. A number of research groups have investigated the electrical and structural properties of Schottky contacts on GaN using different metal materials, including the single elementary metal layers of Ni, Au, Pt, Pd, Ti, Al, Mo, W, Ag, Cr, Cu, Rh, Re, Ir, [11, 13, 20, 21, 24, 27-30, 34, 35, 51, 55, 75, 77, 86, 105, 114, 115, 146-164] the compounds of TiN, W_2B, ZrB_2, WN_x, WSi_x, $TiWN_x$ and WN_x [28,

165-167], as well as multi-layers of Ni/Au, Ni/Pd, Ni/Mo, Pt/Au, Ru/Au, Ir/Pt, Ti/Au, Pd/Mo, Pt/Mo, Ir/Pt, Cu/Au, TiB_2/Ti, Ti/Al, Ti/Au, Au/Ni, Al/Ni/Au, V/Ti/Au, Ni/Mo/Au, Ni/Pd/Au, Ni/Ir/Au, Ni/Pt/Au, W/Pt/Au, Au/Ti/Ai/Ti, Ni/Ti/Pt/Au, Ta/Ti/Ni/Au [48, 58, 59, 61, 74, 99, 111, 113, 164, 168-185]

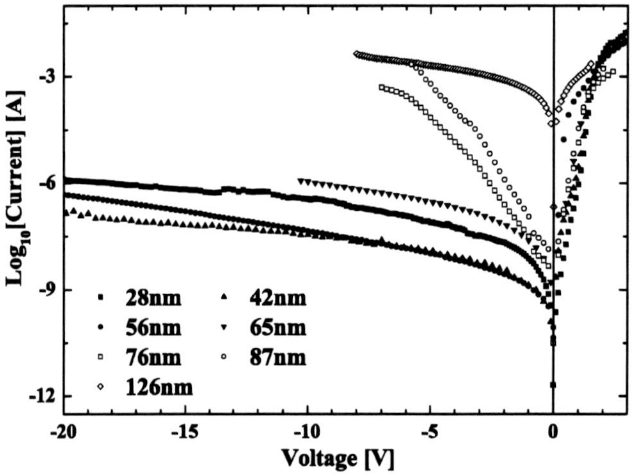

Figure 9. Current-voltage (*I-V*) characteristics of the Au/GaN Schottky diodes having different Au film thicknesses.

Figure 9 shows the *I-V* characteristics of the diodes having different Au film thicknesses. [86] A trend that the Au/GaN Schottky diodes exhibit better rectifying characteristics as the Au thickness decreases when their forward *I-V* curves remain almost unchanged can be clearly identified. In other words, the reverse characteristics of the devices show significant degradation with increasing Au film thickness. For example, in the cases of the contacts having the thinnest Au films of 28 and 42 nm, almost all the diodes are able to keep a leakage current of less than 1 μA when a 20 V reverse bias is applied. As the Au films become thicker, however, a lot of diodes become failure for a 20 V reverse bias. In the case of the 56 nm Au film as shown in Figure 10, 50% of the diodes show good *I-V* characteristics being similar to those obtained for the diodes with 28 and 42 nm Au films while others underwent breakdown below 10 V reverse bias. When the Au film is further increased to 65 nm, only 25% of fabricated diodes can bear a 20 V reverse bias. Furthermore, the leakage currents of the diodes are typically 10-100 times larger than those obtained for the two thinnest Au contacts. The mean reverse breakdown voltage for the diodes is only 4.5 V. When the Au film thickness is beyond 76 nm, no diodes are able to withstand a 20 V bias. Based on the thermionic emission theory, the ideality factors of the diodes are determined, which consistently show degradation behavior as the Au film thickness increases. The barrier heights are found to reduce from 0.78 eV for the thinnest film to 0.69 eV for the thickest film. O'Mahony et al. proposed that larger ideality factor would tend towards the thermal activation of a deep level trap attributed to recombination within the depletion region in the case of wide bandgap semiconductors. [89] The observed degradation of the electrical characteristics of the Au/GaN Schottky contacts with the Au film thickness is likely due to the increase of structural defects in the surface layer of GaN, non-uniformity of interfacial charges, atom inter/out-diffusion. It is worth mentioning that the presence of an insulating or high resistivity thin layer may lead to the modification of the surface potential thus affecting the *I-V* characteristics of the Schottky diodes [120, 186].

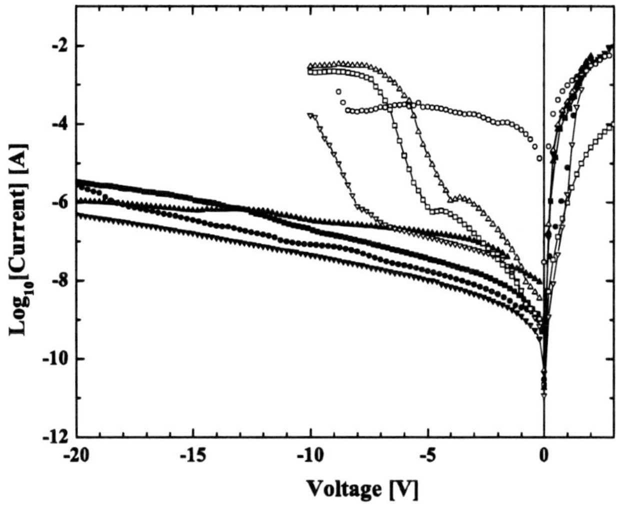

Figure 10. *I-V* characteristics. Solid and hollow symbols represent respectively the good and poor I-V curves of diodes with 56 nm thick Au films.

Figure 11 shows the variation of the atomic concentrations as a function of sputtering time for the sample of the 56 nm Au film on GaN. An interface width may be approximately defined as the distance for atomic concentration transition from 100% Au to 100% Ga+N. The measured interface widths for some Au/GaN contacts are tabulated in Table I. The upper and lower bounds of the interface are plotted as a function of Au thickness in Figure 12. It can be seen that, in the first approximation, the interface thickness scales in proportion to the film thickness. Such a phenomenon is well known for the Au/GaAs interface where its interpretation lies in there being a fixed solubility limit for Ga in the Au overlayer. [42] As the Au deposition amount is increased, more Ga is able to diffuse out of the GaN and into the Au, causing further GaN bond disintegration and interdiffusion and mixing becomes significant. The increased mixing of elemental components and the associated widening of the interfacial region with increased Au coverage can lend some understanding to the observed contact degradation phenomenon. Within the interface and the nearby region, GaN lattice distortion tends to be severe. Since Ga is out-diffusing into the Au film, a high density of V_{Ga} vacancies (and possibly V_{Ga} related centers) will be expected to form in GaN. This is because that Ga vacancies have a relatively low formation energy and are more favorable in the *n*-type GaN. [106, 187, 188] It is known that V_{Ga} causes a deep level at E_V+0.87 eV which is close to the valence band maximum. Such deep level may act an efficient center for carrier recombination. [84, 86, 189] The increase of acceptor' defects near the intersurface region would cause "self-compensation" effect, leading to the widening of the depletion region and, hence, the increase of the barrier height. On the other hand, since recombination centers are also good carrier generation centers a high concentration of V_{Ga} within the GaN will likely produce a major source of the generation current. When the V_{Ga} defect is likely to be a dominant carrier generation center, it is still conceivable that other defect centers may be responsible for additional sites of carrier generation and/or the observed Fermi level pinning at E_C-E_t ~1 eV. For example Au atoms in GaN may bring out some additional carrier generation deep centers. Zou et al. investigated the variation of Ga binding energy and Au-Ga alloy formation by chemical reaction at the interface at very initial coverage stage. [190] Wang et al. observed an enhancement in SBH for Au deposition on *n*-GaN at a lower deposition temperature. [31] Lu et al. and Sreenidhi et al. observed the formation of gallide-based alloy due to interatomic diffusion of deposition metal in GaN. Such alloy might have a different work function and thus have an influence on the barrier height. [120, 168] The barrier-height inhomogeneity at the Au/GaN interface was investigated by Roul et al. [64] Table I shows the values derived from the *I-V* and *C-V* measurements. It can be seen that the drop in the *C-V* measured barrier height is more pronounced. The phenomenon is quite complicated and not yet well understood now. [42, 191] Possible causes include: *(i)* the width widening of the interfacial region; *(ii)* the existence of Schottky barrier height inhomogeneity; *(iii)* non-uniform of the thickness and composition of the contact layer; *(iv)* non-uniform distribution of interfacial charges; and *(v)* the presence of a thin insulating layer between metal and semiconductor. Interfacial layers normally have the effect of lowering the barrier height, the magnitude of which depending on the interface thickness. [42, 69, 86, 190] The dominate current in the *I-V* measurement flows through the regions of lower SBH. The fact that the barrier height measured through the *I-V* measurements is more representative of the generation–recombination current rather than the thermionic emission would explain why it undergoes a lesser reduction with the Au film thickness. The relative steadiness of the ideality factor (n≈2) supports the picture of a dominant generation current across the barrier. In contrast, the *C-V* measured barrier height is affected arithmetically by the depletion region boundary and the charge distribution in the average manner of the SBH inhomogeneity. The barrier height measured by *C-V* is thus close to the average value of the SBHs. [39, 191]

Table I. Summary of characteristics of Au/GaN contacts with different thickness

Au film thickness (nm)		28	42	56	65	76	87	126
Good diode No.		10/10	7/8	4/8	3/12	0/8	0/6	0/5
Reverse Current I_R at	-5 V	29.7nA	12 nA	18.1nA	291nA	Break-down	Break-down	Break-down
	- 10 V	136 nA	35.4nA	74.3nA	1.10μA			
	-20 V	1.22μA	139 nA	6.16μA	50.6μA			
Mean Breakdown Voltage (V)*		> 20	>20	6.1	4.5	4.6	2.8	-
Barrier height (eV) (by C-V)		1.05	1.03	0.96	0.89	0.83	-	-
barrier height (eV) (by I-V)		0.79	0.76	0.78	0.73	0.72	0.69	-
Ideal Factor n (by I-V)		1.18-3.02	1.20-2.76	1.14-3.12	1.27-3.0	1.28-3.02	~3.25	
Interface width (nm) (by AES		-	16.8	19.2	21.6	26.4	31.2	-

[* For those diodes that could not survive a 20V bias].

It should also be concerned that reliability reduced and mean reverse breakdown voltage of the Au/GaN Schottky diodes were lower with increasing Au coverage. Again, there is a wide range of speculation about the origin, but it is likely that the carrier conduction along the threading dislocations is the main cause of these phenomena. When the nanopipes and/or the ditches are present in GaN, the bigger and the deeper these defects are, the larger the number of metal atoms entering these nanopipes is. In the situation, the electronic interactions among Au, Ga, and N atoms should be modified and related to crystal structures and grain sizes [31] as well as the formation of gold gallides or other new phases at the interface. [120] Therefore, the effective barrier height and the reverse characteristics of Schottky diodes on GaN should be largely determined by the presence of dislocation defects and other structural defects as argued in above section. Hierro et al. suggested that the V_{Ga} defects could decorate threaded dislocations, which run through the GaN film. [189] This may lend some further insight into the film thickness degradation since as more Au is deposited, more V_{Ga} defects could be formed in the near surface of the GaN thereby making the dislocations more conducting.

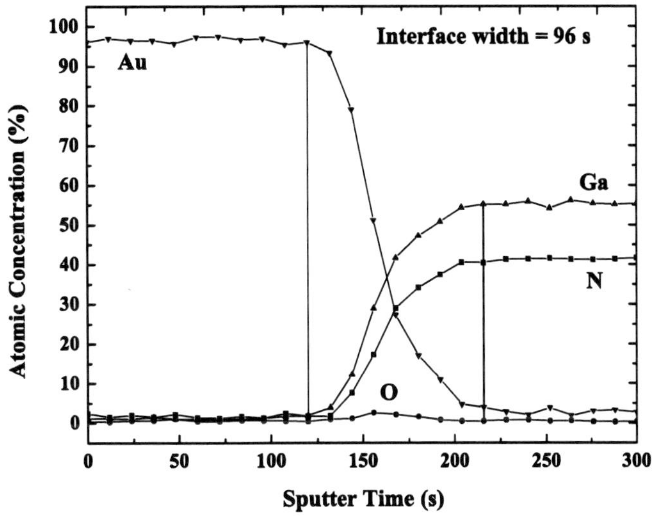

Figure 11. AES depth profile of Au/GaN Schottky contact with 56 nm Au thickness. The distance between the two vertical lines represents the width of the interface.

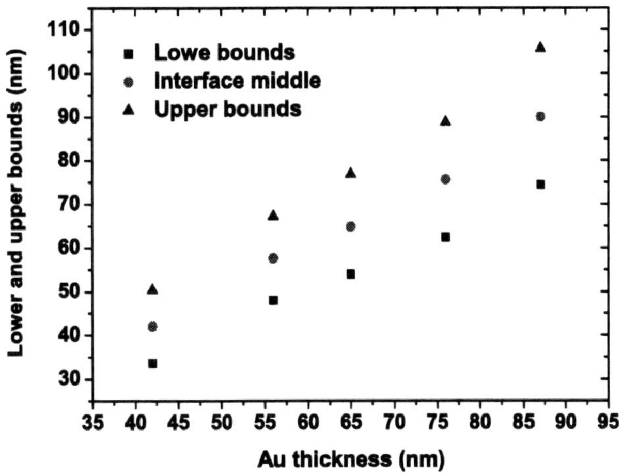

Figure 12. Upper (▲) and lower (■) bounds of the interface as a function of Au thickness.

As emphasized above, regarding the dependence of SBH on the metals, interface states and preparation processes, it is still an open issue. Tracy et al. proposed that the Fermi level which influences the SBH value may not pinned at the GaN surface after investigating the variation of thickness of platinum, gold and silver. [143] Some researchers believed that the Schottky barrier heights on GaN depend directly on the difference of work function between the metal and GaN due to the ionic nature of GaN. [193] Schmitz et al.'s study showed that a monotonic increase in $q\phi_b$ but does not scale as the metal work function increases. [29] However, Guo et al. stated that the SBH is only weakly dependent on the metal work function. [51, 146] Roccaforte et al. ascribed it to a partial pinning of the Fermi level at the metal/GaN interface due to the presence of a high density of surface states. [90] Rennie et al. argued on the Fermi-level pinning. [194] Kampen and Monch suggested that the presence of the metal-induced gap states and the electronegativity influence Schottky contacts on GaN. [195] The possible chemical reactions of the deposited metal with GaN at room temperature may play a dominant role in the determination of the Schottky barrier, as pointed out by Wu et al. [121] Thus, these are not yet a unified consensus on this matter because so many factors may produce no neglected influence on the electrical properties of Schottky-type metal contacts on GaN.

B. Transparent Conducting Oxide Contacts

With the rapid development of optoelectronic devices, transparent conducting oxides (TCO) have been used in a great variety of practical devices. [196-200] Compared with metallic materials which are opaque and thus block more light, transparent conductive electrodes could be used to obtain higher efficiency of optical to electrical energy conversion. Some TCO materials such as tin-doped indium oxide (ITO) and Ga-doped zinc oxide (GZO) have been demonstrated to have both exceptional electrical conductivity and optical transparency. [12, 200-203].

Because the oxide part of TCO materials may chemically interact with the surface atoms of GaN, for example, reacting with Ga to form the insulating solid GaO_x, or replacing N atoms, and/or diffusing into the deeper region to become recombination centers, the preparation techniques and conditions selected for deposition of TCO materials on GaN, especially for the purpose of fabrication of TCO/GaN Schottky contacts should be optimized. The deposition conditions have been shown to strongly affect the microstructures and chemical compositions of the deposited TCO films, [196,204] which in turn modify the electrical and optical properties of the films and thus impact on their specific performance on the GaN devices [196-208].

Figure 13 shows typical room-temperature *I-V* characteristics of the ITO/*n*-GaN Schottky contacts fabricated under different oxygen partial pressures. It can be seen that all the *I-V* curves clearly exhibit good rectifying characteristics. [196] However, it should be noted that the reverse leakage currents of the contacts deposited with oxygen flow rates higher than 3 SCCM are much smaller than those of the contacts deposited with lower or zero oxygen flow rates. Based on the thermionic emission theory, the extracted Schottky parameters are listed in Table II. In sharp contrast to the substantial variation of the reverse leakage current, the values of $q\phi_b$ and n are found to be not very sensitive to the oxygen partial pressure, which is consistent with Hsu *et al*'s results [76].

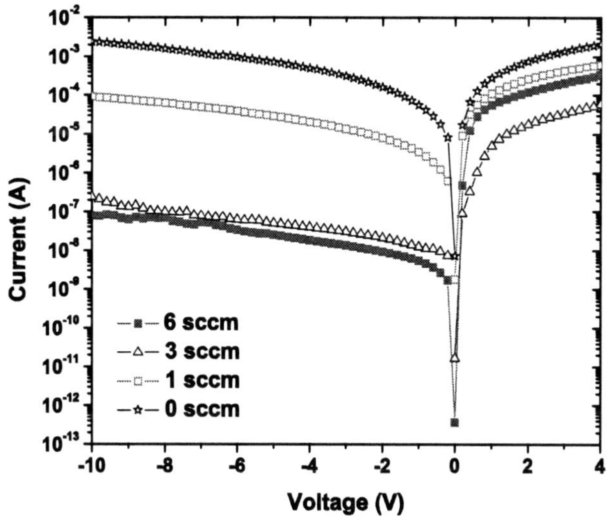

Figure 13. Typical *I-V* characteristics of ITO/*n*-GaN Schottky contacts fabricated under different partial pressures of oxygen.

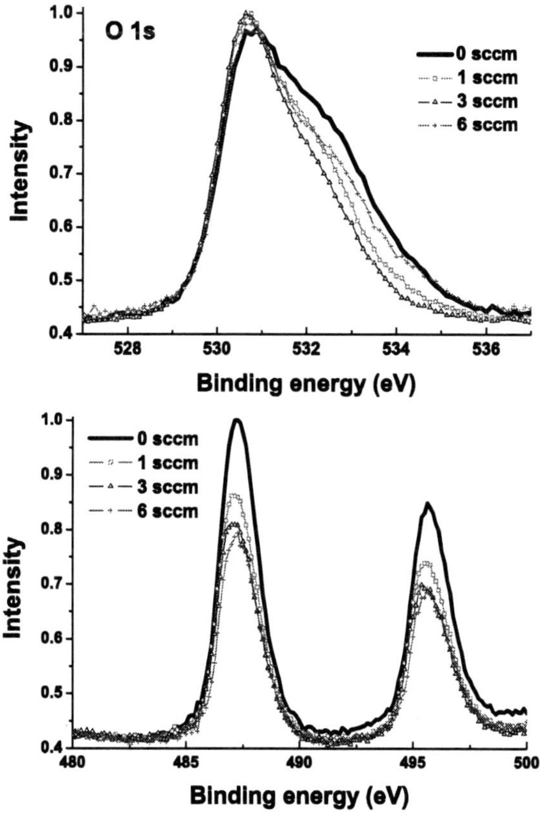

Figure 14. O 1*s* (top) and Sn 3*d* (bottom) normalized XPS curves of the ITO films deposited on GaN wafer under different partial pressures of oxygen.

Table II. Summary of characteristics of ITO/GaN contacts under different preparation conditions

Oxygen flow rate (sccm)	0	1	3	6
Barrier height (eV)	0.63	0.67	0.77	0.77
Ideality factor *n*	2.10	1.75	1.87	1.75
I_R at -10V (mA)	2300	91.5	0.031	0.078
Sn/In ratio (%)	27	15	12	12
Sn/O+In+Sn ratio (%)	6	4	4	3

In order to understand the influence of the oxygen environment on the reverse-bias characteristics of the ITO/GaN Schottky contacts, a detailed XPS study was performed. Figure 14 shows the XPS data of Sn-3*d* and O-1*s* of the ITO films deposited under different oxygen partial pressures. It can be seen that the relative intensity of Sn decreases with increasing the oxygen partial pressure. The integrated intensity ratios of Sn/In and Sn/In+Sn+O, as obtained from the XPS data, are given in Table II. Clearly, these ratios of the films deposited under low or zero oxygen partial pressures are almost twice those of the films deposited under high oxygen partial pressures. It is known that Sn acts as a cationic dopant in ITO film when it substitutes for In and is surrounded by In_2O_3. In this configurational interaction, it provides an electron to the conduction band of In_2O_3 making the ITO film conductive. However, too high a Sn/In ratio may bring about a negative influence on the electrical properties of ITO film. [196, 204-207] A number of speculations have been proposed to account for this behavior. [196, 204-209] Interestingly, it is noted that within a range of the low oxygen partial pressures, the sheet resistance of the ITO films decreases with increasing oxygen partial pressure. This can easily be understood as the primary Sn doping effect. [210] However, for higher oxygen partial pressures, the sheet resistance of ITO films increases with increasing oxygen flow rate. One possible explanation for this effect is the decrease in the concentration of oxygen vacancies. [208] Figure 15 shows the variable energy Doppler broadening spectroscopy (VEDBS) scans taken on the different contacts grown under different partial pressures of oxygen. [196, 211] In general, the low electron momentum *S* parameter is higher in material that contains a high concentration of open volume point defects. This is a result of the positron trapping into open volume defect sites where it predominantly samples low momentum valence electrons. The VEDBS data in the 0-10 keV implantation energy range are indicative of annihilations occurring in the ITO overlayer. At 10 keV, the positron beam is maximally sampling the ITO-GaN interface while at higher energies, the *S* parameter drops to the value characterizing the GaN film. The remarkable feature of this positron annihilation data is that the ITO *S* parameter is significantly higher in the zero oxygen partial pressure case and room-temperature deposition. [211] At the higher deposition temperatures and higher oxygen partial pressures, the *S*-parameter at the interface becomes higher than the bulk, indicating the presence of open volume sites at the interface. Such open volume sites are most likely to be due to the higher lattice mismatch that exists between the GaN and ITO at higher temperatures. [196, 211, 212] This clearly supports the oxygen vacancy model in which the oxygen vacancies are more predominant in the films grown without oxygen and lower deposition temperatures. Also noted, particularly in the zero oxygen case, is the *S* parameter just peaking at the ITO/GaN interface. This indicates the presence of vacancy-type defects and that oxygen has segregated at the interface. This peaking phenomenon seems to reduce under higher oxygen partial pressures, presumably as a result of the point defects being removed under oxygenation. The above evidence for the presence of oxygen vacancy defects both in the ITO film and at the ITO-GaN interface can lend some understanding to the observed pattern of reverse leakage current. However, it must be pointed out that the real picture is likely to be more complicated with other types of point defects making contributions to the generation and tunneling currents. Miller et al. have shown that, at higher temperatures, the reverse leakage current of Ni/*n*-GaN Schottky contact is mainly due to a two-step, trap-assisted tunneling. [23] It is also possible that the increased oxygen content of the ITO film under higher oxygen partial pressures may result in the formation of a GaO_x insulator layer [196, 200, 203] that can reduce tunneling. Further studies are expected to give a more comprehensive understanding to the ITO thin film itself and the ITO/GaN contacts.

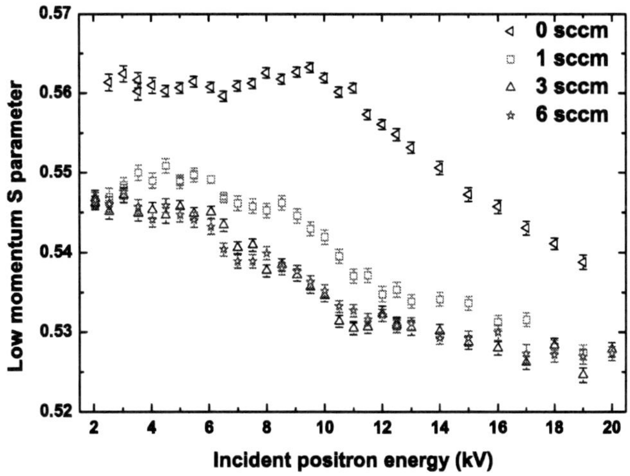

Figure 15. Positron beam VEDBS scans of the ITO/GaN Schottky contacts deposited under different oxygen partial pressures.

4. Applications

To-date, GaN Schottky contacts have been used in a variety of practical areas, such as high temperature/high power electronics [213, 214], microwave rectifiers, sensors for detecting gases, polar liquids and mechanical pressures, and UV light emitting diodes. Depending on the application areas, the detailed structures and contact materials shall be carefully considered and chosen for fabricating GaN-based Schottky devices.

For high temperature/high power electronics application, it has been reported that the critical electric field of the breakdown is estimated to be >4 MV/cm for GaN, which is 10 times larger than conventional Si rectifiers. [215] For example, when the GaN layer with doping concentration $<10^{15}$ cm^{-3} is an ~100 μm thickness, a 20 kV device may be obtained. [216] Two kinds of structures have been adopted in GaN rectifiers, i.e., lateral geometry GaN rectifiers whose reverse breakdown voltages V_{Break} have been achieved to ~9.7 kV [217, 218] and vertical geometry GaN rectifiers usually fabricated on free-standing GaN. [219-221] Furthermore, high frequency rectifiers which can be used as a key element of inverter modules is another particularly promising devices in speed performance due to GaN's excellent electron transport properties, including good mobility, and high saturated drift velocity. [222-225] The on-state resistance R_{on} of the drift region in GaN can be expressed as

$$R_{on} = \frac{W_D}{q\mu_{GaN} N}, \tag{35}$$

where μ_{GaN} is 1000 cm^2 /Vs for GaN and W_D is the drift region thickness. R_{on} of the GaN power Schottky diodes is approximately 400 times lower than that of Si-based devices at a given voltage. It has be reported that the maximum I_{DS} of the GaN based metal-semiconductor field effect transistors can be reached to 270 mA/mm [226] and an output power density of 1.7 W/mm at 8.4 GHz can be yielded for the transistors with a gate length of 0.25 μm. [223] The polarization-doped devices with a gate length of 0.7 μm yield f_T = 19 GHz and f_{max} =46 GHz. [224] These higher power, higher breakdown voltage and faster switching speed characteristics of GaN devices make them very promising and attractive in commercial and industrial applications in power management and control electronics, such as heavy electric vehicles, motor controllers, power supplies, nuclear power, coal and fossil energy applications, automobile, high-frequency wireless base stations and broad-band links, spacecraft, commercial and military radar, and satellite communications. [223, 226]

In the application field of detections, GaN-based Schottky contacts can work as solid-state sensors for detecting chemical gas, polar liquids, pressure, biosystem, X-ray, Alpha particle, and radiation leakage. [13, 227-230] It has been reported that the GaN Schottky diodes exhibit strong changes in current upon exposure to the hydrogen-containing ambient. [13, 14, 231-234] Ali et al. observed that the barrier height was decreased by 123 meV due to the change of atmosphere from synthetic air to 1% H_2 in synthetic air at 280 °C and both the forward and reverse current increase with the detection temperature and H_2 ratio when the Pt-gated GaN Schottky diodes was employed to test in H_2-containing ambient. [14] The demonstrated good sensitivity of the metal/GaN Schottky devices makes them suitable for detecting ethylene (C_2H_4), hydrocarbons, fuel leak detection in spacecraft and release of toxic or corrosive gases, exhaust diagnosis and emissions.

For photodetector (PD) applications, a good number of reports on the GaN based devices using metal and TCO as the contact materials have been published. [197, 235-239] The structures of these PD devices can be either a single Schottky contact or metal-semiconductor-metal (MSM) with lateral, mesa, and vertical geometries. [12, 197, 238-247] Li et al. reported that compared with the device made of symmetric Schottky, the optical responsivity of the device can be improved by using the asymmetric Schottky structure. [239] Saito et al. investigated the composition effect of III-nitrides on the characteristics of Schottky contact based PDs. [237] Sun et al.,'s transient response measurements showed that the rise time and the fall time of the photoresponse were respectively 10 ns and 960 ns for the ultraviolet photoconductive detectors with the Ga-doped ZnO films. [240] Sheu et al. fabricated an ultraviolet band-pass photodetector whose peak responsivity is around 0.10A/W at 365 nm. [242] Figure 16 shows the typical spectral responses of GaN based photodetectors in MSM structures using metal (AuNi) and GZO as the contact materials, namely sample E and F, respectively. [12] It can be seen that both of sample E and F have an efficient response in the UV range. For sample E, there is a peak response at around 365 nm and a sharp cut-off response at longer wavelength side, which reflects the band edge absorption of GaN. For sample F, however, an additional response peak occurs at ~250 nm that is in the deep ultraviolet spectral range. Moreover, this new response peak is quite broad. Since the only difference between sample E and F is the electrode contact material, the observed new deep UV response in sample F is clearly due to the GZO contact material. Figure 17 shows the room-temperature *I-V* characteristics of sample E under the dark conditions and under the illumination of different light wavelengths. Clearly, the current of the device under the light illumination is significantly larger than the dark current due to the extra contribution of photo absorption excited excess carriers (i.e., electrons in conduction band and holes in valence band) to the current. The current-voltage characteristics of the two kinds of devices under the illumination of light with the same wavelength are illustrated in Figure 18 for comparison. Again, sample F shows much stronger photo response to the deep UV light of 250 nm.

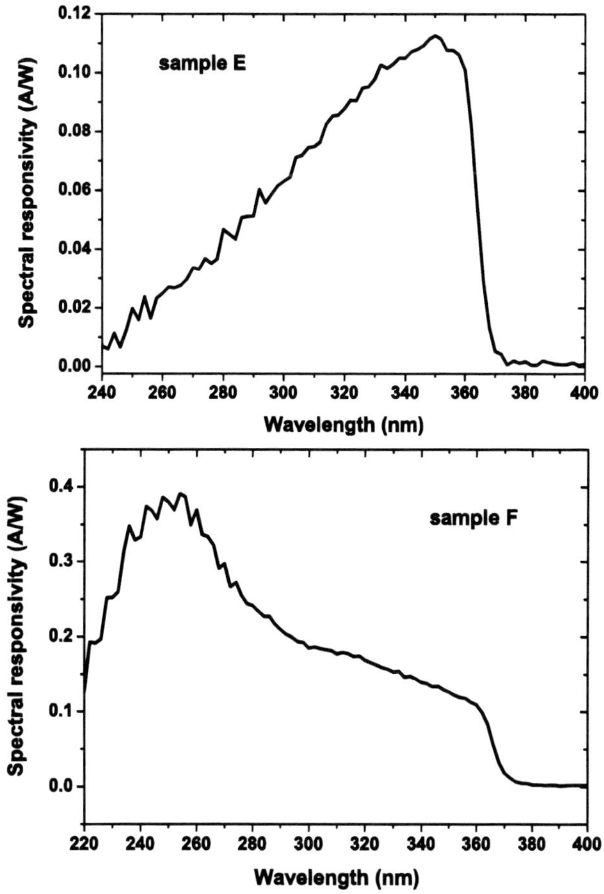

Figure 16. Typical spectral responses of photodetectors in MSM structures of sample E and F.

It is known that most of TCO materials themselves are optoelectronic materials with wide band gaps. Unlike opaque metal materials, TCO materials can have a good optical absorption for incident photons with energy higher than the band gaps of the TCO materials. Therefore, the excess carriers due to the photo absorption in the TCO electrodes may contribute an extra current to the device. Sheu et al. also reported that the efficiency of GaN-based light-emitting diode can be markedly improved when transparent contact layer was used as the electrode material. [248] In sharp contrast to the case of sample E with metal electrodes, sample F with GZO electrodes exhibits two responsivity peaks at around 250 and 350 nm. In particular, the former is 0.4 A/W which is much larger than the latter. From measured value of the responsivity R_{opt}, the derived product of the quantum efficiency η_{opt} and gain g can be estimated by the following equation: [249, 250]

$$R_{opt} = \eta_{opt} g \frac{q\lambda}{hc}, \tag{36}$$

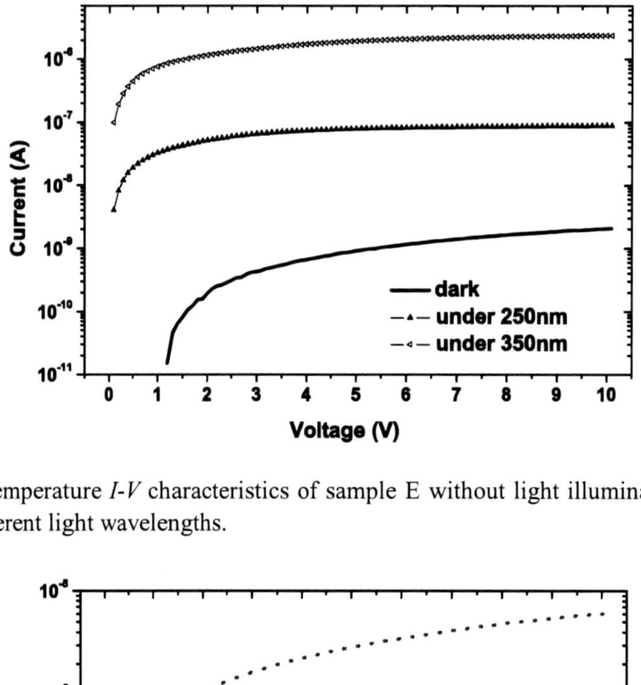

Figure 17. Room-temperature *I-V* characteristics of sample E without light illumination and under the illumination of different light wavelengths.

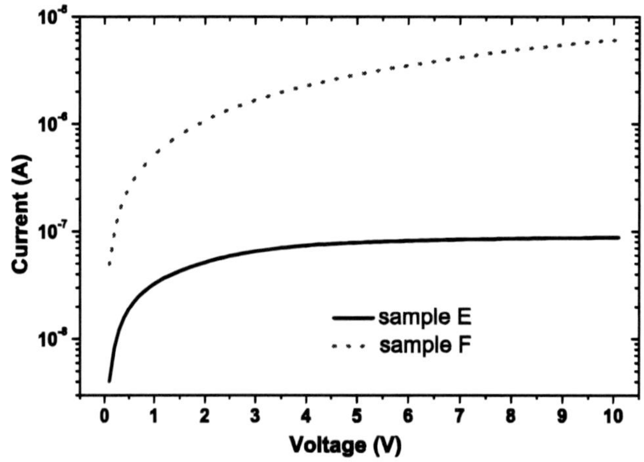

Figure 18. Room-temperature *I-V* characteristics of sample E and F under 250 nm light illumination.

where λ is the wavelength of incident light, and c is the speed of light in vacuum. Clearly, the value of the product of quantum efficiency and the internal gain of sample F is larger than that of sample E, indicating the extra contribution of the efficient photo absorption in the GZO contact layer. For sample E, the efficient photo absorption occurs only in the GaN layer. For sample F, however, efficient photo absorption can occur in both the GZO contact layer and GaN underlying layer. [12] Therefore, the enhanced photo response, in particular in the deep UV region can be understood in sample F. In addition, sample F shows an interesting dependence of the spectral responsivity on the frequency of the incident light, as shown in Figure 19. Further studies are worth conducting for getting an in depth understanding of the TCO/GaN photodetectors.

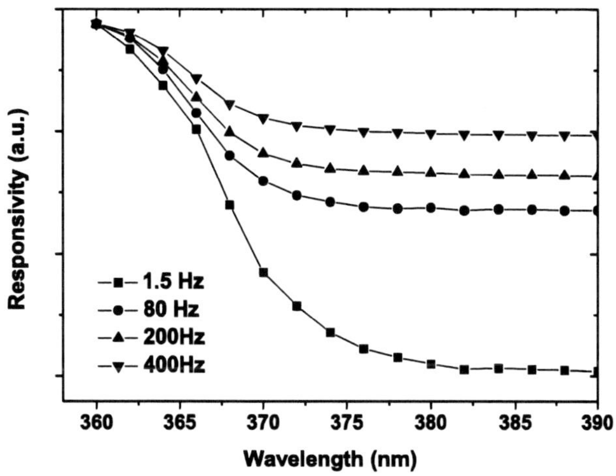

Figure 19. Room-temperature normalized optical responsivity of sample F under different frequency.

Conclusion

In this chapter, a comprehensive review of the GaN based Schottky contacts is given. The physical mechanisms responsible for the electrical characteristics of the devices as a function of the materials quality and the preparation conditions are particularly addressed and discussed in detail. In summary, the *I-V* characteristics of the GaN-based Schottky contacts are not only very sensitive to the as-grown GaN materials quality and the surface status, but also dependent strongly on contact materials. The surface treatment processes of GaN prior to the deposition of contact materials, the selection of contact materials, and the deposition conditions are discussed and summarized since all these practical processes have important influence on the performance of the Schottky contacts. In particular, the influence of various preparation conditions such as the deposition amount of the Au contact material and the deposition atmospheres of ITO thin films on the characteristic of GaN Schottky contacts is described in some detail, and the physical mechanisms responsible for the change in SBH are discussed. Last but not least, some application examples of the GaN Schottky contacts are discussed. It is worth noting that the GaN based Schottky contacts and applications are still a very active and open issue, and many more rigorous studies need to be conducted for obtaining a deeper understanding of the subject matter and for exploiting their applications in more fields.

Acknowledgments

This work was supported by the NSFC grant (No. 60977030) and HK-RGC GRF grant (No. HKU 705812P). The authors would like to thank Mr. K. Wang for his early collaborations to the fabrication and characterization of the Au/GaN Schottky contacts.

References

[1] S. J. Pearton, J. C. Zolper, R. J. Shul, and F. Ren, "GaN: Processing, defects, and devices", *J. Appl. Phys.* 86, 1-78 (1999).
[2] S. N. Mohammad, A. A. Salvador, and H. Morkoc, "Emerging gallium nitride based devices", *Proceedings of the IEEE* 83, 1306-1355 (1995).
[3] S. J. Pearton, F. Renb, A. P. Zhang, and K. P. Lee, "Fabrication and performance of GaN electronic devices", *Materials Science and Engineering* R30, 55-212 (2000).
[4] E. Monroy, F. Calle, J. L. Pau, F. J. Sánchez, E. Muñoz, F. Omnès, B. Beaumont, and P. Gibart, "Analysis and modeling of $Al_xGa_{1-x}N$-based Schottky barrier photodiodes", *J. Appl. Phys.* 88, 2081-2091 (2000).
[5] T. K. Ko, S. J. Chang, Y. K. Su, M. L. Lee, C. S. Chang, Y. C. Lin, S. C. Shei, J. K. Sheu, W. S. Chen, and C. F. Shen, "AlGaN-GaN Schottky-barrier photodetectors with LT GaN cap layers", *J. Cryst. Growth* 283, 68-71 (2005).
[6] X. Zhang, P. Kung, D. Walker, J. Piotrowski, A. Rogalski, A. Saxler, and Razeghi, "Photovoltaic effects in GaN structures with p-n junctions", *Appl. Phys. Lett.* 67, 2028-2030 (1995).
[7] S. Nakamura and G. Fosol, "The Blue Laser Diode", (Springer, Berlin, 1998).
[8] G. J. Sullivan, M. Y. Chen, J. A. Higgins, J. W. Yang, Q. Chen, R. L. Pierson, and B. T. McDermott, "High-power 10-GHz operation of AlGaN HFET's on insulating SiC", *IEEE Electron Device Lett.* 19, 198-199 (1998).
[9] P. Kozodoy, J. P. Ibbetson, H. Marchand, R. T. Fini, S. Keller, J. S. Speck, S. P. DenBaars, and U. K. Mishra, "Electrical characterization of GaN p-n junctions with and without threading dislocations", *Appl. Phys. Lett.* 73, 975-977 (1998).
[10] A. E. Belyaev, N. S. Boltovets, V. N. Ivanov, V. P. Klad'ko, R. V. Konakova, Y. Y. Kudrik, A. V. Kuchuk, V. V. Milenin, Y. N. Sveshnikov, and V. N. Sheremet, "Mechanism of dislocation-governed charge transport in Schottky diodes based on gallium nitride", *Semiconductors* 42, 689-693 (2008).
[11] Y. Q. Wang, S. Alur, Y. Sharma, F. Tong, R. Thapa, P. Gartland, T. Issacs-Smith, C. Ahyi, J. Williams, M. Park, M. Johnson, T. Paskova, E. A Preble, and K. R. Evans, "Ultra-low leakage and high breakdown Schottky diodes fabricated on free-standing GaN substrate", *Semicond. Sci. Technol.* 26, 022002 (2011).
[12] R. X. Wang, L. C. Yang, Y. M. Zhang, K. Fu, S. J. Xu, J. F. Wang, K. Xu, B. S. Zhang, and H. Yang, "The effect of Ga-doped nanocrystalline ZnO electrode on deep-ultraviolet enhanced GaN photodetector", *Appl. Phys. Lett.* 102, 212104 (2013).
[13] J. Schalwig, G. Müller, U. Karrer, M. Eickhoff, O. Ambacher, M. Stutzmann, L. Görgens, and G. Dollinger, "Hydrogen response mechanism of Pt-GaN Schottky diodes", *Appl. Phys. Lett.* 80, 1222-1224 (2002).

[14] M. Ali, V. Cimalla, V. Lebedev, H. Romanus, V. Tilak, D. Merfeld, P. Sandvik, and O. Ambacher, "Pt/GaN Schottky diodes for hydrogen gas sensors", *Sensors and Actuators B* 113, 797-804 (2006).

[15] W. Kruppa, S. C. Binari, and K. Doverspike, "Low-frequency dispersion characteristic of GaN HFETs", *Electron. Lett.* 31, 1951-1952 (1995).

[16] F. Ren, C. R. Abernathy, J. Van Hove, P. P. Chow, R. Hickman, K. B. Jung, H. Cho, T. LaRoche, R. J. Shul, J. Han, and S. J. Pearton, "300℃GaN/AlGaN heterojunction bipolar transistor", *MRS Internet J. Nitride Semicond. Res.* 3, 41 (1998).

[17] F. Ren, M. Hong, S. N. G. Chu, M. A. Marcus, M. J. Schurman, A. G. Baca, S. J. Pearton, and C. R. Abernathy, "Effect of temperature on $Ga_2O_3(Gd_2O_3)$/GaN metal–oxide–semiconductor field-effect transistors", *Appl. Phys. Lett.* 73, 3893-3895 (1998).

[18] M. A. Khan, J. N. Kuznia, A. R. Bhattarai, and D. T. Olsen, "Metal semiconductor field effect transistor based on single crystal GaN", *Appl. Phys. Lett.* 62, 1786-1788 (1993).

[19] M. A. Khan, A. R. Bhattarai, J. N. Kuznia, and D. T. Olsen, "High electron mobility transistor based on a GaN-$Al_xGa_{1-x}N$ heterojunction", *Appl. Phys. Lett.* 63, 1214-1216 (1993).

[20] E. A. Preble, K. M. Tracy, S. Kiesel, H. McLean, P. Q. Miraglia, R. J. Nemanich, R. F. Davis, M. Albrecht, and D. J. Smith, "Electrical, structural and microstructural characteristics of as-deposited and annealed Pt and Au contacts on chemical-vapor-cleaned GaN thin films", *J. Appl. Phys.* 91, 2133-2137 (2002).

[21] Z. J. Lin, J. Z. Zhao, T. D. Corrigan, Z. Wang, Z. D. You, Z. G. Wang, and W. Lu, "The influence of Schottky contact metals on the strain of AlGaN barrier layers", *J. Appl. Phys.* 103, 044503 (2008).

[22] E. D. Readinger, B. R. Luther, S. E. Mohney, and E. L. Piner, "Environmental aging of Schottky contacts to n-AlGaN", *J. Appl. Phys.* 89, 7983-7990 (2001).

[23] E. J. Miller, E. T. Yu, P. Waltereit, and J. S. Speck, "Analysis of reverse-bias leakage current mechanisms in GaN grown by molecular beam epitaxy", *Appl. Phys. Lett.* 84, 535-537 (2004).

[24] L. S. Yu, Q. Z. Liu, Q. J. Xing, D. J. Qiao, S. S. Lau, and J. Redwing, "The role of the tunneling component in the current–voltage characteristics of metal-GaN Schottky diodes", *J. Appl. Phys.* 84, 2099-2104 (1998).

[25] K. J. Duxstad, E. E. Haller, and K. M. Yu, "High temperature behavior of Pt and Pd on GaN", *J. Appl. Phys.* 81, 3134-3137 (1997).

[26] S. N. Mohammad, "Contact mechanisms and design principles for Schottky contacts to group-III nitrides", *J. Appl. Phys.* 97, 063703 (2005).

[27] T. Sawada, N. Kimura, K. Imai, K. Suzuki, and K. Tanahashi, "Interpretation of current transport properties at Ni/n–GaN Schottky interfaces", *J. Vac. Sci. Technol.* B 22, 2051-2058 (2004).

[28] M. Maeda, T. Yamasaki and Y. Takahashi, "Effect of interfacial reaction on electrical conduction across the interface between n-type gallium nitride and contact materials", *J. Phys.: Conference Series* 379, 012020b (2012).

[29] A. C. Schmitz, A. T. Ping, M. A. Khan, Q. Chen, J. W. Yang, and I. Adesida, "Schottky barrier properties of various metals on n-type GaN", *Semicond. Sci. Technol.* 11, 1464-1467 (1996).

[30] A. Ahaitouf, H. Srour, S. O. S. Hamady, N. Fressengeas, A. Ougazzaden, and J. P. Salvestrini, "Interface state effects in GaN Schottky diodes", *Thin Solid Films* 522, 345-351 (2012).
[31] H. T. Wang, S. Jang, T. Anderson, J. J. Chen, B. S. Kang, F. Ren, L. F. Voss, L. Stafford, R. Khanna, B. P. Gila, S. J. Pearton, H. Shen, J. R. LaRoche, and K. V. Smith, "Increased Schottky barrier heights for Au on n- and p-type GaN using cryogenic metal deposition", *Appl. Phys. Lett.* 89, 122106 (2006).
[32] S. Chand and S. Bala, "A comparative study of numerical and analytical methods of simulating inhomogeneous Schottky diode characteristics", *Semicond. Sci. Technol.* 20, 1143-1148 (2005).
[33] C. Merz, M. Kunzer, U. Kaufmann, I. Akasaki, and H. Amano, "Free and bound excitons in thin wurtzite GaN layers on sapphire", *Semicond. Sci. Technol.* 11, 712-716 (1996).
[34] H. Kim, S. N. Lee, Y. Park, J. S. Kwak, and T. Y. Seong, "Metallization contacts to nonpolar -plane n-type GaN", *Appl. Phys. Lett.* 93, 032105 (2008).
[35] S. Oyama, T. Hashizume, and H. Hasegawa, "Mechanism of current leakage through metal/n-GaN interfaces", *Appl. Surf. Sci.* 190, 322-325 (2002).
[36] J. Kotani, T. Hashizume, and H. Hasegawa, "Analysis and control of excess leakage currents in nitride-based Schottky diodes based on thin surface barrier model", *J. Vac. Sci. Technol. B* 22, 2179-2189 (2004).
[37] Y. P. Song, R. L. Van Meirhaeghe, W. H. Laflere, and F. Cardon, "On the difference in apparent barrier height as obtained from capacitance-voltage and current-voltage-temperature measurements on Al/P-InP Schottky barriers", *Solid-State Electron.* 29, 633-638 (1986).
[38] H. Hasegawa and S. Oyama, "Mechanism of anomalous current transport in n-type GaN Schottky contacts", *J. Vac. Sci. Technol.* B 20, 1647-1655 (2002).
[39] H. J. Werner and H. H. Güttler, "Barrier inhomogeneities at Schottky contacts", *J. Appl. Phys.* 69, 1522-1533 (1991).
[40] N. N. K. Reddy and V. R. Reddy, "Barrier characteristics of Pt/Ru Schottky contacts on n-type GaN based on I–V–T and C–V–T measurements", *Bull. Mater. Sci.* 35, 53-61 (2012).
[41] S. M. Sze, "Physics of Semiconductor Devices", 2nd ed. (Wiley, New York, 1981).
[42] E. H. Rhoderick and R. H. Williams, "Metal-Semiconductor Contacts", 2nd ed. (Clarendon, Oxford, 1988).
[43] S. Chand and J. Kumar, "Electron transport and barrier inhomogeneities in palladium silicide Schottky diodes", *Applied Physics A-materials Science & Processing* 65, 497-503 (1997).
[44] R. T. Tung, "Recent advances in Schottky barrier concepts", *Mater. Sci. Eng.* R35, 1-138 (2001).
[45] S. Chand and S. Bala, "Analysis of current-voltage characteristics of inhomogeneous Schottky diodes at low temperatures", *Appl. Surf. Sci.* 252, 358-363 (2005).
[46] N. Yildirim, K. Ejderha, and A. Turut, "On temperature-dependent experimental I-V and C-V data of Ni/n-GaN Schottky contacts", *J. Appl. Phys.* 108, 114506 (2010).

[47] Y. Zhou, D. Wang, C. Ahyi, C.-C. Tin, J. Williams, M. Park, N. M. Williams, A. Hanser, and E. A. Preble, "Temperature-dependent electrical characteristics of bulk GaN Schottky rectifier", *J. Appl. Phys.* 101, 024506 (2007).
[48] G. Parish, R. A. Kennedy, G. A. Umana-Membreno, and B. D. Nener, "Localised defect-induced Schottky barrier lowering in n-GaN Schottky diodes", *Solid-State Electron.* 52, 171-174 (2008).
[49] D. K. Schroder, "Semiconductor Material and Device Characterization", (New York: Wiley) (1990).
[50] P. Hacke, T. Detchprohm, K. Hiramatsu, and N. Sawaki, "Schottky barrier on n-type GaN grown by hydride vapor phase epitaxy", *Appl. Phys. Lett.* 63, 2676-2678 (1993).
[51] J. D. Guo, M. S. Feng, R. J .Guo, F. M. Pan, and C. Y. Chang, "Study of Schottky barriers on n - type GaN grown by low - pressure metalorganic chemical vapor deposition", *Appl. Phys. Lett.* 67, 2657-2659 (1995).
[52] H. Norde, "A modified forward I–V plot for Schottky diodes with high series resistance", *J. Appl. Phys.* 50, 5052-5053 (1979).
[53] M. Mamor, "Interface gap states and Schottky barrier inhomogeneity at metal/n-type GaN Schottky contacts", *J. Phys: Condens. Matter* 21, 335802 (2009).
[54] A. R. Arehart, B. Moran, J. S. Speck, U. K. Mishra, S. P. DenBaars, and S. A. Ringel, "Effect of threading dislocation density on Ni/n-GaN Schottky diode I-V characteristics", *J. Appl. Phys.* 100, 023709 (2006).
[55] N. Yildirim, K. Ejderha, and A. Turut, "On temperature dependent experimental I-V and C-V data of Ni / n -GaN Schottky contacts", *J. Appl. Phys.* 108, 114506 (2010).
[56] A. Kumar, S. Arafin, M. C. Amann, and R. Singh, "Temperature dependence of electrical characteristics of Pt/GaN Schottky diode fabricated by UHV e-beam evaporation", *Nanoscale Research Lett.* 8, 481 (2013).
[57] S. Huang, B. Shen, J. Wang, F. J. Xu, J. Y. Wang, H. Y. Yang, F. Lin, L. Lu, Z. P. Chen, Z. X. Qin, Z. J. Yang, and G. Y. Zhang, "Current transport mechanism of Au/Ni/GaN Schottky diodes at high temperatures", *Appl. Phys. Lett.* 91, 072109 (2007).
[58] K. R. Peta, B. G. Park, S. T. Lee, M. D. Kim, and J. E. Oh, "Temperature-dependent electrical properties of (Pt/Au)/Ga-polarity GaN/Si(111) Schottky diode", *Microelectron. Eng.* 93, 100-104 (2012).
[59] P. K. Rao, B. Park, S. T. Lee, Y. K. Noh, M. D. Kim, and J. E. Oh, " Analysis of leakage current mechanisms in Pt/Au Schottky contact on Ga-polarity GaN by Frenkel-Poole emission and deep level studies*", J. Appl. Phys.* 110, 013716 (2011).
[60] S. Dogan, S. Duman, B. Gurbulak, S. Tuezemen, H. Morkoc, "Temperature variation of current-voltage characteristics of Au/Ni/n-GaN Schottky diodes", Physice-Low-dimensional *Systems & Nanostructures* 41, 646-651 (2009).
[61] M. S. P. Reddy, A. A. Kumar, and V. R. Reddy, "Electrical transport characteristics of Ni/Pd/n-GaN Schottky barrier diodes as a function of temperature", *Thin Solid Films* 519, 3844-3850 (2011).
[62] H. C. Card and E. H. Rhoderick, "Studies of tunnel MOS diodes: I. Interface effects in silicon Schottky diodes", *J. Phys D.: Appl. Phy.* l, 1589-1601 (1971).
[63] S. Chand and J. Kumar, "Evidence for the double distribution of barrier heights in Pd2Si/n-Si Schottky diodes from I–V–T Measurements", *Semicond. Sci. Technol.* 11, 1203-1208 (1996).

[64] Y. B. Roul, T. N. Bhat, M. Kumar, M. K. Rajpalke, A. T. Kalghatgi, and S. B. Krupanidhi, "Analysis of the temperature-dependent current–voltage characteristics and the barrier-height inhomogeneities of Au/GaN Schottky diodes", *Phys. Status Solidi A* 209, 1575-1578 (2012).

[65] N. Yıldırım, A. Turut, and V. Turut, "The theoretical and experimental study on double-Gaussian distribution in inhomogeneous barrier-height Schottky contacts", *Microelectron. Eng.* 87, 2225-2229 (2010).

[66] K. Ejderha, N. Yıldırım, B. Abay, and A. Turut, "Examination by interfacial layer and inhomogeneous barrier height model of temperature-dependent I–V characteristics in Co/p-InP contacts", *J. Alloy. Compd.* 484, 870-876 (2009).

[67] N. Yıldırım and A. Türüt, "A theoretical analysis together with experimental data of inhomogeneous Schottky barrier diodes", Microelectron. Eng. 86, 2270-2274 (2009).

[68] A. Many, Y. Goldstein, and N. B. Grover, "Semiconductor Surface", (North-Holland, Amsterdam, (1965) p. 220.

[69] R. X. Wang, S. J. Xu, S. L. Shi, C. D. Beling, S. Fung, D. G. Zhao, H. Yang, and X. M. Tao, "Probing deep level centers in GaN epilayers with variable-frequency capacitance-voltage characteristics of Au/GaN Schottky contacts", *Appl. Phys. Lett.* 89, 143503 (2006).

[70] S. Suresh, M. Balaji, V. Ganesh, and K. Baskar, "Investigations on the nonidealities in Pd/n-GaN Schottky diodes grown by MOCVD", *J. Alloy. Compd.*, 506, 615-619 (2010).

[71] K. Saarinen, P. Seppälä, J. Oila, P. Hautojärvi, C. Corbel, O. Briot and R. L. Aulombard, "Gallium vacancies and the growth stoichiometry of GaN studied by positron annihilation spectroscopy", *Appl. Phys. Lett.* 73, 3253-3255 (1998).

[72] H. Zhang, E. J. Miller, and E. T. Yu, "Analysis of leakage current mechanisms in Schottky contacts to GaN and$Al_{0.25}Ga_{0.75}N$/GaN grown by molecular-beam epitaxy", *J. Appl. Phys.* 99, 023703 (2006).

[73] J. W. P. Hsu, M. J. Manfra, R. J. Molnar, B. Heying, and J. S. Speck, "Direct imaging of reverse-bias leakage through pure screw dislocations in GaN films grown by molecular beam epitaxy on GaN templates", *Appl. Phys. Lett.* 81, 79-81 (2002).

[74] D. J. Ewing, M. A. Derenge, P. B. Shah, U. Lee, T. S. Zheleva, and K. A. Jones, "Improvement in GaN and AlGaN/GaN Schottky diode performance by reduction in epitaxial film dislocation density", *J. Vac. Sci. Technol.* B 26, 1368-1372 (2008).

[75] J. Suda, K.Yamaji, Y. Hayashi, T. Kimoto, K. Shimoyama, H. Namita, and S. Nagao, "Nearly ideal current–voltage characteristics of Schottky barrier diodes formed on hydride-vapor-phase-epitaxy-grown GaN free-standing substrates", *Applied Physics Express 3,* 101003 (2010).

[76] J. W. P. Hsu, M. J. Manfra, D. V. Lang, S. Richter, S. N. G. Chu, A. M. Sergent, R. N. Kleiman, L. N. Pfeiffer, and R. J. Molnar, "Inhomogeneous spatial distribution of reverse bias leakage in GaN Schottky diodes", *Appl. Phys. Lett.* 78, 1685-1687 (2001).

[77] E. J. Miller, D. M. Schaadt, E. T. Yu, C. Poblenz, C. Elsass, and J. S. Speck, "Reduction of reverse-bias leakage current in Schottky diodes on GaN grown by molecular-beam epitaxy using surface modification with an atomic force microscope", *J. Appl. Phys.* 91, 9821-9826 (2002).

[78] E. Arslan, S. Bütün and E. Ozbay, "Leakage current by Frenkel–Poole emission in Ni/Au Schottky contacts on $Al_{0.83}In_{0.17}N$/AlN/GaN heterostructures", *Appl. Phys. Lett.* 94, 142106 (2009).

[79] E. J. Miller, X. Z. Dang, and E. T. Yu. "Gate leakage current mechanisms in AlGaNGaN heterostructure field-effect transistors", *J. Appl. Phys.* 88, 5951-5958 (2000).

[80] T. Hashizume, J. Kotani, and H. Hasegawa. "Leakage mechanism in GaN and AlGaN Schottky interfaces", *Appl. Phys. Lett.* 84, 4484-4486 (2004).

[81] M. Ozbek and B. J. Baliga, "Tunneling coefficient for GaN Schottky barrier diodes", *Solid-State Electronics* 62, 1-4 (2011).

[82] S. J. Xu, W. Liu, and M. F. Li, "Effect of temperature on longitudinal optical phonon-assisted exciton luminescence in heteroepitaxial GaN layer", *Appl. Phys. Lett.* 77, 3376-3378 (2000).

[83] B. Monemar, "Gallium Nitride (GaN) I, *Semiconductors and Semimetals*" Vol. 50, Chapter 11, (1971).

[84] D. C. Look and J. R. Sizelove, "Dislocation scattering in GaN", *Phys. Rev. Lett.* 82, 1237-1240 (1999).

[85] H. M. Ng, D. Doppalapudi, T. D. Moustakas, N. G. Weimann, and L. F. Eastman, "The role of dislocation scattering in n-type GaN films", *Appl. Phys. Lett.* 73, 821-823 (1998).

[86] K. Wang, R. X. Wang, S. Fung, C. D. Beling, X. D. Chen, Y. Huang, S. Li, S. J. Xu, and M. Gong, "Film thickness degradation of Au/GaN Schottky contact characteristics", *Mater. Sci. Eng.* B 117, 21-25 (2005).

[87] S. Li, J. D. Zhang, C. D. Beling, K. Wang, R. X. Wang, M. Gong, and C. K. Sarkar, "Large lattice relaxation deep levels in neutron-irradiated GaN", *J. Appl. Phys.*, 98, 093517 (2005).

[88] D. G. Zhao, D. S. Jiang, H. Yang, J. J. Zhu, Z. S. Liu, S. M. Zhang, J. W. Liang, X. Li, X. Y. Li, and H. M. Gong, "Role of edge dislocations in enhancing the yellow luminescence of n-type GaN", *Appl. Phys. Lett.* 88, 241917 (2006).

[89] D. O'Mahony,W. Zimmerman, S. Steffen, J. Hilgarth, P. Maaskant, R. Ginige, L. Lewis, B. Lambert and B. Corbett, "Free-standing gallium nitride Schottky diode characteristics and stability in a high-temperature environment", *Semicond. Sci. Technol.* 24, 125008 (2009).

[90] F. Roccaforte, F. Giannazzo, F. Iucolano, J. Eriksson, M. H. Weng, and V. Raineri, "Surface and interface issues in wide band gap semiconductor electronics", *Appl. Surf. Sci.* 256, 5727-5735 (2010).

[91] T. L. Chu, K. Ito, R. K. Smeltzer, and S. S. C. Chu, "Crystal-growth and characteristic of gallium nitride", *J. Electrochem. Soc.* 121, 159-162 (1974).

[92] H. Amano, N. Sawaki, I. Akasaki, and Y. Toyada, "Metalorganic vapor phase epitaxial growth of a high quality GaN film using an AlN buffer layer", *Appl. Phys. Lett.* 48, 353-355 (1986).

[93] Y. Kato, S. Kitamura, K. Hiramatsu, and N. Sawaki, "Selective growth of wurtzite GaN and $Al_xGa_{1-x}N$ on GaN sapphire substrates by metalorganic vapor-phase epitaxy", *J. Cryst. Growth* 144, 133-140 (1994).

[94] H. Marchand, X. H. Wu, J. P. Ibbetson, P. T. Fini, P. Kozodoy, S. Keller, J. S. Speck, S. P. DenBaars, and U. K. Mishra, "Microstructure of GaN laterally overgrown by metalorganic chemical vapor deposition", *Appl. Phys. Lett.* 73, 747-749 (1998).
[95] D. Alquier, F. Cayrel, O. Menard, A. E. Bazin, A. Yvon, and E. Collard, "Recent progresses in GaN power rectifier", *Jpn. J. Appl.Phys.* 51, 01AG08 (2012).
[96] H. Marchand, J. P. Ibbetson, P. T. Fini, P. Kozodoy, S. Keller, S. DenBaars, J. S. Speck, and U. K. Mishra, "Atomic force microscopy observation of threading dislocation density reduction in lateral epitaxial overgrowth of gallium nitride by MOCVD", *Mater. Res. Soc. Int. J. Nitride Semi Res.* 3, 3 (1998).
[97] H. Marchand, J. P. Ibbetson, P. T. Fini, S. Keller, S. P. DenBaars, J. S. Speck, and U. K. Mishra, "Mechanisms of lateral epitaxial overgrowth of gallium nitride by metalorganic chemical vapor deposition", *J. Cryst. Growth* 195, 328-332 (1998).
[98] X. A. Cao, S. J. Pearton, G. Dang, A. P. Zhang, F. Ren, and J. M. Van Hove, "Effects of interfacial oxides on Schottky barrier contacts to n- and p-type GaN", *Appl. Phys. Lett.* 75, 4130-4132 (1999).
[99] H. W. Jang, J. H. Lee, and J. L. Lee, "Characterization of band bendings on Ga-face and N-face GaN films grown by metalorganic chemical-vapor deposition", *Appl. Phys. Lett.* 80, 3955-3957 (2002).
[100] J, W. Jeon, S. H. Park, S. Y. Jung, J. Moon, J. O. Song, G. Namgoong, and T. Y. Seong, "Electrical characteristics of V/Ti/Au contacts to Ga-polar and N-polar n-GaN prepared by different methods", *Electrochemical and Solid-State Letters* 13, H125-H127 (2010).
[101] J. Osvald, J. Kuzmik, G. Konstantinidis, P. Lobotka, and A. Georgakilas, "Temperature dependence of GaN Schottky diodes I-V characteristics", *Microelectron. Eng.* 81, 181-187 (2005).
[102] V. M. Bermudez, "Study of oxygen chemisorption on the GaN(0001)-(1×1) surface", *J. Appl. Phys.* 80, 1190-1200 (1996).
[103] S. Kurtin, T. C. McGill, and C. A. Mead, "Fundamental transition in the electronic nature of solid", *Phys. Rev. Lett.* 22, 1433-1436 (1969).
[104] H. W. Jang, T. Sands, and J. L. Lee, "Effects of KrF excimer laser irradiation on metal contacts to n -type and p -type GaN", *J. Appl. Phys.* 94, 3529-3535 (2003).
[105] H. Ishikawa, S. Kobayashi, Y. Koide, S. Yamasaki, S. Nagai, J. Umezaki, M. Koike, and M. Murakami, "Effects of surface treatments and metal work functions on electrical properties at p-GaN/metal interfaces", *J. Appl. Phys.* 81, 1315-1322 (1997).
[106] T. Kimura and T. Hashizume "Effect of carbon incorporation on electrical properties of n -type GaN surfaces". *J. Appl. Phys.* 105, 014503 (2009).
[107] H. W. Choi, S. J. Chua, and X. J. Kang, "Surface modification and ohmic contact formation to n and p-type GaN", *Phys. Stat. Sol. A* 188, 399-402 (2001).
[108] A. Shintani and S. Minagawa, "Etching of GaN using phosphoric-acid", *J. Electrochem. Soc.* 123, 706-713 (1976).
[109] E. J. Miller, D. M. Schaadt, E. T. Yu, P. Waltereit, C. Poblenz, and J. S. Speck, "Reverse-bias leakage current reduction in GaN Schottky diodes by electrochemical surface treatment", *Appl. Phys. Lett.* 82, 1293-1295 (1998).

[110] H. G. Kim, S. H. Kim, P. Deb, and T. Sands, "Effect of KOH treatment on the Schottky barrier height and reverse leakage current in Pt/n-GaN", *J. Electron. Mater.* 35, 107-112 (2006).

[111] M. S. P. Reddy, V. R. Reddy, I. Jyothi, and C. J. Choi, "Electrical, structural and morphological characteristics of rapidly annealed Ni/Pd Schottky rectifiers on n-type GaN", *Surf. Interface Anal.* 43, 1251-1256 (2011).

[112] J. Spradlin, S. Dogan, M. Mikkelson, D. Huang, L. He, D. Johnstone, H. Morko, and R. J. Molnar, "Improvement of n-GaN Schottky diode rectifying characteristics using KOH etching", *Appl. Phys. Lett.* 82, 3556-3558 (2003).

[113] N. Miura, T. Nanjo, M. Suita, T. Oishi, Y. Abe, T. Ozeki, H. Ishikawa, T. Egawa, and T. Jimbo, "Thermal annealing effects on Ni/Au based Schottky contacts on n-GaN and AlGaN/GaN with insertion of high work function metal", *Solid-State Electron.* 48, 689-695 (2004).

[114] D. W. Kim, J. C. Bae, W. J. Kim, H. K. Baik, J. M. Myoung, and S. M. Lee, "The Improvement of electrical properties of Pd-based contact to p-GaN by surface treatment", *J. Electron. Mater.* 30, 183-187 (2001).

[115] L. L. Smith, S.W. King, R. J. Nemanich, and R. F. Davis, "Cleaning of GaN surfaces", *J. Electron. Mater.* 25, 805-810 (1996).

[116] S. W. King, L. L. Smith, J. P. Barnak, J.-H. Ku, J. A. Christman, M. C. Benjamin, M. D. Bremser, R. J. Nemanich, and R. F. Davis, "Ex situ and in situ methods for oxide and carbon removal from AlN and GaN surfaces", *Mater. Res. Soc. Symp. Proc.* 395, 739-744 (1996).

[117] Q. Z. Liu and S. S. Laua, "Review of the metal-GaN contact technology", *Solid-State Electron.* 42, 677-691 (1998).

[118] G. L. Martinez, M. R. Curiel, B. J. Skromme, and R. J. Molnar, "Surface recombination and sulfide passivation of GaN", *J. Electron. Mater.* 29, 325-331 (2000).

[119] Y. J. Lin, W. X. Lin, C. T. Lee, and F. T. Chien, "Changes in optical and electrical properties and surface recombination velocity of n-type GaN due to $(NH_4)_2S_x$ treatment", *Solid State Communications* 137, 257-259 (2006).

[120] T. Sreenidhi, K. Baskar, A. DasGupta, and N. DasGupta, "Reactive ion etching of GaN in SF_6 + Ar and $SF_6 + N_2$ plasma", *Semicond. Sci. Technol.* 23, 125019 (2008).

[121] S.-H. Su, C.-C. Hou, M.-H. Lin, M. Yokoyama, S.-M. Chen, and H. Kuan, "Reduction of ohmic contact resistance on n-GaN by surface treatment using Cl_2 inductively coupled plasma following laser lift", *Jpn. J. Appl. Phys.* 46, 965-967 (2007).

[122] C. I. Wu and A. Kahn, "Investigation of the chemistry and electronic properties of metal/gallium nitride interfaces", *J. Vac. Sci. Technol.* B 16, 2218-2223 (1998).

[123] X. A. Cao, H. Cho, S. J. Pearton, G. T. Dang, A. P. Zhang, F. Ren, R. J. Shul, L. Zhang, R. Hickman, and J. M. Van Hove, "Depth and thermal stability of dry etch damage in GaN Schottky diodes", *Appl. Phys. Lett.* 75, 232-234 (1999).

[124] A. S. Usikov, W. L. Lundin, U. I. Ushakov, B. V. Pushnyi, N. M. Schmidt, Y. M. Zadiranov, and T. V. Shubtra, "Electronic and optical properties of the device structures based on GaN after dry etching", *Proc. Electrochem. Soc.* 98-14, 57-62 (1998).

[125] R. W. Hunt, L. Vanzetti, T. Castro, K. M. Chen, L. Sorba, P. I. Cohen, W. Gladfelter, J. M. Van Hove, J. N. Kuznia, M. A. Khan, and A. Franciosi, "Electronic-structure, surface-composition and long-range order in GaN", *Physica B* 185, 415-421 (1993).

[126] C. Y. Lee, H. Sekiguchi, H. Okada, and A. Wakahara, "Plasma-induced damage and recovery on Au/n-GaN Schottky diode in different processes", *Jpn. J. Appl. Phys.* 51, 076503 (2012).

[127] Q. Fan, S. Chevtchenko, X. F. Ni, S. J. Cho, F. Yun, and H. Morkoç, "Reactive ion etch damage on GaN and its recovery", *J. Vac. Sci. Technol.* B 24, 1197-1201 (2006).

[128] A. T. Ping, A. C. Schmitz, I. Adesida, M. A. Khan, Q. Chen, and J. W. Yang, "Characterization of reaction ion etching-induced damage to n-GaN surface using Schottky diodes", *J. Electron. Mater.* 26, 266-271 (1997).

[129] Z. Fan, S. N. Mohammed, W. Kim, O. Aktas, A. E. Botchkarev, and H. Morkoç, "Very low resistance multilayer Ohmic contact to n-GaN", *Appl. Phys. Lett.* 68, 1672-1674 (1996).

[130] A. T. Ping, Q. Chen, J. W. Yang, M. A. Khan, and I. Adesida, "The effects of reactive ion etching-induced damage on the characteristics of ohmic contacts to n-Type GaN", *J. Electron. Mater.* 27, 261-265 (1997).

[131] D. Basak, M. Verdu, M. T. Montojo, M. A. Sanchez-Garcia, F. J.Sanchez, E. Munoz, and E. Calleja, "Reactive ion etching of GaN layers using SF_6", *Semicond. Sci. Technol.* 12, 1654-1657 (1997).

[132] B. Rong, R. Cheung, W. Gao, M. M. Alkaisi, and R. J. Reeves, "Effects of reactive ion etching on the electrical characteristics of GaN", *J. Vac. Sci. Technol. B* 18, 3467-3470 (2000).

[133] J. M. Lee, K. M. Chang, S. W. Kim, C. Huh, I. H. Lee, and S. J. Park, "Dry etch damage in n-type GaN and its recovery by treatment with an N_2 plasma", *J. Appl. Phys.* 87, 7667-7670 (2000).

[134] S. Tripathy, A. Ramam, S. J. Chua, J. S. Pan, and A. Huan, "Characterization of inductively coupled plasma etched surface of GaN using Cl_2/BCl_3 chemistry", *J. Vac. Sci. Technol. A* 19, 2522-2532 (2001).

[135] S. J. Cho, S. Doğan, S. Sabuktagin, M. A. Reshchikov, D. K. Johnstone, and H. Morkoc¸ "Surface band bending in as-grown and plasma-treated n-type GaN films using surface potential electric force microscopy", *Appl. Phys. Lett.* 84, 3070-3072 (2004).

[136] H. W. Choi, S. J. Chua, A. Raman, J. S. Pan, and A. T. S. Wee, "Plasma-induced damage to n-type GaN", *Appl. Phys. Lett.* 77, 1795-1797 (2000).

[137] B. Rong, E. Van Der Drift, R. J. Reeves, W. G. Sloof, and R. Cheung, "Inductively coupled plasma etching of GaN and its effect on electrical characteristics", *J. Vac. Sci. Technol.* B 19, 2917-2920 (2001).

[138] P. S. Chen, T. H. Lee, L. W. Lai, and C. T. Lee, "Schottky mechanism for Ni Au contact with chlorine-treated n -type GaN layer", *J. Appl. Phys.* 101, 024507 (2007).

[139] X. A. Cao, A. P. Zhang, G. T. Dang, F. Ren, S. J. Pearton, R. J. Shul, and L. Zhang, "Schottky diode measurements of dry etch damage in n- and p-type GaN", *J. Vac. Sci. Technol. A* 18, 1144-1148 (2000).

[140] H. W. Jang, C. M. Jeon, J. K. Kim, and J. L. Lee, "Room-temperature ohmic contact on n-type GaN with surface treatment using Cl_2 inductively coupled plasma", *Appl. Phys. Lett.* 78, 2015-2017 (2001).

[141] H. S. Kim, G. Y. Yeom, J. W. Lee, and T. I. Kim, "A study of GaN etch mechanisms using inductively coupled Cl_2/Ar plasmas", *Thin Solid Films* 341, 180-183 (1999).

[142] Y. J. Hsn, S. Xue, W. P. Guo, C. Z. Sun, Z. B. Hao, and Y. Luo, "Characteristics of n-GaN after Cl_2/Ar and Cl_2/N_2 inductively coupled plasma etching", *Jpn. J. Appl. Phys.* 42, 6409-6412 (2003).
[143] K. M. Tracy, P. J. Hartlieb, S. Einfeldt, R. F. Davis, E. H. Hurt, and R. J. Nemanich, "Electrical and chemical characterization of the Schottky barrier formed between clean n-GaN (0001) surfaces and Pt, Au, and Ag", *J. Appl.Phys.* 94, 3939-3948 (2003).
[144] L. Dobos, B. Pécz, L. Tóth, Zs. J. Horváth, Z. E. Horváth, A. Tóth, and M.-A. Poisson, "Annealed Ti/Cr/Al contacts on n-GaN", *Vacuum* 100, 46-49 (2014).
[145] L. Dobos, B. Pe´cz, L. To´th, Zs. J. Horva´th, Z. E. Horva´th, B. Beaumont, and Z. Bougrioua, "Structural and electrical properties of Au and Ti/Au contacts to n-type GaN", *Vacuum* 82, 794-798 (2008).
[146] J. D. Guo, F. M. Pan, M. S. Feng, R. J. Guo, P. F. Chou, and C. Y. Chang, "Schottky contact and the thermal stability of Ni on n-type GaN", *J. Appl. Phys.* 80, 1623-1627 (1996).
[147] O. Menard, F. Cayrel, E. Collard, and D. Alquier, "Ni based planar Schottky diodes on gallium nitride (GaN) grown on sapphire", *Phys. Status Solidi C* 7, 112-115 (2010).
[148] M. R. H. Khan, T. Detchprohm, P. Hacke, K. Hiramatsu, and N. Sawaki, "The barrier height and interface effect of a Au-n-GaN Schottky diode", *J. Appl. D: Appl. Phys.* 28, 1169-1174 (1995).
[149] H. T. Wang, S. Jang, T. Anderson, J. J. Chen, B. S. Kang, F. Ren, L. F. Voss, L. Stafford, R. Khanna, B. P.Gila, S. J. Pearton, H. Shen, J. R. LaRoche, and K. V. Smith, "Increased Schottky barrier heights for Au on n- and p-type GaN using cryogenic metal eposition", *Appl. Phys. Lett.* 89, 122106 (2006).
[150] K. J. Reddy, V. R. Reddy, and P. N. Reddy, "Thermal annealing behavior on Schottky barrier parameters and structural properties of Au contacts to n-type GaN", *J. Mater. Sci.: Mater Electron* 19, 333-338 (2008).
[151] L. Wang, M. I. Nathan, T.-H. Lim, M. A. Khan, and Q. Chen, "High barrier height GaN Schottky diodes: Pt/GaN and Pd/GaN", *Appl. Phys. Lett.* 68, 1267-1269 (1996).
[152] Y. Kribes, I. Harrison, T. S. Cheng, and C. T. Foxon, "Investigation of Au Schottky contacts on GaN grown by molecular beam epitaxy", *Semicond. Sci. Technol.* 12, 913-917 (1997).
[153] B. Akkal, Z. Benamara, H. Abid, A. Talbi, and B. Gruzza, "Electrical characterization of Au/n-GaN Schottky diodes", *Mater. Chem. Phys.* 85, 27-31 (2004).
[154] A. T. Ping, A. C. Schmitz, M. A. Khan, and I. Adesida, "Characterisation of Pd Schottky barrier on n-type GaN", *Electron. Lett.* 32, 68-70 (1996).
[155] A. C. Schmitz, A. T. Ping, M. A. Khan, Q. Chen, J. W. Yang, and I. Adesida, "High temperature characteristics of Pd Schottky contacts on n-type GaN", *Electron. Lett.* 32, 1832-1833 (1996).
[156] Y. J. Lin, "Electronic transport and Schottky barrier heights of Pt/n-type GaN Schottky diodes in the extrinsic region", *J. Appl. Phys.* 106, 013702 (2009).
[157] F. K. Yam and Z. Hassan, "Investigation of the effects of porous layer on the electrical properties of Pt/n-GaN Schottky contacts", *Physica B* 403, 3105-3109 (2008).
[158] H. Kim, S. H. Phark, K.-M. Song, and D. W. Kim, "Schottky contacts to polar and nonpolar n-type GaN", *J. Korean Phys. Soc.* 60, 104-107 (2012).

[159] K. Çnar, N. Yildirim, C. Cokun, and A. Turut, "Temperature dependence of current-voltage characteristics in highly doped Ag/p–GaN In Schottky diodes", *J. Appl. Phys.* 106, 073717 (2009).

[160] L. Dobos, B. Pécz, L. Tóth, Zs. J. Horváth, Z. E. Horváth, A. Tóth, E. Horváth, B. Beaumont, and Z. Bougrioua, "Metal contacts to n-GaN", *Appl. Surf. Sci.* 253, 655-661 (2006).

[161] C. K. Ramesh, V. R. Reddy, and C.-J. Choi, "Electrical characteristics of molybdenum Schottky contacts on n-type GaN", *Mater. Sci. Eng.* B 112, 30-33 (2004).

[162] C. K. Ramesh, V. R. Reddy, and K. S. R. K. Rao, "Effect of annealing temperature on electrical characteristics of ruthenium-based Schottky contacts on n-type GaN", *J. Mater. Sci.-Mater. Electron* 17, 999-1004 (2006).

[163] H. S. Venugopalan and S. E. Mohney, "Thermally stable rhenium Schottky contacts to n-GaN", *Appl. Phys. Lett.* 73, 1242-1244 (1998).

[164] R. Mehandru, S. Kang, S. Kim, F. Ren, I. Kravchenko,W. Lewis, and S. J. Pearton, "Effect of deposition conditions and annealing on W Schottky contacts on n-GaN", *Mater. Sci. Semicond. Process.* 7, 95-98 (2004).

[165] F. Tian and E. F. Chor, "Rhodium-based Schottky contacts on n-doped gallium nitride", *Phys. Status Solidi* C 5, 1953-1955 (2008).

[166] R. Khanna, S. J. Pearton, F. Ren, I. Kravchenko, C. J. Koa, and G. C. Chi, "W_2B-based rectifying contacts to n-GaN", *Appl. Phys. Lett.* 87, 052110 (2005).

[167] T. N. Order, P. Martin, J. Y. Lin, H. X. Jiang, J. R. Williams, and T. Issacs-Smith, "Thermally stable Schottky contacts on n-type GaN using ZrB_2" , *Appl. Phys. Lett.* 88, 183505 (2006).

[168] C. Z. Lu, X. L. Zhang, X. S. Xie, S. W. Feng, I. Diagne, A. Khan, and S. N. Mohammad, "Interface states mediated reverse leakage through metal/$Al_xGa_{1-x}N$/GaN Schottky diodes". *J. Vac. Sci. Technol.* B 26, 1987-1992 (2008).

[169] C. Y. Hu, Z. X. Qin, Z. X. Feng, Z. Z. Chen, Z. B. Ding, Z. J. Yang, T. J. Yu, X. D. Hu, S. D. Yao, and G. Y. Zhang, "Temperature dependent diffusion and epitaxial behavior of oxidized Au/Ni/p-GaN ohmic contact", *Mater. Sci. and Eng. B* 128, 37-43 (2006).

[170] J. Dumont, E. Monroy, E. Munoz, R. Caudano, and R. Sporken, "Investigation of metal–GaN and metal–AlGaN contacts by XPS depth profiles and by electrical measurements", *J. Cryst. Growth* 230, 558-563 (2001).

[171] M. H. Kim, S. N. Lee, C. Huh, S. Y. Park, J. Y. Han, J. M. Seo, and S. J. Park, "Interfacial reaction and Fermi level movement induced by sequentially deposited metals on GaN: Au/Ni/GaN", *Phys. Rev. B* 61, 10966-10971 (2000).

[172] I. Jyothi, V. R. Reddy, M. S. P. Reddy, C. J. Choi, and J.-S. Bae, "Structural and electrical properties of rapidly annealed Ni/Mo Schottky barriers on n-type GaN", *Phys. Status Solidi A* 207, 753-759 (2010).

[173] V. R. Reddy and P. K. Rao, "Annealing temperature effect on electrical and structural properties of Cu/Au Schottky contacts to n-type GaN", *Microelectron. Eng.* 85, 470-476 (2008).

[174] V. R. Reddy, P. K. Rao, and C. K. Ramesh, "Annealing effects on structural and electrical properties of Ru/Au on n-GaN Schottky contacts", *Mater. Sci. Eng. B* 137, 200-204 (2007).

[175] C. L. Yu, C. H. Chen, S. J. Chang, and P. C. Chang, "GaN metal-semiconductor-metal ultraviolet photodetectors with Ir/Pt contact electrodes", *J. Elect. Chem. Soc.*154, J71-J72 (2007).
[176] V. R. Reddy, M. Ravinandan, and P. K. Rao, "Electrical and structural properties of rapidly annealed Pd/Mo Schottky contacts on n-type GaN", *Semicond. Sci. Technol.* 23, 095026 (2008).
[177] V. R. Reddy, M. Ravinandan, P. K. Rao, and C.-J. Choi, "Effects of thermal annealing on the electrical and structural properties of Pt/Mo Schottky contacts on n-type GaN", *J. Mater. Sci. Mater. Electron.* 20, 1018-1025 (2009).
[178] M. Ravinandan, P. K. Rao, and V. R. Reddy, "Analysis of the current-voltage characteristics of the Pd/Au Schottky structure on n-type GaN in a wide temperature range", *Semicond. Sci. Technol.* 24, 035004 (2009).
[179] Y. H. Jung, M. A. Mastro, J. Hite, C. R. Eddy, and J. Kim, "Post-annealing behavior of Ni/Au Schottky contact on non-polar a-plane GaN", *Thin Solid Films* 518, 5810-5812 (2010).
[180] I. Jyothi and V. R. Reddy, "Electrical, structural and surface morphological properties of thermally stable low-resistance W/Ti/Au multilayer ohmic contacts to n-type GaN", *Appl. Surf. Sci.* 257, 67-71 (2010).
[181] F. Liu, T. Wang, B. Shen, S. Huang, F. Lin, N. Ma, F. J. Xu, P. Wang, and J. Q. Yao, "Thermal annealing behaviour of Al/Ni/Au multilayer on n-GaN Schottky contacts", *Chin. Phys. B* 18, 1618-1621 (2009).
[182] R. Khanna, S. J. Pearton, F. Ren, and I. Kravchenko, "Annealing temperature dependence of TiB_2 Schottky barrier contacts on n-GaN", *J. Electron. Mater.* 35, 658-662 (2006).
[183] L. Wang, F. M. Mohammed, and I. Adesida, "Characterization of Au and Al segregation layer in post-annealed thin Ti/Al/Mo/Au Ohmic contacts to n-Ga N", *J. Appl. Phys.* 98, 106105 (2005).
[184] A. E. Belyaev, N. S. Boltovets, V. N. Ivanov, R. V. Konakova, Ya. Ya. Kudryk, P. M. Lytvyn, V. V. Milenin, and Yu. N. Sveshnikov, "Thermal stability of multilayer contacts on gallium nitride", *Tech. Phys. Lett.* 31, 1078-1081 (2005).
[185] A. Motayed, K. A. Jones, M. A. Derenge, M. C. Wood, D. N. Zakharov, Z. Liliental-Weber, D. J. Smith, A. V. Davydov, W. T. Anderson, A. A. Iliadis, and S. N. Mohammad, "Electrical, microstructural, and thermal stability characteristics of Ta/Ti/Ni/Au contacts to n-GaN", *J. Appl. Phys.* 95, 1516-1524 (2004).
[186] S. Mou, J. V. Li, and S. L. Chuang, "Surface channel current in InAs/GaSb type-II superlattice photodiodes", *J. Appl. Phys.* 102, 066103 (2007).
[187] S. Limpijumnong and C. G. Van de Walle, "Diffusivity of native defects in GaN", *Phys. Rev. B* 69, 035207 (2004).
[188] M. A. Reshchikov and H. Morkoc, "Luminescence properties of defects in GaN", *J. Appl. Phys.* 97, 061301(2005).
[189] A. Hierro, M. Hansen, J. J. Boeckl, L. Zhao, J. S. Speck, U. K. Mishra, S. P. DenBaars, and S. A. Ringel, "Carrier trapping and recombination at point defects and dislocations in MOCVD n-GaN", *Phys. Stat. Sol. B* 228 937-946 (2001).
[190] C. W. Zou, B. Sun, G. D. Wang, W. H. Zhang, P. S. Xu, H. B. Pan, and F. Q. Xu, "Initial interface study of Au deposition on GaN(0 0 0 1)", *Physica B* 370, 287-293 (2005).

[191] T. P. Chen, T. C. Lee, C. C. Ling, C. D. Beling, and S. Fung, "Current transport and its effect on the determination of the Schottky-barrier height in a typical system–gold on silicon", *Solid-State Electron.* 36, 949-954 (1993).
[192] J. P. Sullivan, R. T. Tung, M. R. Pinto, and W. R. Graham, "Electron transport of inhomogenous Schottky barriers: a numerical study", *J. Appl. Phys.* 70, 7403-7424 (1991).
[193] J. S. Foresi and T. D. Moustakas, "Metal contact to gallium nitride", *Appl. Phys. Lett.* 62, 2859-2861 (1993).
[194] J. Rennie, M. Onomura, S.-Y. Nunoue, G.-I. Hatakoshi, H. Sugawara, and M. Ishikawa, "Effect of metal type on the contacts to n-type and p-type GaN", *J. Cryst. Growth* 189-190, 711-715 (1998).
[195] T. U. Kampen and W. Monch, "Barrier heights of GaN Schottky contacts", *Appl. Surf. Sci.* 117, 388-393 (1997).
[196] R. X. Wang, S. J. Xu, A. B. Djurišić, C. D. Beling, C. K. Cheung, C. H. Cheung, S. Fung, D. G. Zhao, H. Yang, and X. M. Tao, "Influence of ITO thin-film quality on reverse leakage current of ITO/n-GaN Schottky contacts", *Appl. Phys. Lett.* 89, 033503 (2006).
[197] T. K. Ko, S. J. Chang, J. K. Sheu, S. C. Shei, Y. Z. Chiou, M. L. Lee, C. F. Shen, S. P. Chang, and K. W. Lin, "AlGaN/GaN Schottky-barrier UV-B bandpass photodetectors with ITO contacts and LT-GaN cap layers", *Semicond. Sci. Technol.* 21, 1064-1068 (2006).
[198] C. H. Chen, S. J. Chang, Y. K. Su, G. C. Chi, J. Y. Chi, C. A. Chang, J. K. Sheu, and J. F. Chen, "GaN Metal–Semiconductor–Metal Ultraviolet Photodetectors With Transparent Indium–Tin–Oxide Schottky Contacts", *IEEE Photonics Technol. Lett.* 13, 848-850 (2001).
[199] J. D. Hwang, C. C. Lin, and W. L. Chen, "Electrical properties of sputtered-indium tin oxide film contacts on n -type GaN", *J. Appl.Phys.* 100, 044908 (2006).
[200] J. K. Sheu, Y. K. Su, G. C. Chi, M. J. Jou, and C. M. Chang, "Effects of thermal annealing on the indium tin oxide Schottky contacts of n-GaN", *Appl. Phys. Lett.* 72, 3317-3319 (1998).
[201] P. K. Rao and V. R. Reddy, "Effect of annealing temperature on electrical and structural properties of transparent indium tin oxide electrode to n-type GaN", *Mater. Chem. Phys.* 114, 821-826 (2009).
[202] K.-H. Shim, M. C. Paek, B. T. Lee, C. Kim, and J. Y. Kang, "Preferential regrowth of indium–tin oxide (ITO) films deposited on GaN(0001) by rf-magnetron sputter", *Appl. Phys. A* 72, 471-474 (2001).
[203] T. Han, Y. Shi, H. Wu, and C. Liu, "Fabrication and characterization of aluminum-doped zinc oxide Schottky diodes on n-GaN", *Curr. Appl. Phys.* 12, 1536-1540 (2012).
[204] J. K. Sheu, K. H. Chang, M. L. Lee, J. F. Huang, K. S. Kang, W. L. Wang, W. C. Lai, and T. H. Hsueh, "Characterization of gallium-doped zinc oxide contact on n-type gallium nitride epitaxial layers", *J. Electrochem. Soc.* 156, H679-H683 (2009).
[205] M. Rottmann, H. Hennig, B. Ziemer, R. Kalahna, and K. H. Heckner, "Variations in microstructure and composition of indium tin oxide films with the deposition technique", *J. Mater. Sci.* 31, 6465-6500 (1996).
[206] F. O. Adurodija, H. Izumi, T. Ishihara, H. Yoshioka, and M. Motoyama, "Effect of Sn doping on the electronic transport mechanism of indium–tin–oxide films grown by

pulsed laser deposition coupled with substrate irradiation", *J. Appl. Phys.* 88, 4175-4180 (2000).

[207] R. X. Wang, C. D. Beling, S. Fung, A. B. Djurišić, C. C. Ling, C. Kwong, and S. Li, "Influence of annealing temperature and environment on the properties of indium tin oxide thin films", *J. Phys. D* 38, 2000-2005 (2005).

[208] R. X. Wang, C. D. Beling, S. Fung, A. B. Djurišić, C. C. Ling, and S. Li, "Influence of gaseous annealing environment on the properties of indium tin oxide thin films", *J. Appl. Phys.* 97, 033504 (2005).

[209] W. G. Haines and R. H. Bube, "Effects of heat treatment on the optical and electrical properties of indium–tin oxide films", *J. Appl. Phys.* 49, 304-307 (1978).

[210] H. Kim, C. M. Gilmore, A. Piqué, J. S. Horwitz, H. Mattoussi, H. Murata, Z. H. Kafafi, and D. B. Chrisey, "Electrical, optical, and structural properties of indium–tin–oxide thin films for organic light-emitting devices", *J. Appl. Phys.* 86, 6451-6461 (1999).

[211] S. Noguchi and H. Sakata, "Electrical-properties of Sn-doped In_2O_3 films prepared by reactive evaporation", *J. Phys. D* 14, 1523 (1981).

[212] C. K. Cheung, R. X. Wang, C. D. Beling, A. B. Djurišić and S. Fung, "Positron beam study of indium tin oxide films on GaN", *J. Phys. Condensed Matter* 19, 086204 (2007).

[213] T. P. Chow and R. Tyagi, "Wide bandgap compound semiconductors for superior high-voltage unipolar power devices", *IEEE Trans Electron Dev.* 41, 1481-1483 (1994).

[214] J. I. Chyi, C. M. Lee, C. C. Chuo, X. A. Cao, G. T. Dang, A. P. Zhang, F. Ren, S. J. Pearton, S. N. G. Chu, and R. G. Wilson, "Temperature dependence of GaN high breakdown voltage diode rectifiers", *Solid-State Electronics* 44, 613-617 (2000).

[215] J. S. Pearton, R. C. Abernathy, and F. Ren, "Gallium Nitride Processing for Electronics, Sensors and Spintronics", Springer-Verlag London Limited (2006).

[216] Z. Z. Bandic, P. M. Bridger, E. C. Piquette, T. C. McGill, R. P. Vaudo, V. M. Phanse, and J. M. Redwing "High voltage (450 V) GaN Schottky rectifiers", *Appl. Phys. Lett.* 74, 1266-1268 (1998).

[217] A. P. Zhang, J. W. Johnson, F. Ren, J. Han, A. J. Polyakov, N. B. Smirnov, A. V. Govorkov, J. M. Redwing, K. P. Lee, and S. J. Pearton, "Lateral $Al_xGa_{1-x}N$ power rectifiers with 9.7 kV reverse breakdown voltage", *Appl. Phys. Lett.* 78, 823-825 (2001).

[218] A. P. Zhang, G. Dang, F. Ren, J. Han, A. Y. Polyakov, N. B. Smirnov, A. V. Govorkov, J. M. Redwing, X. A. Cao, and S. J. Pearton, "Al composition dependence of breakdown voltage in $Al_xGa_{1-x}N$ Schottky rectifier", *Appl. Phys. Lett. 76*, 1767-1769 (2000).

[219] J. W. Johnson, J. R. LaRoche, F. Ren, B. P. Gila, M. E. Overberg, C. R. Abernathy, J. I. Chyi, C. C. Chuo, T. E. Nee, C. M. Lee, K. P. Lee, S. S. Park, J. I. Park, and S. J. Pearton, "Schottky rectifiers fabricated on free-standing GaN substrates", *Solid-State Electron.* 45, 405-410 (2001).

[220] J. W. Johnson, A. P. Zhang, W. B. Luo, F. Ren, S. J. Pearton, S. S. Park, Y. J. Park, and J. I. Chyi, "Breakdown voltage and reverse recovery characteristics of free-standing GaN Schottky rectifiers", *IEEE Trans. Electron. Dev.* 49, 32-36 (2002).

[221] J. W. Johnson, B. Luo, F. Ren, D. Palmer, S. J. Pearton, S. S. Park, and Y. J. Park, "1.6 A GaN Schottky rectifiers on bulk GaN substrates", *Solid-State Electron.* 46, 911-913 (2002).

[222] Q. Chen, J. Yang, R. Gaska, M. Khan, M. Shur, G. Sullivan, A. Sailor, J. Higgins, A. Ping, and I. Adesida, "High-power microwave 0.25μm gate doped-channel GaN/AlGaN heterostructure field effect transistor", *IEEE Electron Device Lett.* 19, 44-46 (1998).

[223] R. Quay, "Gallium Nitride Electronics", Springer Series in materials science Editors: R. Hull R.M. Osgood, Jr. J. Parisi H.Warlimont, Springer-Verlag Berlin Heidelberg (2008).

[224] S. Rajan, H. Xing, S. DenBaars, U. Mishra, and D. Jena, "AlGaN/GaN polarization-doped field-effect transistor for microwave power applications", *Appl. Phys. Lett.* 84, 1591-1593 (2004).

[225] R. Gaska, J. W. Yang, A. Osinsky, Q. Chen, M. A. Khan, A. O. Orlov, G. L. Snider, and M. S. Shur, "Electron transport in AlGaN–GaN heterostructures grown on 6H–SiC substrates", *Appl. Phys. Lett.* 72, 707-709 (1998).

[226] L. Zhang, L. Lester, A. Baca, R. Shul, P. Chang, C. Wilson, U. Mishra, S. DenBaars, and J. Zolper, "Epitaxially-grown GaN junction field effect transistors", *IEEE Trans. Electron Devices* 47, 507-511 (2000).

[227] H. Morkoç: "Nitride Semiconductors and Devices", Springer Series in materials science Editors: R. Hull, R. M. Osgood Jr., H. Sakaki. A. Zunger, Springer-Verlag Berlin Heidelberg (1999).

[228] S. J. Pearton, B. S. Kang, S. Kim, F. Ren, B. P. Gila, C. R. Abernathy, J. Lin, and S. N. G. Chu, "GaN-based diodes and transistors for chemical, gas, biological and pressure sensing", J. Phys.: Condens. Matter 16, R961-R994 (2004).

[229] B. S. Kang, S. Kim, and F. Ren, J. W. Johnson, R. J. Therrien, P. Rajagopal, J. C. Roberts, E. L. Piner, and K. J. Linthicum, S. N. G. Chu, K. Baik, B. P. Gila, C. R. Abernathy, and S. J. Pearton, "Pressure-induced changes in the conductivity of AlGaN/GaN high-electron mobility-transistor membranes", *Appl. Phys. Lett.* 85, 2962-2964 (2004).

[230] A. Y. Polyakov, N. B. Smirnov, A. V. Govorkov, A. V. Markov, E. A. Kozhukhova, I. M. Gazizov, N. G. Kolin, D. I. Merkurisov, V. M. Boiko, and A. V. Korulin, "Alpha particle detection with GaN Schottky diodes", *J. Appl. Phys.* 103708 (2009).

[231] B. P. Luther, S. D. Wolter, and S. E. Mohney, "High temperature Pt Schottky diode gas sensors on n-type GaN", *Sens. Actuator B-Chem.* 56, 164-168 (1999).

[232] J. Schalwig, G. Müller, O. Ambacher, and M. Stutzmann "Group-III-nitride based gas sensing devices", *Phys. Status Solidi A* 185, 39-45 (2001).

[233] R. Neuberger, G. Müller, O. Ambacher, and M. Stutzmann, "High-electron-mobility AlGaN/GaN transistors (HEMTs) for fluid monitoring applications", *Phys. Status Solidi A* 185, 85-89 (2001).

[234] G. Steinhoff, M. Hermann, W. J. Schaff, L. F. Eastmann, M. Stutzmann, and M. Eickhoff, "pH response of GaN surfaces and its application for pH-sensitive field-effect transistors", *Appl. Phys. Lett.* 83, 177-179 (2003).

[235] F. Xie, H. Lu, D. J. Chen, F. F. Ren, R. Zhang, and Y. D. Zheng, "Bias-selective dual-operation-mode ultraviolet Schottky-barrier photodetectors fabricated on high-resistivity homoepitaxial GaN", *IEEE Photonics Technol. Lett.* 24, 2203-2205 (2012).

[236] P. Deb, H. Kim, Y. X. Qin, R. Lahiji, M. Oliver, R. Reifenberger, and T. Sands, "GaN nanorod Schottky and p-n junction diodes", *Nano Lett.* 6, 2893-2898 (2006).

[237] T. Saito, T. Hitora, H. Hitora, H. Kawai, I. Saito, and E. Yamaguchi, "UV/VUV photodetectors using group III-nitride semiconductors", *Phys. Status Solidi* C 6, S658-S661 (2009).

[238] J. C. Carrano, T. Li, P. A. Grudowski, C. J. Eiting, R. D. Dupuis, and J. C. Campbel, "Current transport mechanisms in GaN-based metal–semiconductor–metal photodetectors", *Appl. Phys. Lett.* 72, 542-544 (1998).

[239] D. B. Li, X. J. Sun, H. Song, Z. M. Li, H. Jiang, Y. R. Chen, G. Q. Miao, and B. Shen, "Effect of asymmetric Schottky barrier on GaN-based metal-semiconductor-metal ultraviolet detector", *Appl. Phys. Lett.* 99, 261102 (2011).

[240] J. Sun, F. J. Liu, H.-Q. Huang, J. W. Zhao, Z. F. Hu, X. Q. Zhang, and Y. S. Wang, "Fast response ultraviolet photoconductive detectors based on Ga-doped ZnO films grown by radio-frequency magnetron sputtering", *Appl. Surf. Sci.* 3, 921-924 (2010).

[241] D. Walker, E. Monroy, P. Kung, J. Wu, M. Hamilton, F. J. Sanchez, J. Diaz, and M. Razeghi, "High-speed, low-noise metal–semiconductor–metal ultraviolet photodetectors based on GaN", *Appl. Phys. Lett.* 74, 762-764 (1999).

[242] J. K. Sheu, K. H. Chang, and M. L. Lee, "Ultraviolet band-pass photodetectors formed by Ga-doped ZnO contacts ton-GaN", *Appl. Phys. Lett.* 92, 113512 (2008).

[243] S. S. Shinde and K. Y. Rajpure, "Fast response ultraviolet Ga-doped ZnO based photoconductive detector", *Mater. Res. Bull.* 46, 1734-1737 (2011).

[244] S. J. Chang, M. L. Lee, J. K. Sheu, W. C. Lai, Y. K. Su, C. S. Chang, C. J. Kao, G. C. Chi, and J. M. Tsai, "GaN metal–semiconductor–metal photodetectors with low-temperature-GaN cap layers and ITO metal contacts", *IEEE Electron Device Lett.* 24, 212-214 (2003).

[245] Q. Chen, J. W. Yang, A. Osinsky, S. Gangopadhyay, B. Lim, M. Z. Anwar, M. A. Khan, D. Kuksenkov, and H. Temkin, "Schottky barrier detectors on GaN for visible–blind ultraviolet detection", *Appl. Phys. Lett.* 70, 2277-2279 (1997).

[246] E. Munoz, E. Monroy, J. L. Pau, F. Calle, F. Omnes, and P. Gibart, "III nitrides and UV detection", *J. Phys.: Condens. Matter.* 13, 7115-7137 (2001).

[247] H. C. Liu, "Photoconductive gain mechanism of quantum-well intersubband infrared detectors", *Appl. Phys. Lett.* 60, 1507-1509 (1992).

[248] J. K. Sheu, Y. S. Lu, M. L. Lee, W. C. Lai, C. H. Kuo, and C.-J. Tun, "Enhanced efficiency of GaN-based light-emitting diodes with periodic textured Ga doped ZnO transparent contact layer", *Appl. Phys. Lett.* 90, 263511 (2007).

[249] R. X. Wang, L. C. Yang, S. J. Xu, X. D. Zhang, X. Dong, Y. C. Zhao, K. Fu, B. S. Zhang, H. Yang, "Bias-voltage dependent ultraviolet photodetectors prepared by GaO_x + ZnO mixture phase nanocrystalline thin films", *J. Alloy. Compd.* 566, 201-205 (2013).

[250] W. W. Gartner, "Depletion-layer photoeffects in semiconductors", *Phys. Rev.* 116, 84-87 (1959).

In: Gallium Nitride
Editor: Kiera Olivia Peak

ISBN: 978-1-63321-387-6

Chapter 6

GaN MOSFET on AlGaN/GaN Heterostructure

Jin-Ping Ao
The University of Tokushima, Japan

Abstract

GaN metal-oxide-semiconductor field-effect transistors (MOSFETs) with recessed gate on AlGaN/GaN heterostructure are described in which the drain and source Ohmic contacts were fabricated on the AlGaN/GaN heterostructure and the electron channel was formed on the GaN buffer layer by removing the AlGaN barrier layer. General consideration to develop an enhancement-mode (E-mode) GaN MOSFET and the basic evaluation methods are described. GaN MOSFETs with a silane-based SiO_2 insulator deposited by plasma-enhanced chemical vapor deposition (PECVD) were demonstrated on an AlGaN/GaN heterostructure. E-mode operation up to a gate voltage of 15 V was confirmed at an ignorable gate leakage current. To investigate the effects of the fabrication process, including the dry recess process, the dielectric, the annealing, the dielectric thickness and the surface treatment, on the device performance, devives were fabricated and evaluated. We found that the electron mobility and the interface state density were improved by a thermal treatment at 1000 °C in nitrogen ambient. Trenching effect at the bottom near the sidewall was observed when positive photoresist was utilized and the inductively coupled plasma (ICP) power was low. The recess profile was improved by adopting SiO_2 as the etching mask and increasing the ICP power. It indicates that the nitrogen vacancies are mainly responsible for the negative threshold voltages rather than the silicon contamination. The maximum field-effect mobility of about 150 cm^{-2} V^{-1} s^{-1} and the minimum interface state density of about 1×10^{11} cm^{-2} eV^{-1} were obtained from the optimized process conditions. It demonstrates that the optimization of the ICP recess conditions and the removal of the etching damage to improve the surface condition are still necessary to realize enhancement-mode GaN MOSFETs on AlGaN/GaN heterostructure.

1. Introduction

The metal-oxide-semiconductor (MOS) structure is the most important structure among the semiconductor structures. The metal-oxide-semiconductor field-effect transistor (MOSFET) is by far the most common transistor in both digital and analog circuits.

Enhancement-mode (E-mode) MOSFET, also called as normally-off MOSFET, is the common switching elements in most MOS logic families. In this operation mode, the MOSFET is in off-state at zero gate–source voltage, and can be turned on by pulling the gate voltage in the direction of the drain voltage. For the application in the field of power electronics, the E-mode is the key element for realizing safe operation and reducing power consumption due to the normally-off operation. E-mode devices are also attractive for developing single-power systems and realizing low-power direct-coupled logic by the combined use of E- and D-mode devices.

Till now, almost all the MOSFETs are fabricated upon silicon technologies. Depending on the type of the substrate, the channel can contain electrons on a p-type substrate (called an n-channel), or holes on an n-type substrate (called a p-channel). Due to the limitation on material properties, silicon technologies are still limited in some specific application, such as high-power, high-frequency and high-voltage. Gallium nitride (GaN) has been regarded as one of the candidates to overcome the limitation of silicon physical properties due to its unique properties, such as wide band-gap and high electron conductivity of the two-dimensional electron gas (2DEG) in the AlGaN/GaN heterostructure [1-4]. However, the conventional AlGaN/GaN high electron mobility transistors (HEMTs) operate in normally-on mode. For power electronics application, an E-mode operation is necessary to achieve safe operation and low power consumption.

2. General Consideration to Develop an E-Mode GaN MOSFET

The MOSFET is a four-terminal device with a source (S), a gate (G), a drain (D), and sometimes a substrate electrode if the substrate is conductive (figure 1). The issues needed to be developed when we try to develop a GaN MOSFET include: (a) a high-quality channel with high electron mobility and saturation velocity, which can be GaN and some related materials like InGaN; (b) a kind of dielectric with low leakage current, high interface carrier mobility and low interface state; (c) Ohmic contact structure to fabricate the source and drain electrodes.

To develop an n-channel GaN MOSFET, a p-type or intrinsic GaN can be selected as the channel layer. In this structure, Si ion implantation process is necessary to form the n^+- GaN layer for Ohmic contact formation. After the ion implantation, activation at high temperature (normally over 1000°C) should be conducted to reduce the sheet resistance of the implanted layer. To avoid the ion implantation and high temperature process, a GaN MOSFET structure on AlGaN/GaN heterostructure had been proposed, as shown in figure 2. In this structure, the 2DEG layer in the AlGaN/GaN heterostructure is used as the drain and source electrodes. Normally, the sheet resistance of the 2DEG layer is around 500 Ω/square. However, in this case, dry recess process is necessary to remove the AlGaN layer above the channel region. The effects of the dry etching process to the device performance should be evaluated carefully.

Four types of MOSFET design can be used to evaluate the transistor performance, which are (a) long-channel ring-type, (b) short-channel ring-type, (c) long-channel bar-type, and (d) short-channel bar-type (figure 3). The long-channel device can be used freely because the series resistance can be ignored compared to the high resistance of the channel. The long-channel structure is also necessary to measure the capacitance precisely owing to the large area. The bar-type transistor is useful because it is very easy to define the dimension of the device. However, device isolation process is necessary to fabricate the bar-type structure. The effects, such as the leakage current from the isolation region, cannot be ignored.

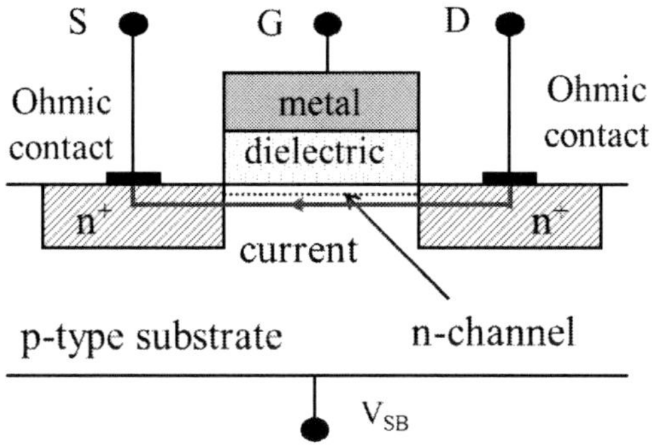

Figure 1. An illustration of an n-channel MOSFET.

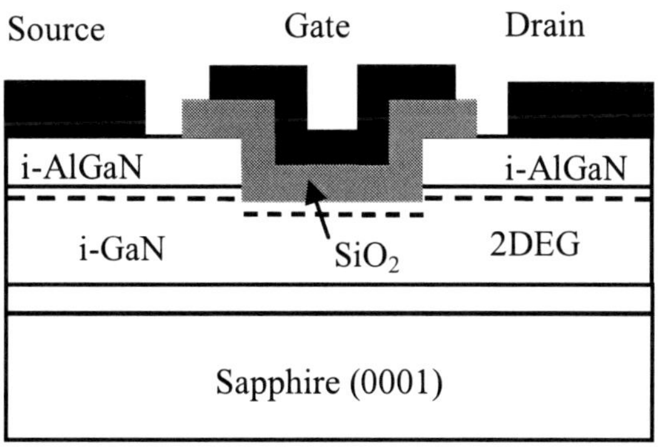

Figure 2. The device structure of a GaN MOSFET on an AlGaN/GaN heterostructure.

The field-effect electron mobility can be determined by a gate capacitance-transconductance method. According to the basis of the gradual channel approximation of MOSFETs, field-effect mobility μ_{FE} could be calculated as

$$\mu_{FE} = \frac{G_m L}{C_{ox} W V_d}, \tag{1}$$

where C_{ox} is the oxide capacitance per unit area, G_m the transconductance, L the gate length, W the gate width, and V_d the drain voltage in the linear region, for instance, at a drain voltage of 0.1 V. For the long-channel ring-type MOSFET, the gate length L is defined as the distance from the outer radius to the inner radius. The effective width W_{eff} can be calculated from the formula below with the channel length, inner radius r_{in} and outer radius r_{out}:

$$W_{eff} = \frac{2\pi L}{\ln r_{out} - \ln r_{in}}. \tag{2}$$

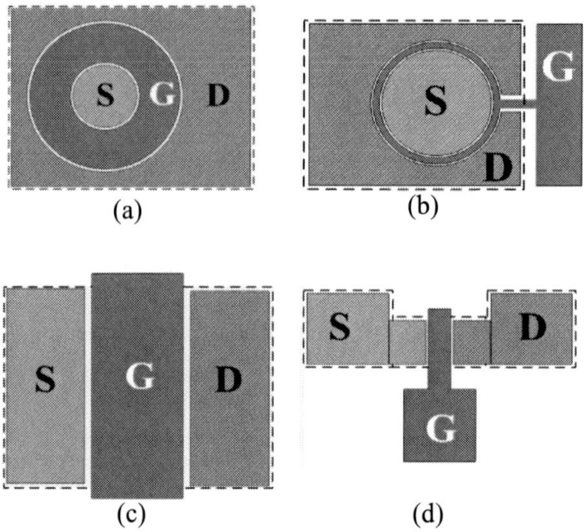

Figure 3. Four types of MOSFET for device evaluation. (a) long-channel ring-type. (b) short-channel ring-type. (c) long-channel bar-type. (d) short-channel bar-type.

To investigate the interface characteristics of the SiO_2-GaN MOS structure, subthreshold swing and interface state can be estimated. The subthreshold swing (S), which is the gate voltage (V_G) needed to reduce the drain current (I_{DS}) by one decade, is defined as:

$$S = \frac{dV_G}{d \log I_{DS}}. \tag{3}$$

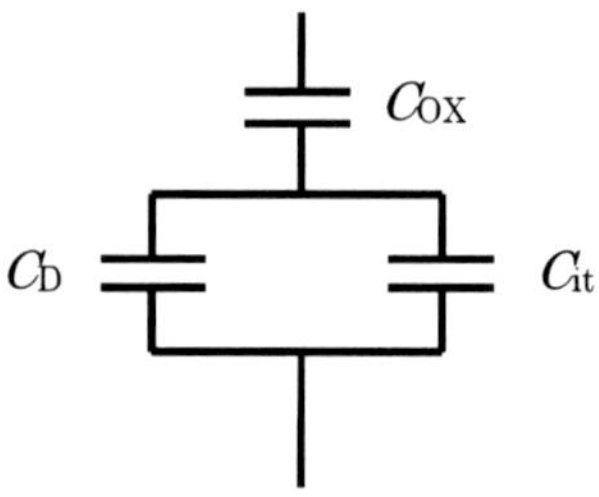

Figure 4. The equivalent circuit of the GaN MOSFET on an AlGaN/GaN heterostructure.

The equivalent circuit of a MOS structure can be expressed as the oxide capacitance C_{OX} connected in series with a parallel connection of the depletion capacitance C_D and the interface-related capacitance C_{it}, as shown in figure 4. In this case, the subthreshold swing can be expressed as

$$S = (\ln 10)\left(\frac{kT}{q}\right)\left(1 + \frac{C_D}{C_{OX}} + \frac{C_{it}}{C_{OX}}\right), \tag{4}$$

where k is the Boltzmann constant, T is the absolute temperature, and q is the electron charge. If the subthreshold swing is known, using Equ. (4) and assuming that the depletion capacitance is zero, we can estimate the density of the interface state using

$$D_{it} = C_{it} / q^2 . \tag{5}$$

3. The Development of GaN MOSFET on AlGaN/GaN Heterostructure

We have reported the performance and characteristics of a GaN MOSFET with tetraethylorthosilicate (TEOS) SiO_2 as the gate insulator [5]. Device operation up to a gate voltage of 10 V was realized at a low gate leakage current. Unfortunately, the maximum field-effect mobility was only approximately 45 $cm^2\ V^{-1}\ s^{-1}$ at an interface state density of $1.02 \times 10^{13}\ cm^{-2}\ eV^{-1}$. After that, we reported the performance and characteristics of a GaN MOSFET with a gate oxide deposited by silane-based plasma-enhanced chemical vapor deposition (PECVD) and an AlGaN/GaN heterostructure as the source and drain structure [6]. We found that the performance on mobility and interface density had been improved by thermal treatment at high temperature.

The GaN MOSFET was developed on an AlGaN/GaN HFET structure grown on a sapphire (0001) substrate, including a buffer layer, a 3 μm i-GaN layer, and a 24 nm u-AlGaN layer with Al composition ratio of 25 %. The sheet resistance of the wafer is about 490 Ω/□. In the MOSFET structure, the two-dimensional gas (2DEG) layer was used as the ohmic contact layer for drain and source electrodes, and the semi-insulating buffer layer was used as the channel layer by removing the AlGaN layer, as shown in figure 2. The device fabrication process was started from device isolation by inductively coupled plasma (ICP) etching for an approximate 100 nm depth with $SiCl_4$ and Cl_2 gases. The AlGaN layer in the channel region was also recessed with the same system using a very slow etching rate to avoid etching damage. The etching depth was 40 nm with an etching rate of 1.2 nm/min. After these dry etching processes, the samples were cleaned using chemicals such as acetone and methanol, followed by a wet treatment with HNO_3:BHF = 1:1 solution. Next, a SiO_2 insulator with thickness of around 100 nm was deposited. Three kinds of sample were then thermally treated at 850 or 1000 °C for 10 min in N_2 or O_2 ambient. After the gate insulator patterning, Ohmic contact was formed using Ti/Al/Ti/Au with annealing temperature of 850 °C for 1 min in N_2 ambient. Finally, Ni/Au was deposited as the gate electrode.

Measurements showed that the sheet resistance and the contact resistance had no obvious degradation after process for the devices only treated in Ohmic annealing process (marked as N_2/850/1), at 1000 °C for 10 min in N_2 ambient (marked as N_2/1000/10), and 850 °C for 10 min in O_2 ambient (marked as O_2/850/10), respectively. Degradation in both sheet resistance and contact resistance was found for the device treated at 1000 °C for 10 min in O_2 ambient (marked as O_2/1000/10). Capacitance-voltage measurement and step profile measurement by atomic force microscope (AFM) showed that the thickness and permittivity of the oxide were 112.3 nm and 4.44, respectively. The gate leakage current is suppressed below 10^{-10} A with gate bias from -30 V to 15 V for the N_2/1000/10 device.

Figure 5 shows the current-voltage (*I-V*) characteristics of a device which was treated in N_2 at 1000 °C for 10 minutes. The gate length and the gate width of the device are 3 μm and 50 μm, respectively, with the drain-source distance of 21 μm. It should be mentioned that the gate length is defined as the edge-to-edge distance in the recess region. The overlap lengths of SiO_2 and gate metal on AlGaN are 6 μm and 3 μm, respectively. Operation of drain current enhancement was confirmed up to a gate voltage of 15 V. Maximum drain current density of about 240 mA/mm and transconductance of about 15 mS/mm were obtained at drain voltage of 20 V and gate voltage of 15 V from the transfer characteristics. The threshold voltage is about -3.5 V.

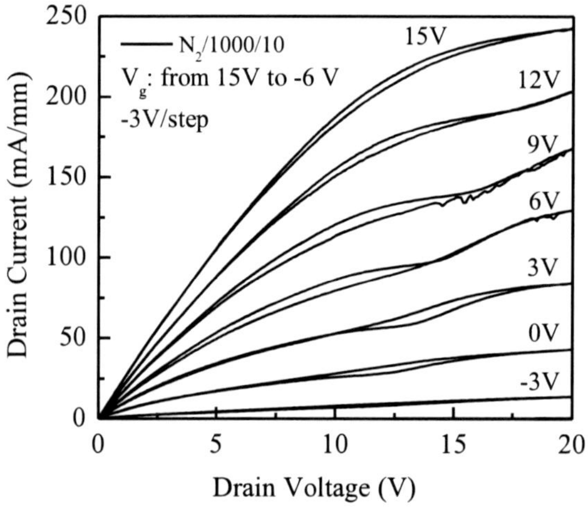

Figure 5. Current-voltage characteristics of the device at drain voltages from 0 to 20 V and gate voltages from 0 to 15 V [6].

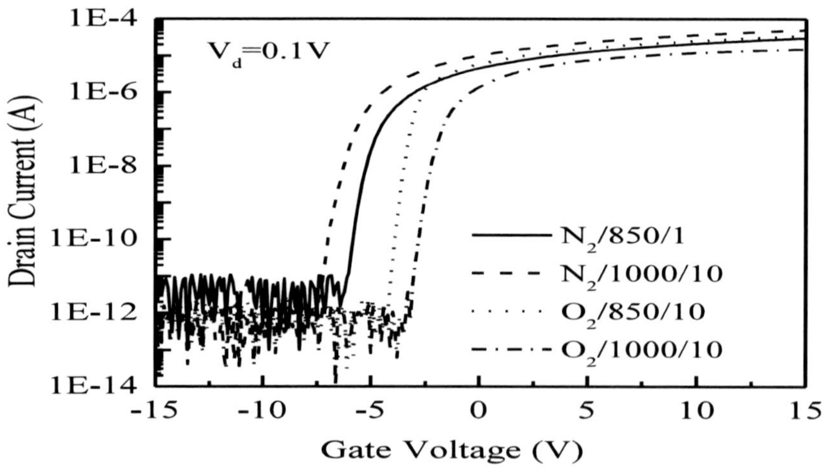

Figure 6. Subthreshold characteristics of the devices with different thermal treatment conditions [6].

Figure 6 shows the subthreshold characteristics of four kinds of ring-type transistors with inner electrode radius of 56 μm and the gate length of 100 μm. For all the devices, operation of drain current enhancement was confirmed up to a gate voltage of 15 V. The threshold voltages of the samples which were treated in O_2 ambient were found to have positive shift. Especially, the threshold voltage of the sample treated in O_2 at 1000 °C is about 1.0 V. The negative threshold voltage is considered to be mainly due to the existing interface state considering that a semi-insulating GaN layer is used as the surface channel. To move the threshold to the positive side, an improved interface with low interface state density is desired. Furthermore, a p-type GaN substrate is necessary to get a positive threshold voltage.

The measured subthreshold swings were 0.249 V/dec (N_2/850/1), 0.212 V/dec (N_2/1000/10), 0.189 V/dec (O_2/850/10), 0.209 V/dec (O_2/1000/10), respectively. Using equation (2) and assuming that the depletion capacitance is zero, we estimated that the densities of the interface state between SiO_2 and GaN were 3.32×10^{11} (N_2/850/1), 2.21×10^{11} (N_2/1000/10), 1.57×10^{11} (O_2/850/10), and 2.16×10^{11} (O_2/1000/10) cm^{-2} eV^{-1}, respectively. As shown in Equ. (4), the subthreshold swing will increase by the existence of the interface-related capacitance C_{it}. If the interface state is below 10^{10} cm^{-2} eV^{-1}, the affect can be ignored in the above calculation case. In our GaN MOSFET case, the depletion capacitance is assumed as zero considering that the carrier concentration is extremely low in the semi-insulating substrate, so the effect on subthreshold swing comes mainly from the interface state. In this experiment, the subthreshold swing and interface state density were improved by thermal treatment. A low subthreshold swing of 0.189 V/dec and a low interface state density of 1.57×10^{11} cm^{-2} eV^{-1} were obtained even though the thickness of the oxide is 112.3 nm. A very good SiO_2-GaN interface was achieved.

Capacitance and conductance measurement showed that the field-effect mobility values were 108 and 79 cm^2/Vs, respectively, for the devices treated at 850 °C and 1000 °C in oxygen. For the devices treated at Ohmic annealing condition and 1000 °C in nitrogen ambient, the measured mobility values were 110 and 137 cm^2/Vs, respectively. Obviously, channel mobility and interface characteristics were improved through thermal treatment in N_2 ambient. The mobility was degraded by treating in O_2 ambient even though the interface was improved. This is considered to be due to the oxygen diffusion into the channel.

4. The Influence of the Dry Recess Process

As we discussed in the above, to realize an E-mode GaN MOSFET, a AlGaN/GaN heterostructure could be utilized as the source and drain electrodes to avoid ion implantation and activation at high temperature. In this structure, dry recess process is necessary to remove the AlGaN layer for forming the channel region on the GaN layer. In the dry recess process, ICP process with low power is generally required for precise controlling by slowing down the etching speed and reducing etching damage [5-7]. However, the recess profile could probably become non-flat if the ICP power and bias power are too low, sometimes even leading to failure in the MOSFETs operation. Good recess profile with flat surface and uniform depth is desired. In this situation, to get good recess profiles and avoid etching damage, trade-off among the ICP conditions is also necessary. In this part, dry recess experiment with different ICP etching condition was firstly done to find an optimum ICP etching condition by comparing the recess profiles [8].

A dry recess experiment for removing the AlGaN layer is done on AlGaN/GaN heterostructure using ICP system to find a proper ICP etching condition. The AlGaN/GaN heterostructure was grown on sapphire substrate with AlGaN thickness of 30 nm and i-GaN thickness of 2 μm. During the experiments, the following set of conditions was used as standard etching conditions: etching gas of $SiCl_4$, pressure of 0.25 Pa, gas flow rate of 3 sccm, and etching time of 33 min. Before dry etching, the samples were cleaned with chemicals such as acetone and methanol in sequence to remove the possible contamination on the AlGaN surface. Photolithography technology was used to fabricate all etching pattern. Etching mask of positive photoresist (Fujifilm, HPR1183L) and SiO_2 were used for comparison. The SiO_2 film with thickness of about 500 nm was deposited by tetraethylorthosilicate (TEOS)-based PECVD and patterned with BHF wet etching. During dry etching process, ICP power from 50 to 200 W and bias from 20 to 60 W were adjusted to compare the etching rate and recess profile. After dry etching, etching mask of positive photoresist and SiO_2 were removed with acetone and BHF respectively. Finally, etching depths and profiles are evaluated by AFM.

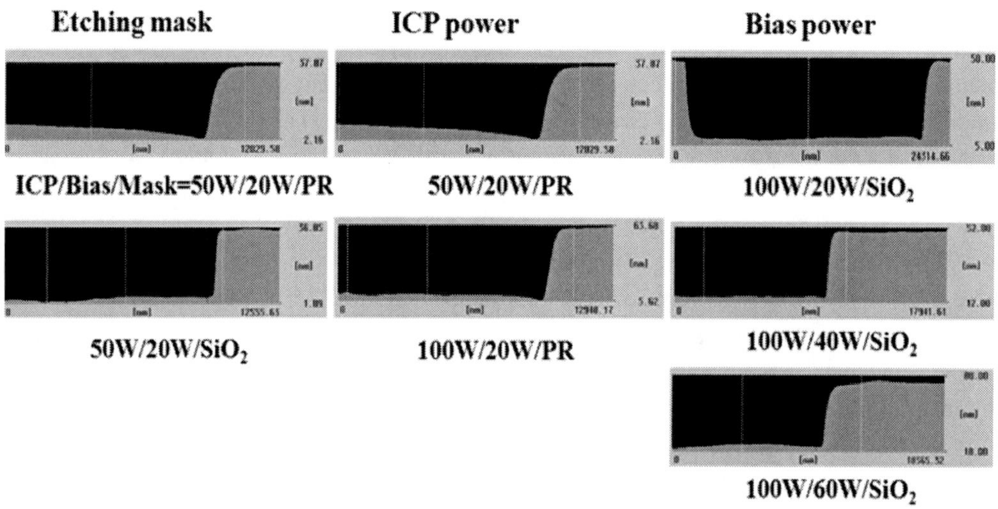

Figure 7. Recess profiles of different ICP etching conditions scaned by AFM [8].

Figure 7 shows the recess profiles of different ICP etching conditions scanned by AFM. The samples with photoresist mask and low ICP power (50 W) got a hill-like profile in the trench, showing trenching effect at the bottom of the sidewall. However, SiO_2 mask had a better recess profile than that of photoresist even with the same ICP power. The recess profile was also improved by increasing the ICP power. When the ICP power was fixed at 100 W, good profiles were also obtained by increasing the bias power. The trenching effect at the bottom of the sidewall is a serious problem to fabricate GaN MOSFET on AlGaN/GaN heterostructure. Similar effect had also been observed in ion-milled samples resulting from enhanced sputtering by the energetic ions which was reflected at low angles from the sidewalls [9, 10]. Also, a deposition effect would probably lower the etching rate in the center of the recess [11]. Carbon from the photoresist layer also possibly accelerates the etching rate near the sidewall to form such a profile. Furthermore, other effects, such as the surface charging effect, may also change the profile [12, 13]. But it often shows a bowling or barreling shape which are quite different form profile with the trenching effect. The possible reasons for the trenching effect in our dry recess experiment are considered to be the combination of the plasma concentration resulting from ion reflection from the sidewalls and the lower etching rate in the center resulting from the deposition effect. It is also found that by increasing ICP power from 50 to 100 W, the trenching effect at the bottom of the side wall become weak. Two reasons may account for it. The first one is that the higher etching rate resulting from higher ICP power reduced the deposition effect. So the profile in the center becomes deeper. The second is that higher plasma density resulting from higher ICP power let the plasma concentration at the side wall not so obvious compared with that in the center, so the etching rate become much uniform. In general, etching condition of SiO_2 mask with proper higher ICP power would get a good recess profile with uniform depth.

Based on this experiment, GaN MOSFETs on AlGaN/GaN heterostructure were fabricated using different recess conditions to evaluate the device performances. To form the channel, the 2DEG layer in the channel region was recessed for 40 nm by controlling etching time to 50, 30, 10, and 6 min for the ICP/bias conditions of 50/20, 100/20, 100/40, and 100 W/60 W, respectively, according to the etching speed obtained from the dry etching experiment. After that, samples were treated in HNO_3 : HF = 1 : 1 solution to remove the possible Si contamination on the etched surface considering that $SiCl_4$ was used as the etching gas. Then SiO_2 insulator with thickness of 97.7 nm was deposited using silane-based PECVD. The ohmic contact was formed using Ti/Al/Ti/Au (50 /200/40/40 nm) by annealing at 850°C for 1 min in N_2 ambient. Finally, Ni/Au (70/30 nm) was deposited as the gate metal.

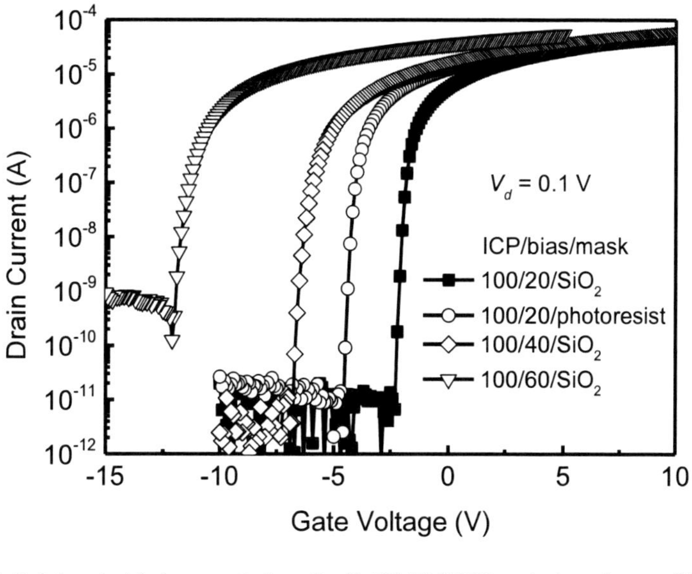

Figure 8. Subthreshold characteristics of a GaN MOSFET at drain voltage of 0.1 V [8].

Figure 8 shows the transfer characteristics of the devices in the sub-threshold region, which were measured at a drain voltage of 0.1V. Using Equ. (4) and assuming the depletion capacitance as zero under a positive gate bias, we estimated that the averaged density of the interface states between SiO_2 and GaN were 1.39×10^{11}, 1.61×10^{11}, 3.17×10^{11}, and 4.38×10^{11} cm^{-2} eV^{-1} for samples with ICP condition of 100 W/20 W/SiO_2, 100 W/20 W/photoresist, 100 W/40 W/SiO_2, and 100 W/60 W/SiO_2, respectively. It showed that higher bias power would bring more interface states and lower threshold voltage. It was considered to be due to the much serious surface damage and Si contamination occurring at higher bias power. A leakage current of about 1×10^{-9} A was obtained in the sample with

ICP condition of 100 W/60 W/SiO_2. It was much higher than that of the other samples implying the higher Si contamination and surface damage. On the other hand, samples with SiO_2 mask have lower interface states and higher threshold voltage than that of photoresist. It is mainly because that sample with SiO_2 mask had a better profile, little contamination and interface states than that of photoresist mask.

It should be mentioned that, theoretically, the threshold voltage of an ideal MOSFET formed on i-GaN surface should be around 0 V. However, in our experiment, a threshold voltage of -1.1 V was observed for the optimized sample and other samples showed negative-shift threshold voltages. The negative threshold voltages may mainly result from two reasons. The first is the unavoidable silicon contamination from the etching gas of $SiCl_4$ and nitrogen-vacancy resulting from surface damage in the dry recess process that work as n-type dopants in GaN. By this reason, the GaN substrate becomes weak n-type [14]. The second is the possible positive charges existing in the non-ideal SiO_2 layer leading to a negative shift on the threshold voltage [15]. In figure 8, the deeper threshold voltage at higher bias power possibly demonstrates the first reason because higher bias power will bring much surface damage and silicon contamination. Since the threshold voltage of the MOSFETs can be tuned by substrate doping, the positive threshold voltage will be realized easily by p-type doping in the substrate.

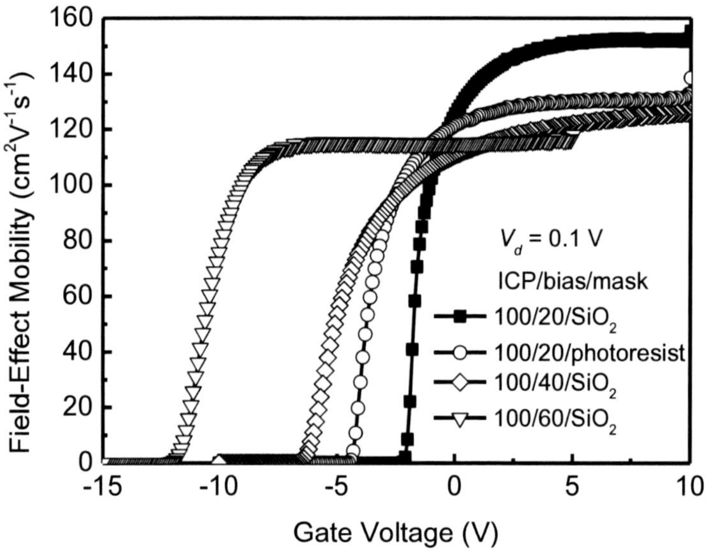

Figure 9. ICP dependency of field-effect mobility [8].

The measured field-effect mobility of the devices is shown in figure 9. Electron field-effect mobility reached 152.8 $cm^2\ V^{-1}\ s^{-1}$ for a device with SiO_2 etching mask under bias power of 20 W and ICP power of 100 W. The mobility degraded at increased bias power. Also, MOSFET with the SiO_2 etching mask showed a higher mobility than that of photoresist mask. Like the situation of interface state density, high bias power and the use of photoresist etching mask will bring much surface damage and contamination which will act as the scattering center in the channel leading to a decrease on the channel mobility.

5. The Process Dependence of the Threshold Voltage

Not only the dry recess process, but also the other fabrication process, such as the type of the dielectric, the thickness of the dielectric, can affect the channel mobility, the threshold voltage and the interface state density. Especially, to investigate the reasons of the negative threshold voltages, GaN MOSFETs with different oxide thickness, etching gas and bias power of ICP were fabricated [16].

The fabrication process is based on the standard photolithography and lift-off technologies. After the device isolation by ICP etching with $SiCl_4$ gas, the 2DEG layer in the channel region was recessed for 40 nm. In this process, different etching gas ($SiCl_4$, BCl_3, and Cl_2) and bias power (20-60 W) were used to investigate the dry recess process dependency on the performance of the devices. After that, surface treatment with HNO_3/HF (10 min) was given to remove the possible Si contamination on the etched surface. Then SiO_2 insulator with thickness of 30-100 nm was deposited using plasma-enhance chemical vapor deposition (PECVD) with silane and TEOS source, respectively. Then, the samples were annealed at 1000 °C for 10 min in N_2 ambient which was reported as an optimized annealing condition to minimize the interface state density [6, 17]. The ohmic contacts were formed using Ti/Al/Ti/Au (50 /200 /40 /40 nm) and annealed at 850°C for 1 minute in N_2 ambient. The evaluation of transmission line model (TLM) showed that the contact resistance was around 0.3 Ω·mm. Finally, Ni/Au (70 /30 nm) was deposited as the gate metal.

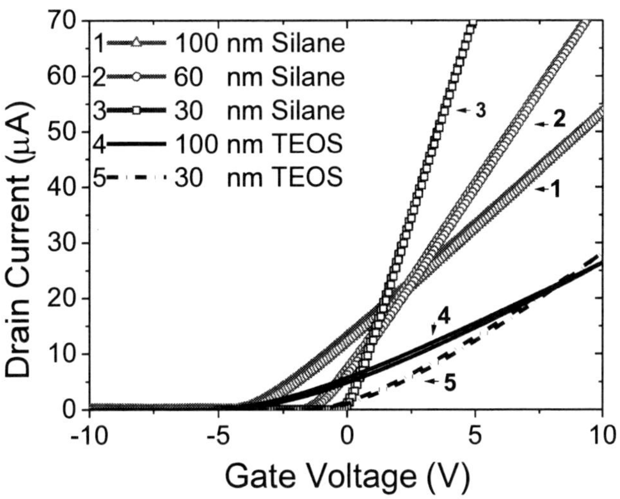

Figure 10. The transfer characteristics of the devices with different oxide layer [16].

Three types of GaN MOSFETs were designed for device evaluation (figure 2), which were long-channel ring-type (W_{eff}=819 μm, L=94μm), long-channel bar-type (W= 200 μm, L=100μm), and short-channel bar-type (W= 56 μm, L=20μm) MOSFETs. Device operation up to gate voltage of 10 V was confirmed for all the types of device. Figure 10 shows the transfer characteristics at a drain voltage of 0.1 V. The dry recess conditions of these devices were with ICP/Bias power of 100/20 W, etching gas of $SiCl_4$ and surface treatment of HNO_3/HF (10 min). Threshold voltages of MOSFETs with silane-based oxide of 100 nm, 60, nm and 30 nm are -3.0 V, -1.1 V, and 0.0 V, respectively. For devices with TEOS-based oxide of 30 nm and 100 nm, the threshold voltages are about -3.0 and 0.0 V, respectively. The capacitance-voltage (*C-V*) characteristics showed similar trends of threshold shifts with the transfer characteristics in figure 10.

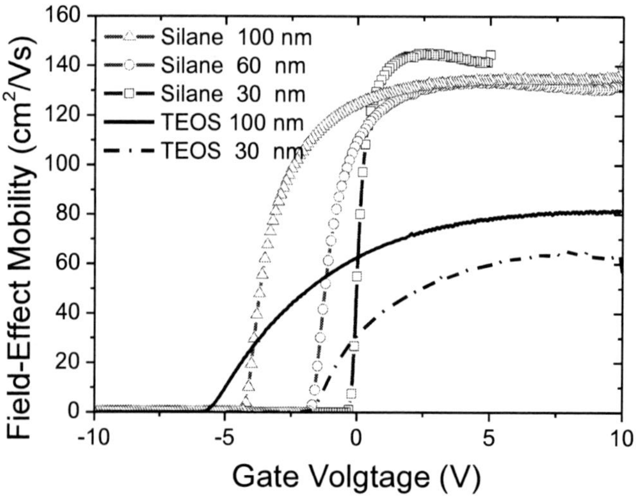

Figure 11. The field-effect mobility characteristics [16].

The measured field-effect mobility values of the devices are shown in figure 11. Electron field-effect mobility values of around 140 $cm^2V^{-1}s^{-1}$ are obtained from the silane-based devices while 65-80 $cm^2V^{-1}s^{-1}$ are obtained from the TEOS-based devices. The lower mobility values of TEOS-based devices may be mainly caused by the poor SiO_2/GaN interface with high interface state density.

For better comparison, subthreshold swing between a drain current from 10^{-10} to 10^{-9} A was used to evaluate the interface state density in our experiment. Figure 12 shows the subthreshold characteristics of the devices with different oxide thickness, which were measured at a drain voltage of 0.1 V. In figure 12, the subthreshold swings of the silane-based devices are around 0.08-0.09 V/decade, while for the TEOS-based devices the subthreshold swings are around 0.12-0.13 V/decade.

We estimated that the average interface state densities between the SiO_2 and the GaN layer were 1.44×10^{11}, 1.63×10^{11}, and 1.84×10^{11} /cm^2eV for the devices with silane-based oxide of 100, 60, and 30 nm thick, respectively. They were 5.14×10^{11} and 4.11×10^{11} /cm^2eV for the devices with TEOS-based oxide of 100 and 30 nm thick, respectively. The interface state densities of the TEOS-based devices are much higher, which may be the possible reason for the lower field-effect mobility values. Similar results have also been reported in [5, 6]. It demonstrates that the oxide deposition process is very important even with the same oxide such as SiO_2.

Theoretically, the threshold voltages of an ideal MOSFET formed on SI-GaN would be positive for the low electron concentration in the substrate and gate metal with higher work function on the oxide. However, in our experiment as shown in figure 10, negative threshold voltages are commonly observed. Also, it is interesting that positive threshold voltage shift was observed when thinner oxide layers were used with both silane- and TEOS-based oxide.

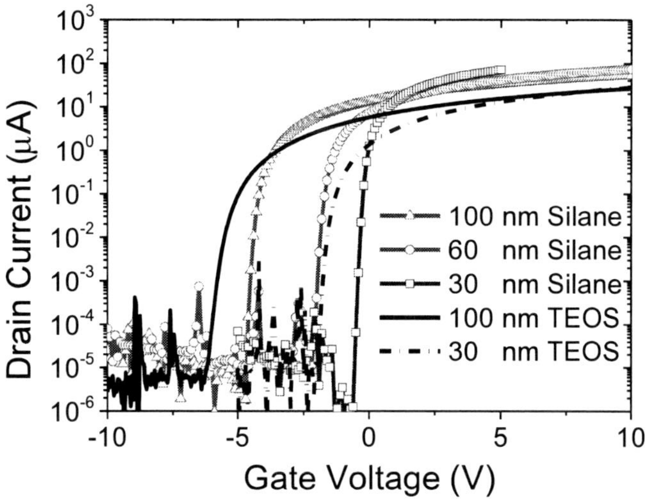

Figure 12. The subthreshold characteristics of the devices with different oxide thickness [16].

Generally, the negative threshold voltage is caused by the positive charges near the interface, including the charges in the oxide layer, the fixed charges at the interface, and also the ionized space charges from the donors. Thus, the relationship between the threshold voltage shift and the thickness difference can be expressed as Equ. 6.

$$\Delta V_T = -\frac{Q_p}{\varepsilon_r \varepsilon_0} \Delta T_{ox}, \qquad (6)$$

where ΔV_T is the threshold voltage shift, ΔT_{ox} the thickness difference, Q_p the positive charge, ε_r the relative permittivity of GaN and ε_0 the permittivity in vacuum. It is estimated that the positive charges Q_p with a density of around 1×10^{12} cm^{-2} exist near the interface in both the silane- and TEOS-based devices. Considering the possible negative bulk charges in the SI-GaN, this value would be a little larger [18]. Possibly, besides the charges in the oxide layer, they may also come from the ionized shallow donors (Si contamination or N vacancy) brought by dry recess process. Si could come from the etching gas of $SiCl_4$ while N vacancies may come from the surface damage [19, 20]. Also, it can be estimated that the threshold voltage would be 1.36 V when the oxide thickness is close to 0. It is related to the barrier height of the Ni/SI-GaN Schottky contact with the typical values from 1.0 to 1.5 eV [21, 22].

To investigate the reasons of the positive charges, dry recess etching with etching gases of $SiCl_4$, BCl_3 and two-step etching of $SiCl_4$ followed by Cl_2 have also been done. Figure 13 shows the transfer characteristics of the devices recessed using different etching gases but nearly the same level of bias power (20-25 W). Small threshold voltage shifts can be much clearly observed in the *I-V* characteristics. It shows that the threshold voltages of the devices with gate recess by Cl_2 and BCl_3 are a little lower than that of $SiCl_4$. It would be contradictory if Si contamination dominates the threshold voltages shift. Thus, it is considered that the Si contamination is not so serious probably owing to the post-treatment in HNO_3:HF=1:1 solution.

The influence of the etching damage was investigated by adjusting the bias power of the ICP system considering that higher bias power may bring much ion bombardment damage.

Figure 14 shows the transfer characteristics of the devices recessed by bias power from 20 to 60 W. Deeper threshold voltage is observed at higher bias power. The threshold voltage shows strong dependence on the bias power of ICP system. It demonstrates that the ion bombardment damage may be the dominant reason for the negative threshold voltage. Possibly, the ion bombardment damage may relate to the N vacancy and some deep donor levels in the SI-GaN layer.

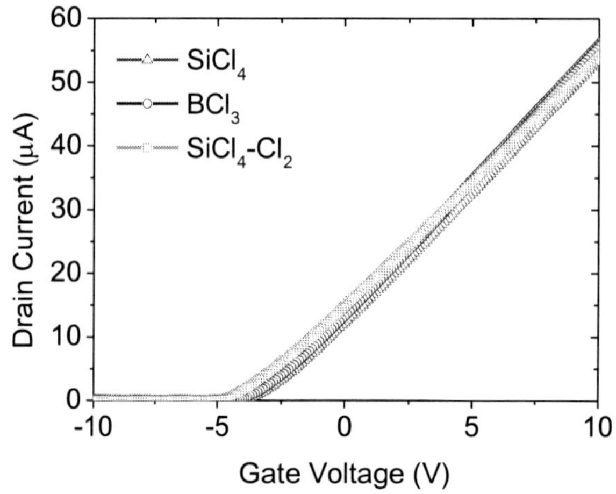

Figure 13. The transfer characteristics of the devices with different etching gas [16].

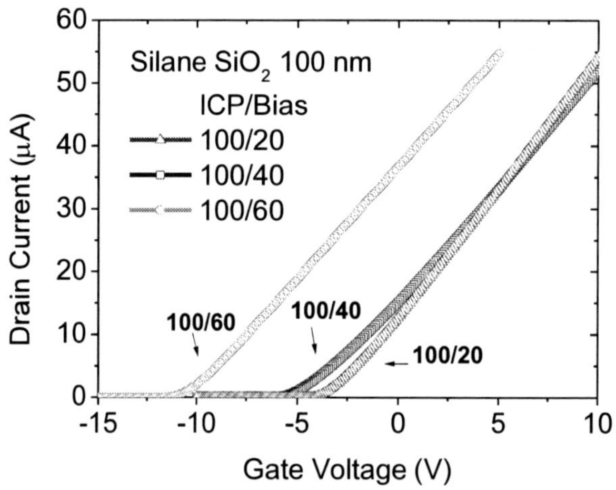

Figure 14. The transfer characteristics of the devices with different etching bias power [16].

Furthermore, experiment of X-ray photoelectron spectroscopy (XPS) was done for the dry-etched samples. It demonstrates that the V_N caused by the dry etching damage may be the much important reason for the negative threshold voltage. The Si contaminates exist even in the samples with $SiCl_4$-Cl_2 two steps etching process when surface treatment is not done. However, it could be easily removed by HNO_3/HF treatment. Also the Si contamination is more likely to be in the form of SiO_2 and the native oxide of Ga_2O_3 seems inevitable as long as the surface is exposed in the air. As a result, it can be referred that the Si contamination in the dry recess process are not an important factor for the negative threshold voltage for that all of the GaN MOSFETs samples were given the HNO_3/HF treatment after the dry recess process [16]. We can also refer that the higher leakage current at the off-state of the bar-type MOSFET may be caused by the etching-damage-induced n-type surface layer. So, to obtain E-mode operation and higher on-off ration in a recessed GaN MOSFET on AlGaN/GaN heterostructure, methods to remove the dry etching damage would be very important.

Conclusion

GaN MOSFETs with recessed gate on AlGaN/GaN heterostructure can be realized by fabricating the drain and source Ohmic contacts on the AlGaN/GaN heterostructure and forming the electron channel on the GaN buffer layer by removing the AlGaN barrier layer. The field-effect mobility values of devices with silane-based oxide are much higher than that of TEOS-based devices due to the lower interface state density. The electron mobility and the interface state density can be improved by a thermal treatment at 1000 °C in nitrogen ambient. The recess profile can be improved to eliminate the trenching effect at the bottom near the sidewall by adopting SiO_2 as the etching mask and increasing the ICP power. Negative threshold voltages were commonly observed in all the devices and it was found that positive charges of around 1×10^{12} /cm^2 exist near the interface at the threshold condition in both the silane- and TEOS-based devices. Nitrogen vacancies (V_N) caused by the dry etching damage may be the main reason for the positive charges and negative threshold voltages. Maximum field-effect mobility of about 150 cm^{-2} V^{-1} s^{-1} and the minimum interface state density of about 1×10^{11} cm^{-2} eV^{-1} were obtained from the optimized process conditions. It demonstrates that to realize E-mode and high on-off ratio GaN MOSFETs with recess gate, optimization of the ICP recess conditions and removal of the etching damage to improve the surface condition are still necessary for the GaN MOSFETs on AlGaN/GaN heterostructure.

References

[1] Akatas, O.; Fan, Z. F.; Botchkarev, A.; Mohammad, S. N.; Roth, M.; Jenkins, T.; Kehias, L.; Morkoc, H. *IEEE Electron Device Lett.* 1997, 18, 293-295.

[2] Ohno, Y., Kuzuhara, M. *IEEE Trans. Electron Devices,* 2001, ED-48, 517-523.

[3] Keller, S.; Wu, Y. -F.; Parish, G.; Zhiang, N.; Xu, J. J.; Keller, B. P.; DenBaars, S. P.; Mishra, U. K. *IEEE Trans. Electron Devices.* 2001, 48, 552-559.

[4] Ao, J. -P.; Kikuta, D.; Kubota, N.; Naoi, Y.; Ohno, Y. *IEEE Electron. Device Lett.* 2003, 24, 500-502.

[5] Ao, J. -P.; Nakatani, K.; Ohmuro, K.; Sugimoto, M.; Hu, C.-Y.; Sogawa, Y.; Ohno, Y. *Japanese J. of Appl. Phys.* 2010, 49, 04DF09.

[6] Ao, J.-P.; Nakatani, K.; Sogawa, Y.; Akamatsu, S.; Kim, Y. H., Miyashita, T., Motoyama, S. -I.; Ohno, Y. *Phys. Status Solidi C,* 2011, 2, 457–460.

[7] Matsuura, K.; Kikuta, D.; Ao, J.-P.; Ogiya, H.; Hiramoto, M.; Kawai, H.; Ohno Y. *Jpn. J. Appl. Phys.* 2007, 46, 2320-2324.

[8] Wang, Q.; Tamai, K.; Miyashita, T.; Motoyama, S. -I.; Wang, D.; Ao, J. -P.; Ohno, Y. *Jpn. J. Appl. Phys.* 2013, 52, 01AG02.

[9] Flemish, J. R.; Xie, K.; Zhao, J. H. *Appl. Phys. Lett.* 1994, 64, 2315-2317.

[10] Williams, R. Modern GaAs Processing Methods, Artech House: London, 1990; 192-193.
[11] Rossnagel, S. M.; Cuomo, J. J.; Westwood, W. D. Handbook of Plasma Processing Technology, Noyes Publications: New York, 1990; 321-323.
[12] Murakawa, S.; Fang, S.; McVittie, J. P. IEDM Tech. Dig. 1992, 57-60.
[13] Murakawa, S.; McVittie, J. P. *Jpn. J. Appl. Phys.* 1994, 33, 2184-2188.
[14] Lee, J.-M.; Chang, K.-M.; Kim, S.-W.; Huh, C.; Lee, I.-H.; Park, S.-J. *J. Appl. Phys.* 2000, 87, 7667-7669.
[15] Gaind, A. K.; Kasprzak, L. A. *Solid-State Electron.* 1979, 22, 303-309.
[16] Wang, Q.; Jiang, J.; Miyashita, T.; Motoyama, S. -I.; Li, L.; Wang, D.; Ohno, Y.; Ao, J. -P. *Solid-State Electron.* 2014 (accepted).
[17] Huang, W.; Khan, T.; Chow, T. P. *J. of electronic materials*, 2006, 35, 726-732.
[18] Ohno, Y.; Kio, Y.; Ikawa, Y.; Ao, J. -P. *Jpn. J. Appl. Phys,* 2013, 52, 08JN28.
[19] Hashizume, T.; Nakasaki, R. *Appl. Phys. Lett,* 2002, 80, 4564-4566.
[20] Lee, J.-M.; Lee, K.-S.; Park, S.-J. *J. Vac. Sci. Technol. B.* 2004, 22, 479-482.
[21] Zhan, A. P.; Dang, G. T.; Fan, R.; Hyun, C.; Lee, K. P.; Pearton, S. J.; Chyi, J. I.; Nee, T. Y.; Chuo, C. -C. *IEEE Trans. Electron Devices,* 2001, 48, 407-411.
[22] Arehart, A. R.; Moran, B.; Speck, J. S; Mishra, U. K.; DenBaars, S. P.; Ringel, S. A. *J. Appl. Phys,* 2006, 100, 023709.

In: Gallium Nitride
Editor: Kiera Olivia Peak
ISBN: 978-1-63321-387-6

Chapter 7

STRUCTURAL AND DIELECTRIC PROPERTIES OF THE GALLIUM NITRIDE AND SILICA NANOPARTICLES INVESTIGATED BY MOLECULAR DYNAMICS

Alexander Y. Galashev*
Institute of Industrial Ecology, Ural Division,
Russian Academy of Sciences, Russia

Abstract

The dielectric, structural, and kinetic properties of $(GaN)_{54}$ $(SiO_2)_{50}$ nanoparticles of the following two types are studied using the method of molecular dynamics: particles of GaN covered by a SiO_2 layer and particles of SiO_2 with a GaN layer deposited onto it. The absorption and reflection spectra of these nanoparticles are considerably different in the respective frequency ranges $0 \le \omega \le 1500$ cm^{-1} and $\omega > 700$ cm^{-1}. The Raman spectrum of $(SiO_2)_{86}$ had three broad bands only, whereas the Raman spectrum of $(GaN)_{129}$ contained a large number of overlapping bands. The shape of Raman spectra for four-component particles depends strongly on the way the GaN -, and SiO_2 -components are located in the nanoparticle. The nanoparticle with a GaN core had a continuous fairly smooth spectrum over the frequency range $0 \le \omega \le 600$ cm^{-1}, whereas the spectrum of the nanoparticle with a SiO_2 core contained well-defined bands caused by vibrations of groups of atoms of different kinds and atoms of the same kind. The single-centered functions of the radial distribution of the nanoparticles differ in their peak positions and intensities, including the "tail" part for $r > 0.9$ nm. The structure of nanoparticles with the SiO_2 core is completely tetrahedral, while the nearest six-atom environment is dominant in the structure of a particle with the GaN core. The mobility of gallium atoms in the nanoparticles is lower by two orders than that of other elements (N, Si, and O).

* E-mail address: galashev@ecko.uran.ru.

Keywords: Gallium nitride; Infrared and Raman spectra; Molecular dynamics; Nanoparticles; Silicon dioxide; Structure

1. Introduction

Nano-scale materials have stimulated great interest due to their importance in basic scientific research and potential technological applications. These materials exhibit unique chemical and physical properties, differing substantially from those of the corresponding bulk solids because of their small size and extremely large surface-to-volume ratio. Blue is one of the primary colors used in white light and hence materials that are sources of blue light are technologically important. There is an interest in using blue and ultraviolet light emitting diodes (LED) as light sources. There are a few reports of blue emission from GaN-based nanomaterials [1, 2]. Decomposition of different mixtures incorporated into a polymer results in the formation of GaN nanoparticles which exhibit blue emission near 426 and 475 nm, respectively. Blue photoluminescence (~450 nm) has also been demonstrated from $InN+SiO_2$ nanomaterials [3]. Unfortunately known devices have low efficiencies and poor stability. The effectiveness of radiation increases as the size of the GaN nanoparticles decreases. The semiconductor nanoparticles can be coated with isolator for a variety of purposes, such as to improve their dispersibility. In addition, such nanoparticles are very stable. However the resulting spectroscopic behaviour is highly dependent on arrangement of semiconductor or isolator components. It is important to understand the correlation between the structure and the optical/electronic properties of these nanomaterials. Better understanding of this correlation will enable to rationally design superior materials for next generation technologies.

It is difficult to grow large GaN crystals using high-temperature methods [4] because of the thermal instability of the materials, which leads to their destruction before the instant of melting ($T_m \sim 2773$K). As a result, GaN is mainly used in the form of thin films. Epitaxial GaN films play an important role in the development of photoelectronic devices such as diodes (blue/green ones) and laser diodes. The advantage of GaN is that it can produce effective luminescence even in the presence of a high density of crystal defects. This contradicts the general consensus that lattice defects behave as nonradiative recombination centers in semiconductors of the groups III–V. Recombination centers decrease the efficiency and the lifetimes of electronic devices. A lack of perfect substrates is the main reason hindering the progress of fabricating highly efficient GaN-based devices. Various single crystals (Al_2O_3, SiC, GaAs, and Si) have been intensively used as substrates for hetero-epitaxial GaN growth. Using the X-ray diffractometry method, the authors of [5] showed that it is possible to control the growth of GaN layers for small angles of disorientation of the crystallites forming these layers. Despite the large difference between the thermal expansion coefficients and lattice parameters of GaN and Si, an epitaxial GaN layer on a Si substrate is especially attractive for devices whose fabrication is being developed as a result of the integration of GaN into Si microelectronics.

The optical and electronic properties of GaN layers growing on both Si substrates and GaN crystals were studied in detail using Raman [6, 7] and infrared (IR) [8] spectroscopy.

Raman scattering is a powerful and widely used method for studying admixtures and defects in materials. This technique is used to rapidly identify materials. Because of high sensitivity of vibrational frequencies and scattering intensities to small differences in crystal structure, this method is successfully used to reveal polymorphism of materials. Raman spectroscopy cannot however be used to determine the type of vibrations (longitudinal or transverse) to which a band is related. The structural and spectral properties of high-purity GaN films were studied using X-ray diffraction and Brillouin, Raman, and IR spectroscopy methods [9]. It was established that the moduli of elasticity C_{13} and C_{44} are more sensitive to residual stress than optical-phonon frequencies. The high-frequency (ε_∞ = 5.14) and static (ε_0 = 9.04) dielectric constants were determined. Short-wavelength devices are miniaturized using nanomaterials, including individual nanoparticles. Information on the properties of the nanomaterials of applied semiconductors is required in this case. Quartz (SiO_2) is a material suitable for the deposition of a GaN layer. The mechanism for the fixing of GaN onto a SiO_2 nanoparticle can be studied in computer experiments. The behavior of the nanoparticle surface tension under three-dimensional tensile strain was studied in [10]. The elastic properties and the stability of deformed Si nanoparticles were considered in [11, 12]. Interband radiative recombination can occur in Si nanocrystals with sizes of approximately 1 nm and is suppressed strongly in the Si bulk [13]. The optical properties of SiO_2 nanoparticles covered by GaN layers and of particles with inverse structures are, in fact, unknown.

The aim of this paper is to study the structural and kinetic properties of $(GaN)_{54}(SiO_2)_{50}$ nanoparticles of two types (a GaN particle covered by a SiO_2 layer and a SiO_2 particle with a GaN layer deposited onto it) and determine the frequency dependence of their complex permittivity and IR absorption and reflection spectra. Besides, our main concern is obtaining the Raman spectra of these nanoparticles as well as $(SiO_2)_{86}$ and $(GaN)_{129}$ nanoformations.

2. Computer Model

The capability of the Tersoff-type potential is limited when describing distinctions between chemical interactions; however, it describes the chemical distinctions between the valence s and p electronic properties. This potential is an effective tool for modeling new materials. The Tersoff potential is based on the concept of bond order. The interatomic potential energy for two neighboring atoms i and j is written as [14]

$$V_{ij} = f_C(r_{ij})\left[Aa_{ij}\exp(-\lambda^{(1)}r_{ij}) - \chi Bb_{ij}\exp(-\lambda^{(2)}r_{ij})\right], \tag{1}$$

$$f_C(r_{ij}) = \begin{cases} 1, & r < R-D \\ \frac{1}{2} - \frac{1}{2}\sin\left[\frac{\pi}{2}(r-R)/D\right], & R-D < r < R+D, \\ 0 & r > R+D \end{cases} \tag{2}$$

where b_{ij} is the manyparticle bond-order parameter describing the creation of the bond formation energy (the attractive part of V_{ij}) for the local atomic position because of the presence of other neighboring atoms (k atoms). The potential energy is a manyparticle function of the positions of atoms i, j, and k and is determined by the following parameters:

$$b_{ij} = (1 + \xi_{ij}^{n_i})^{-1/(2n)}, \tag{3}$$

$$\xi_{ij} = \sum_{k \neq i,j} f_C(r_{ij}) \beta_i g(\theta_{ijk}) \exp\left[\lambda_3^3 (r_{ij} - r_{ik})^3\right], \tag{4}$$

$$g(\theta) = 1 + \frac{c^2}{d^2} - \frac{c^2}{\left[d^2 + (h - \cos\theta)^2\right]}, \tag{5}$$

$$a_{ij} = (1 + q^n \eta_{ij}^n)^{-1/(2n)}, \tag{6}$$

$$\eta_{ij} = \sum_{k \neq i,j} f_c(r_{ij}) \exp\left[\lambda_3^3 (r_{ij} - r_{ik})^3\right], \tag{7}$$

$$\lambda_{ij}^{(k)} = (\lambda_i^{(k)} + \lambda_j^{(k)})/2;$$

$$A_{ij} = (A_i A_j)^{1/2}; \quad B_{ij} = (B_i B_j)^{1/2}; \tag{8}$$

$$R^{(1)} = R - D; \quad R^{(2)} = R + D;$$

$$R_{ij}^{(1)} = (R_i^{(1)} R_j^{(1)})^{1/2}; \; R_{ij}^{(2)} = (R_i^{(2)} R_j^{(2)})^{1/2}.$$

where ξ is the effective coordination number, $g(\theta)$ is the function of the angle between r_{ij} and r_{ik} which stabilizes the tetrahedral structure, λ_3 and q are set to be zero.

The potential defined by expressions (1)–(7) differs from the corresponding potential for a one-component system [15] by the introduction of an additional parameter χ. This parameter increases or decreases the heteropolar bonds with respect to the value obtained by simple interpolation. Thus, the chemical interaction is included in this parameter or is incorporated when choosing the interpolation formula. Here, $\chi_{ii} = 1$ and $\chi_{ij} = \chi_{ji}$ so only one independent parameter is required for a pair of atoms of two types. The parameter β in (4) is included to achieve additional flexibility for a pair of atoms of different types.

The Tersoff-potential parameters for Si, O, Ga, and N are given in Table 1 [16, 17]. Optimized parameters are used in the potentials for Si, O, and N. The procedure chosen to optimize the potential parameters included simultaneous fitting of the energy and forces to the values obtained using calculations in accordance with density functional theory rather than traditional fitting of the potential to the physical properties [16]. The parameters $\lambda_{ij}^{(k)}$, A_{ij}, B_{ij}, and $R_{ij}^{(k)}$ of the potential describing the interaction between atoms of the types i and j were calculated using arithmetic and geometric averages (of Eq. 8). The fitting procedure for the other parameters using the original Tersoff potential was given in [18]. The Tersoff potential can be used for bonding orbitals; the parameters fitted to the sp^3 hybridization can be used to describe interactions in materials with the sp^2 hybridization [19].

Table 1. Parameters of the optimized Tersoff potential

Parameter	Silicon*	Oxygen*	Gallium**	Nitrogen*
A, eV	1803.79	3331.06	993.88	6368.21
B, eV	471.195	260.476	136.123	511.205
λ_1, Å^{-1}	2.62392	3.78336	2.50842	5.60181
λ_2, Å^{-1}	1.88891	3.34402	1.49082	3.16170
R, Å	2.76581	2.78681	3.5	2.08389
D, Å	0.31772	0.52612	0.1	0.33133
b	1.0999×10^{-6}	1.00270	0.23586	4.4422×10^{-3}
n	0.78766	3.98638	3.47290	2.42635
h	–0.56239	–0.52909	0.71459	–0.52909
c	1.0039×10^5	0	0.07630	2.2955×10^4
d	16.21701	1	19.7967	24.78674
χ	1	1	1	1

* The parameter values for Si, O, and N were obtained in [16], and ** those for Ga are taken from [17].

The initial configurations of the nanoparticles were created by cutting spheres and spherical layers from a GaN crystal with the wurtzite structure and from an α-quartz crystal and by the subsequent surrounding of the spheres with layers. The preliminarily created GaN crystal had the following lattice parameters: a = b = 0.3181 nm and c = 0.5184 nm [20]. The arrangement of the SiO_4 tetrahedrons to obtain an α-quartz crystal with the parameters a = b = 0.5082 nm and c = 0.55278 nm [21] was constructed using the GRINSP program-generator of inorganic crystalline structures [22]. To obtain the first-type nanoparticle, a GaN sphere was embedded into a spherical SiO_2 layer with the positions of their centers overlapping. The second-type nanoparticle was constructed similarly by fixing a SiO_2 sphere to a spherical GaN layer. In both cases, spherical-layer atoms located at distances to any sphere atoms that were less than a certain chosen value r_m were eliminated in the region including the sphere surface and the lower boundary of the spherical layer. As a result, after the nanoparticles were collected, the following minimum distances between atoms of the different types were obtained: 3.23 (Ga - Si), 2.67 (Ga - O), 2.22 (N - Si), and 2.00 Å (N - O). The sphere size was chosen so that it contained 50 SiO_2 structural units or 54 GaN atomic units. A similar number of structural units of each type were left in the external environment; atoms located furthest away from the center of mass of the created nanoparticle were eliminated in this case. The physical properties were calculated using a classical molecular dynamics ensemble that was a particular case of the microcanonical ensemble. The equations of motion were integrated by the fourth-order Runge–Kutta method with the time step $\Delta t = 10^{-16}$ s. The atom velocities were corrected to balance the systems at a temperature of 300 K during the preliminary stage of calculation with a duration of 10^5 Δt. The main calculation was carried out without any correction and has a duration of 10^6 Δt. As a result, $T = 304.5$ K was established for the nanoparticle with the GaN nucleus, and $T = 303.3$ K, for the nanoparticle with the "inverse" structure.

3. Dielectric Properties

The infrared spectra of the clusters can be calculated using the autocorrelation function of the total dipole moment. In the model, each atom can be regarded as a polarizable point dipole located at the atom nucleus. The interaction of the atom i having permanent polarizability $\boldsymbol{\alpha}_{i,0}$ with neighboring atoms produces the induced dipole moment $\mathbf{d}_i$ [23]

$$\mathbf{d}_i = \mathbf{d}_{i,0} + \boldsymbol{\alpha}_{i,0} \sum_{j \neq i} \mathbf{T}_{ij} \mathbf{d}_j . \qquad (9)$$

Here, $\mathbf{T}_{ij}$ is the dipole–dipole interaction tensor

$$\mathbf{T}_{ij} = \frac{1}{\left| r_{ij} \right|^3} \left(3\hat{\mathbf{r}}_{ij}\hat{\mathbf{r}}_{ij} - \mathbf{1} \right). \qquad (10)$$

In Eq. (9), $\hat{\mathbf{r}}_{ij}$ is the unit vector in the direction ($\mathbf{r}_i - \mathbf{r}_j$), where $\mathbf{r}_i$ and $\mathbf{r}_j$ are the positions of the nuclei of atoms i and j, and $\mathbf{1}$ denotes the unit (3 × 3) tensor. We used the following values for the atomic polarizability: 3.75 (Si), 0.793 (O), 8.1 (Ga), and 1.1 Å^3 (N) and the following values for the electric charges (in units of elementary electric charge): 4 (Si), –2 (O), 3 (Ga), and –3 (N) [24]. The standard iterative procedure [25] was used to calculate the induced dipole moments. The determination accuracy for $\mathbf{d}_i$ was given in the range 10^{-5}–10^{-4} D.

The static dielectric constant was calculated via fluctuations of the total dipole moment $\mathbf{d}$ [26]:

$$\varepsilon_0 = 1 + \frac{4\pi}{3VkT} \left[\left\langle \mathbf{d}^2 \right\rangle - \left\langle \mathbf{d} \right\rangle^2 \right], \qquad (11)$$

where V is the nanoparticle volume and k is the Boltzmann constant. The permittivity $\varepsilon(\omega)$ as a function of the frequency ω was represented by the complex quantity $\varepsilon(\omega) = \varepsilon'(\omega) - i\varepsilon''(\omega)$ the following equation was used to determine it [26, 27]:

$$\frac{\varepsilon(\omega)-1}{\varepsilon_0-1}=-\int_0^\infty \exp(-i\omega t)\frac{dF}{dt}dt=1-i\omega\int_0^\infty \exp(-i\omega t)F(t)dt \quad (12)$$

where the function $F(t)$ is the normalized autocorrelation function of the total dipole moment of the nanoparticle:

$$F(t)=\frac{\langle \mathbf{d}\ (t)\cdot \mathbf{d}\ (0)\rangle}{\langle \mathbf{d}^2\rangle}. \quad (13)$$

The absorption cross section for IR radiation is defined as

$$\sigma(\omega)=\left(\frac{2}{\varepsilon_v c\hbar\eta}\right)\omega\ \tanh(\frac{\hbar\omega}{2kT})\ \mathrm{Re}\int_0^\infty dte^{i\omega t}\langle \mathbf{M}(t)\cdot\mathbf{M}(0)\rangle, \quad (14)$$

where η is the refractive index, ε_v is the permittivity of vacuum, and c is the speed of light.

The reflection coefficient R is defined as the ratio of the average energy flux reflected from the surface to the incident flux. In the case of normal incidence of the plane monochromatic wave, the reflection coefficient is given by the formula [28]

$$R=\left|\frac{\sqrt{|\varepsilon_1|}-\sqrt{|\varepsilon_2|}}{\sqrt{|\varepsilon_1|}+\sqrt{|\varepsilon_2|}}\right|^2. \quad (15)$$

It is assumed here that the wave propagates from a transparent medium (medium 1) to the medium that can be transparent and nontransparent, i.e., the absorbing and scattering one (medium 2). The subscripts for the permittivity in expression (15) denote the medium.

The Raman spectra of clusters can be calculated via the autocorrelation function of polarizability fluctuations.

Each atom in a model can be treated as a polarizable point dipole situated in the nucleus of the atom. As a result of interactions with neighboring atoms, polarizability $\boldsymbol{\alpha}_i$ is created in an atom with permanent polarizability $\boldsymbol{\alpha}_{i,0}$ [21]

$$\boldsymbol{\alpha}_i=\boldsymbol{\alpha}_{i,0}+\boldsymbol{\alpha}_{i,0}\sum_{j\neq i}\mathbf{T}_{ij}\boldsymbol{\alpha}_j. \quad (16)$$

The $J(\omega)$ Raman spectrum excited by not polarized light is given by the equation [21] (if $\omega<0$, we have Stokes and, if $\omega>0$, anti-Stokes spectrum lines)

$$J(\omega)=\frac{\omega}{(\omega_L-\omega)^4}\left(1-e^{-\hbar\omega/kT}\right)\ \mathrm{Re}\int_0^\infty dte^{i\omega t}\langle \Pi_{xz}(t)\Pi_{xz}(0)\rangle, \quad (17)$$

where

$$\Pi(t) \equiv \sum_{j=1}^{N} \left[\boldsymbol{\alpha}_j(t) - \langle \boldsymbol{\alpha}_j \rangle \right], \tag{18}$$

ω_L is the exciting laser frequency, Π_{xz} is the xz component of the $\Pi(t)$ value, the x axis is directed along the bisectrix of the $O-Si-O$ angle of the SiO_2 structural unit or along the $Ga-N$ bond determining the induced dipole direction. Simulation was performed using ω_L = 19436.3 cm^{-1} (argon laser green line, $\lambda = 514.5$ nm).

4. Infrared Spectra

The configurations of nanoparticles with GaN in the central region and SiO_2 at the periphery obtained at a time instant of 100 ps are shown in Figure 1. The SiO_2 arrangement (especially for the external arrangement of these structural units in the nanoparticle) corresponds to the amorphous state to a greater extent than to α-quartz. In both cases, individual O atoms escape from Si atoms and approach Ga atoms. In the case where GaN is located on the cluster surface, N atoms are tightly pressed against Ga atoms and do not penetrate into the cluster.

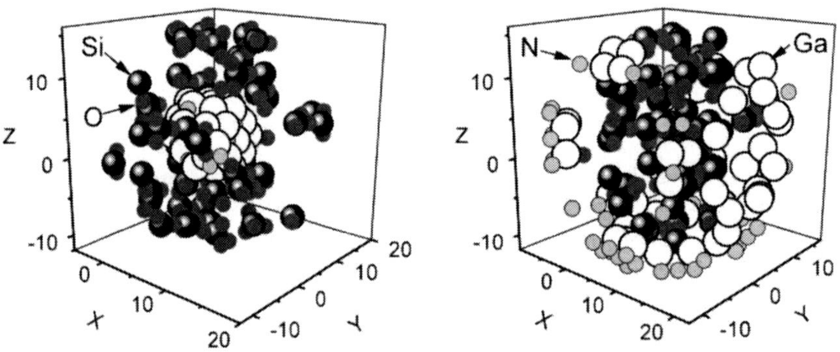

(a) (b)

Figure 1. Configuration of the $(GaN)_{54}$ $(SiO_2)_{50}$ nanoparticles corresponding to a time instant of 100 ps; the nanoparticle cores are (a) GaN and (b) SiO_2.

Ga atoms can be in contact with Si and O atoms. The SiO_2 skeleton is retained except for some losses of O atoms. The inner SiO_2 core has dense and sparse regions. Ga atoms adjoin this core rather closely. If GaN is located in the central region of the nanoparticle, then the surface layer is strongly inhomogeneous and has ruptures. Only individual O atoms approach the GaN core, coming into contact with Ga atoms. The GaN core retains its sphere-like shape during the entire calculation. The N and Ga atoms are tightly bound to one another. During the calculation, these bonds are not broken. Moreover, N atoms tend enter inside the GaN core.

Figure 2 gives a more exact representation of atomic packing in the GaN core. It can be seen that, on the whole, Ga and N atoms form alternating rows. Some rows of Ga and N atoms are greatly curved and are even broken. However, atoms of different types do not mix completely. Ga atoms mainly form the core surface.

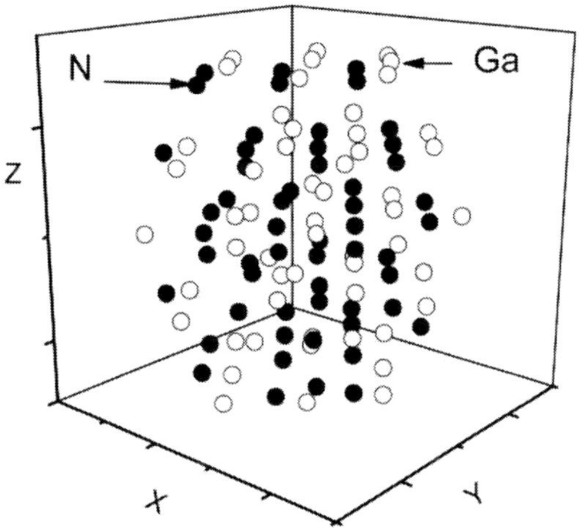

Figure 2. Configuration of the nanoparticle GaN core that corresponds to a time instant of 100 ps.

Figure 3 shows the real and imaginary parts of the $(GaN)_{54}(SiO_2)_{50}$ nanoparticle permittivity. The frequency dependences of ε' and ε'' of the nanoparticles with the core of GaN and SiO_2 are significantly different. The functions $\varepsilon'(\omega)$ and $\varepsilon''(\omega)$ for the nanoparticle with the GaN core have smoother forms. The maximum ε' is at a frequency of 1208 cm^{-1} for the particle with the GaN core and at a frequency of 1240 cm^{-1} for the particle with the SiO_2 core. In both cases, the minimum is observed near 1048 cm^{-1} at a nonzero frequency. The maximum value of the function $\varepsilon''(\omega)$ corresponds to a frequency of 1095 cm^{-1} for the nanoparticle with the GaN core and to 1449 cm^{-1} for the particle whose central region is formed by SiO_2. The minimum of this function for a nonzero frequency is at 91 cm^{-1} for the first-type particle and at 85 cm^{-1} for the second-type particle. The static permittivities of the bulk GaN and SiO_2 are shown in Figures 3a and 3b by the straight dash-dot lines. Knowing the frequency dependence of the complex permittivity makes it possible to determine a series of dielectric properties.

The IR absorption spectra of the $(GaN)_{54}(SiO_2)_{50}$ nanoparticles are shown in Figure 4 (curves *1* and *2*). The spectrum $\sigma(\omega)$ of the nanoparticle with the SiO_2 core is characterized by several well-resolved peaks, while the particle with the GaN core has a rather smooth frequency spectrum $\sigma(\omega)$, in which a series of modes can also be singled out. The IR radiation absorption spectrum of the nanoparticle whose internal part consists of SiO_2 is characterized by lines with higher intensities. For comparison, this figure also shows the experimental IR absorption spectrum of a GaN film grown on a SiO_2/Si substrate [29]. Such a structure is of considerable interest for the development of devices based on the integration of GaN into Si microelectronics. The experimental IR absorption spectrum for the GaN nucleus with island heights of $h = 50$ nm has intense bands at 480 and 1100 cm^{-1}, which are due to the presence of Si atoms [8].

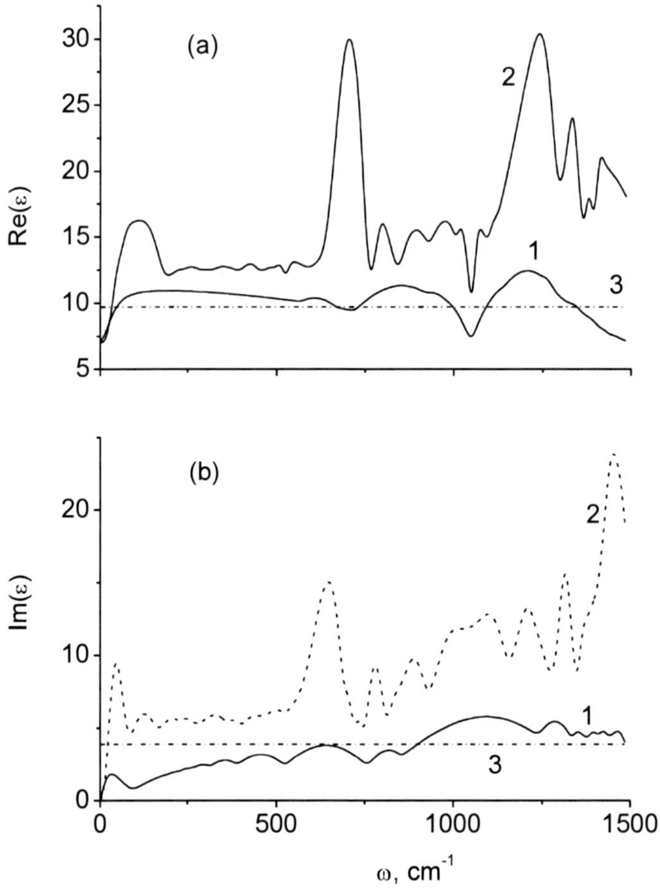

Figure 3. (a) Real and (b) imaginary parts of the permittivity of the $(GaN)_{54}(SiO_2)_{50}$ nanoparticles with a cores formed of (*1*) GaN and (*2*) SiO_2; (*3*) the experimental static dielectric permittivity: (a) GaN and (b) SiO_2.

A series of bands at 472, 680, 824, 1128, and 1289 cm^{-1} are observed in the IR spectrum of the $(GaN)_{54}(SiO_2)_{50}$ nanoparticle with the GaN core. The majority of them are, probably, due to the superposition of closely located absorption signals. The low-intensity bands at 680 and 824 cm^{-1} can be compared with the bands of the experimental IR spectrum of α-quartz: 685 and 789 cm^{-1}, and the strong band at 1128 cm^{-1}, with the high-intensity band of α-quartz at 1063 cm^{-1} [30].

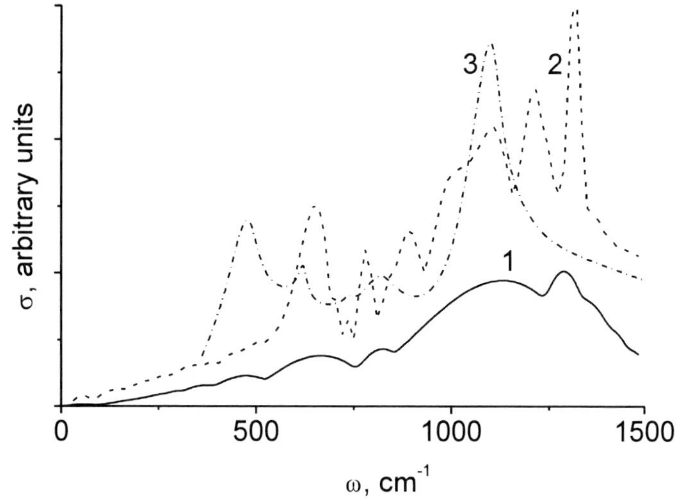

Figure 4. Infrared absorption spectra for the $(GaN)_{54}$ $(SiO_2)_{50}$ nanoparticles with a cores formed from (*1*) GaN and (*2*) SiO_2; (*3*) IR absorption spectrum for the GaN / SiO_2 / Si nanocrystals with an average size of h = 50 nm [29].

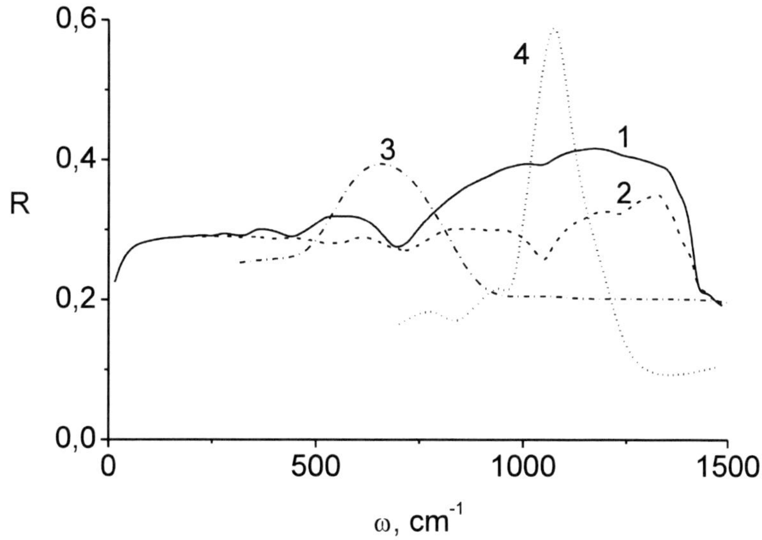

Figure 5. Infrared reflection spectra for the nanoparticles with a cores formed from (*1*) GaN and (*2*) SiO_2; (*3*) IR reflection spectrum obtained by averaging the corresponding spectra for the N and Ga polar layers grown by the method of hydride vapor-phase epitaxy [33]; and (*4*) the reflection spectrum for silica-gel glass at T = 453 K [32].

When going from bulk α -quartz to silica-gel, the asymmetric transverse optical mode (1063 cm^{-1}) shifts to the frequencies 1100, 1150, 1140, and 1120 cm^{-1} for 3-, 4-, 5-, and 6-link rings of SiO_2, respectively [31]. A change in the silica-gel density from 1 to 2 g/cm^3 shifts this band from 1078 to 1122 cm^{-1}. The mode localization also varies strongly with temperature. The band at 1289 cm^{-1} can be compared with the asymmetric valence mode

(1200 cm^{-1}) in silica-gel glass [32]. This mode belonging to optically active oxygen and describing asymmetric stretching was experimentally observed for thin films of amorphous SiO_2 in the form of peaks in the range 1076–1256 cm^{-1}. Bands at 651, 781, 889, 1107, 1212, and 1313 cm^{-1} are pronounced in the IR spectrum of the nanoparticle with the SiO_2 core. The first, second, and fourth bands can be regarded as bands corresponding to α -quartz. However, it can also be assumed that the fourth band is due to the presence of GaN. It is likely that the last two bands are caused by the interaction between the GaN and SiO_2 subsystems.

The calculated reflection spectra of $(GaN)_{54}(SiO_2)_{50}$ nanoparticles with GaN and SiO_2 cores differ considerably from one another, because the surface structures are different (Figure 5). The spectrum $R(\omega)$ for the particle with the GaN core is characterized by high intensities in the frequency range $\omega >$ 687 cm^{-1}. The minimum of the function is observed at a frequency of 687 cm^{-1}.

The frequency 1176 cm^{-1} corresponds to the maximum of the function $R(\omega)$. The deepest minimum in the dependence for the nanoparticle with the SiO_2 core is at a frequency of 1052 cm^{-1}, and the most intense band of this spectrum corresponds to a frequency of 1329 cm^{-1}. The Ga and N polar faces of the GaN crystal with the wurtzite structure suggest a difference between their adsorption and optical properties. Hydride epitaxy in the vapor phase was performed on the Ga and N polar faces of the GaN single crystal at high pressure [33]. The GaN films had layered atomic structures.

The IR reflection spectra of the GaN films grown on the Ga and N polar sides differ considerably from one another. The spectrum for the N polar layer turned out to be dominant because of reflection from a free-electron plasma. The IR reflection spectrum for the Ga polar layer had an intense band produced by a longitudinal phonon in the range 500–750 cm^{-1}.

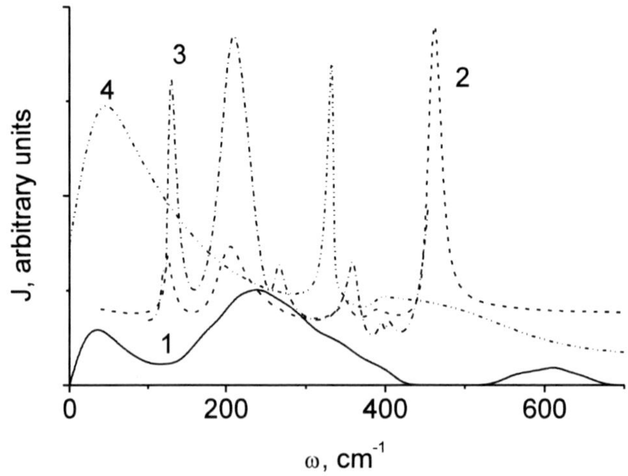

Figure 6. Raman spectra of (*1*) $(SiO_2)_{86}$ nanoparticle, (*2*) α -quartz, (*3*) natural chalcedony from Arizona [36], and (*4*) SiO_2 film grown on sapphire [37].

Curve *3* in Figure 5 corresponds to the combined IR reflection spectrum obtained in the form of the arithmetical mean of the corresponding spectra related to the Ga and N sides of the GaN crystal. This spectrum has one broad intense band in the vicinity of the frequency 663 cm^{-1}. Weakly pronounced bands at 564 (the GaN core) and 614 cm^{-1} (the SiO_2 core) are the nearest ones to it. The silica-gel reflection spectrum has a high-intensity band at 1084 cm^{-1} [31] (Figure 5, curve *4*).

Thus, when transitioning from a crystalline structure to an amorphous one, the most intense band in the reflection spectrum shifts to the blue spectral region and approaches the positions of intense bands in the reflection spectrum of $(GaN)_{54}(SiO_2)_{50}$ nanoparticles.

5. Raman Spectra

The totality of microcrystalline quartz kinds are traditionally divided into two types on the basis of crystalline habitus determined using optical microscopy [34]. Granular diversities include chert and fine-grained sandstone, whereas fibrous varieties of fine-crystalline quartz are combined under a name of chalcedonies. The axis in chalcedonies is most often perpendicular to the long fiber axis. Although the majority of chalcedonies grow with just this orientation, special chalcedonies with the axis parallel to fibers or directed at an angle of 30° to them are also numerous [35].

The Raman spectra of the $(SiO_2)_{86}$ (α-quartz) nanoparticle, natural chalcedony from Arizona [36], and a SiO_2 film [37] grown on sapphire are shown in Figure 6. The spectra of crystalline quartz were measured at 297 $\pm$ 3 K. The low-frequency region with $\omega <$ 50 cm^{-1} was not considered in [37] because, according to the authors, strong spurious scattering was observed at these frequencies. The strongest bands of crystalline SiO_2 were situated at 128, 206, and 464 cm^{-1}. In addition, two bands at 696 and 808 cm^{-1} were observed. The strongest band situated at 400–530 cm^{-1} corresponds to $O-Si-O$ symmetrical stretching-bending modes. The bands at 128 and 206 cm^{-1} correspond to torsional and $O-Si-O$ bending modes. Recent measurements of Raman scattering from a silicon dioxide cell showed the presence of bands at 495.8 and 748.5 cm^{-1} [38]. The bands at 128 and 206 cm^{-1} are strongest in the Raman spectrum of chalcedony. Weak peaks of the same origin were observed in the J spectrum of chalcedony at 262 and 355 cm^{-1}.

The SiO_2 film at $T = 293$ K consisted of the amorphous phase, but crystalline fragments were also present. The structure of SiO_2 glass is an amorphous network of Si atoms tetrahedrally surrounded by O atoms. Tetrahedra are linked with each other through oxygen atoms, and $Si-O-Si$ bridges with interbond angles close to 150° are formed. In other words, each Si atom in the amorphous network is surrounded by four O atoms, and, in turn, each oxygen atom is linked with two silicon atoms. The Raman spectrum of a SiO_2 film grown on a sapphire substrate is shown in Figure 6, curve *4*. A broad band at about 400 cm^{-1} corresponds to $Si-O$ tetrahedron rocking vibrations. A peak at 800 cm^{-1} was also observed. This peak was related to periodic bends of chemical bonds in $Si-O-Si$ bridges. According to [38], an intense narrow peak at 378 cm^{-1} corresponds to plasma radiation in an Ar-laser gas discharge tube. The stoichiometry of SiO_2 is disturbed at elevated temperatures. Nonstoichiometric SiO_2 films were obtained after annealing at 773 K for 15 min. The Raman spectrum of such a film had a broad band centered at approximately 490 cm^{-1} which corresponded to Si - Si bond vibrations. This means that amorphous silicon clusters formed in the film. The first peak in the J spectrum of a low-temperature film situated at ~50 cm^{-1} was a wing of the Raleigh line.

A well-defined low-frequency peak at 34 cm^{-1} was also observed in the Raman spectrum of the $(SiO_2)_{86}$ nanoparticle. Its appearance can also be related to the presence of Raleigh scattering. The most intense broad band appears in the vicinity of a 248 cm^{-1} frequency. This band is caused by torsional vibrations and O - Si - O bending modes. As distinct from crystalline SiO_2, a large number of vibrations close in energy is observed in the $(SiO_2)_{86}$ nanoparticle in the vicinity of the 248 cm^{-1} band. Their superposition forms a broad intense Raman spectrum band. A weak burst in the J spectrum of the nanoparticle is also observed close to the 620 cm^{-1} frequency. This burst is an overtone of modes responsible for the intense Raman spectrum band at 248 cm^{-1}. The positions of the main Raman spectrum bands of the $(SiO_2)_{86}$ nanoparticle, α-quartz, and SiO_2 film are limited by the frequency range $0 \le \omega \le 500$ cm^{-1}. Two new Raman lines at 295 and 380 cm^{-1} and a weak mode at 605 cm^{-1} were recently observed in the Raman spectrum of amorphous silicon dioxide [39]. The authors relate the origin of the new branches to vibrations in five- and more-membered SiO_2 rings.

Because of thermal instability of gallium nitride crystals, GaN semiconductors are mainly produced and used in the form of thin films. The Raman spectra of the GaN film grown on sapphire substrate and on GaAs as well as calculated J spectrum of the $(GaN)_{129}$ nanoparticle are shown in Figure 7. The spectrum of the GaN film grown on sapphire has a noticeable peak at 247 cm^{-1} [40]. In [41], the origin of this peak was related to As impurities in GaN films. This mode is present at 293 K and does not disappear up to 500 K. This is the temperature at which the beginning of the electronic transition with an increase in the number of donors is expected. The peak at 558 cm^{-1}, which appears in the Raman spectrum, corresponds to the existence of stresses in GaN films with the wurtzite structure. The E_2 additional active Raman mode is responsible for the appearance of this peak. When stresses appear, the peak shifts toward higher frequencies; its localization at 566 cm^{-1} was mentioned. The longitudinal asymmetric vibration mode is observed at 733 cm^{-1}. This mode has an obvious “tail” extended toward high frequencies. It is present in the form of a prominent signal at low temperatures. The coalescing lines at 410 and 420 cm^{-1} are acoustic overtones [40]. The lines at 417 and 750 cm^{-1} related to sapphire were excluded from the spectrum. The size of the nanoparticle, as well as the number of layers in the super thin films may have some influence on the Raman spectra of samples of identical composition. The primary change is an increase in intensity of the spectra with increasing particle size and number of layers in the film. Reducing the number of layers in the film may also cause negligible (a few cm^{-1}) red or blue shift of the peaks of Raman spectrum. In the presence of strong coupling of the film to the substrate the reduction of the film layer number can sometimes lead to the emergence of a new peak in the Raman spectrum of the bulk material.

The most striking observation in [42] was that the additional peaks were present only in spectra taken from samples grown on GaAs. Samples grown on sapphire did not exhibit these features. For example, quite pronounced mode at 95 cm^{-1} as well as modes at 60, 102.5, 125, and 250 cm^{-1} were found in the Raman spectrum of a GaN layer grown on GaAs. The theory cannot explain all additional Raman peaks observed at low energies in GaN grown on GaAs. Either N impurities in GaAs or As defects in GaN are likely to be their origin.

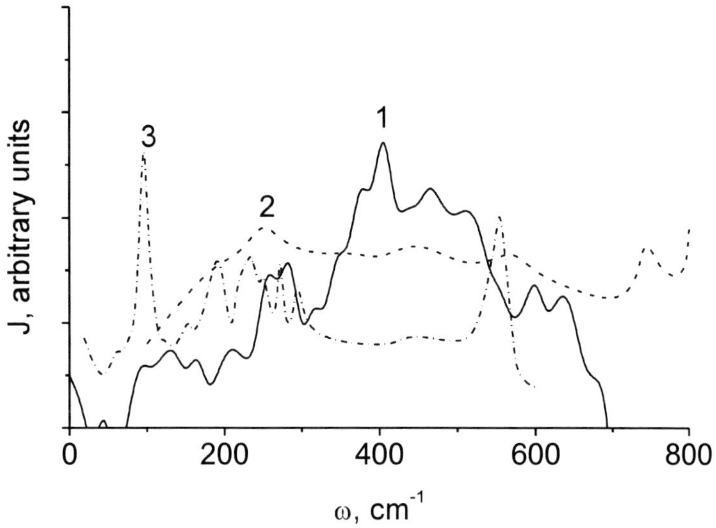

Figure 7. Raman spectra of *(1)* $(GaN)_{129}$ nanoparticle as well as GaN film grown on *(2)* the sapphire substrate [40] and *(3)* the GaAs substrate [42].

The Raman spectrum of the $(GaN)_{129}$ nanoparticle is continuous and contains several bands localized at $0 \le \omega \le 650$ cm^{-1}. The calculated Raman spectrum of the $(GaN)_{129}$ nanoparticle has a noticeable peak of bosons (at 131 cm^{-1}) and two accompanying subpeaks at 98 and 164 cm^{-1} (Figure 7). The boson peak is a universal special feature of the Raman spectra of most of glasses. This peak is equally strong in many glasses and weakly changes as glass connectedness decreases.

This observation was used in [43] to draw the conclusion that the boson peak appeared because of vibrations local in character. These vibrations can be represented by a harmonic oscillator. On the whole, the form of the spectrum of the nanoparticle corresponds to the form of the spectrum of a GaN film over the frequency range $100 \le \omega \le 650$ cm^{-1}. The Raman spectrum of the film has a smoother relief. The mean frequency value of nanoparticle spectrum peaks at 210, 260, and 283 cm^{-1} is 251 cm^{-1}, which closely agrees with the position of the Raman peak observed in a GaN film (247 cm^{-1}).

The most intense band in the J spectrum of the nanoparticle at 408 cm^{-1} can be treated as an acoustic overtone.

Taking anharmonicity into account, the peaks at 464, 510, 599, and 635 cm^{-1} can be assigned to overtones of modes with a 251 cm^{-1} mean frequency. The continuous spectrum of the $(GaN)_{129}$ nanoparticle extends to ~694 cm^{-1}, whereas the J spectrum of the

$(SiO_2)_{86}$ nanoparticle extends to a 415 cm^{-1} frequency only; at 415 cm^{-1}, a break of the spectrum is observed. A large number of active frequencies in the Raman spectrum of the $(GaN)_{129}$ nanoparticle is caused by its small size and a small time interval during which observations were made.

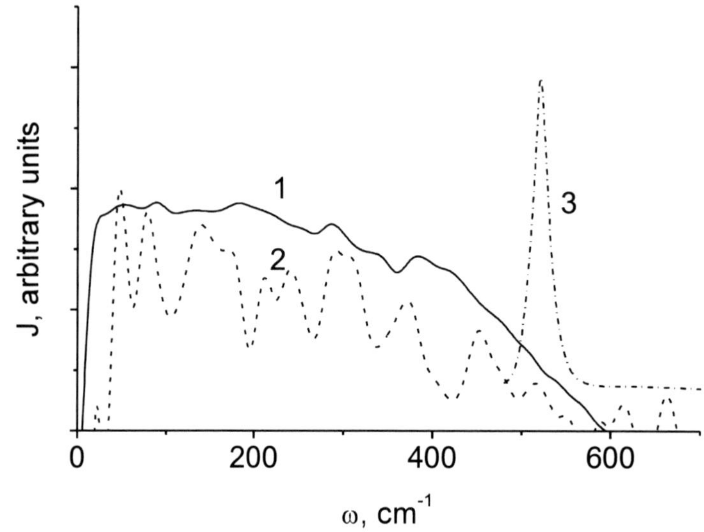

Figure 8. Raman spectra of $(GaN)_{54}$ $(SiO_2)_{50}$ nanoparticles with (*1*) GaN and (*2*) SiO_2 cores and (*3*) Raman spectrum of GaN / SiO_2 / Si nanocrystals with a mean size of h = 50 nm [29].

The $(GaN)_{54}$ $(SiO_2)_{50}$ nanoparticle with a GaN core has a fairly smooth continuous Raman spectrum over the frequency range $0 \leq \omega \leq 600$ cm^{-1} (Figure 8). The band in the vicinity of a 400 cm^{-1} frequency is not most intense, as with the $(GaN)_{129}$ nanoparticle. The band situated close to a 254 cm^{-1} frequency is most intense. The intensity of the J spectrum linearly decreases as the frequency increases from 425 to 600 cm^{-1}. The Raman spectrum of the $(GaN)_{54}$ $(SiO_2)_{50}$ nanoparticle with a SiO_2 core is characterized by a substantially cut relief. This spectrum has a somewhat higher extension compared with the J spectrum of the $(GaN)_{54}$ $(SiO_2)_{50}$ particle with a GaN core. At $\omega >$ 559 cm^{-1}, the Raman spectrum of the particle with a SiO_2 core is not continuous. Over the frequency range $0 \leq \omega \leq 700$ cm^{-1}, the J spectrum of this particle has 11 bands. Some of these peaks (3–5) are split. The fourth peak has two subpeaks, and the third and fifth peaks have subpeaks and shoulders on the right. The first two peaks at 47 and 79 cm^{-1} should be treated as boson peaks. The position of a broad peak at 139 cm^{-1} is in agreement with the low-frequency additional active Raman mode E_2 = 144 cm^{-1} experimentally observed in GaN films [40]. The bands at 212, 241, and 291 cm^{-1}, which have the mean frequency value 248 cm^{-1}, on the one hand, correspond to the main mode (248 cm^{-1}) of the Raman spectrum of the $(SiO_2)_{86}$ nanoparticle and, on the other hand, are close to modes with the mean frequency 251 cm^{-1} observed in the J spectrum of the $(GaN)_{129}$ nanoparticle. The band at 291 cm^{-1} is also close to a 295 cm^{-1}

frequency mode in amorphous silicon [39]. In exactly the same way, the bands at 452 and 516 cm^{-1} correspond to 464 and 510 cm^{-1} modes in the spectrum of the $(GaN)_{129}$ nanoparticle. In addition, the position of the band at 516 cm^{-1} coincides with the localization of the Raman spectrum band for a GaN nanocrystal with size 50 nm grown on a SiO_2 substrate [29]. The Raman spectrum of this nanocrystal contains only one peak (516 cm^{-1}) caused by Si atom vibrations. The position of a well-defined peak at 372 cm^{-1} is close to that of a sharp Raman spectrum peak of a SiO_2 film (378 cm^{-1}) [37] and amorphous SiO_2 (380 cm^{-1}) [39]. The other peaks at 615 and 664 cm^{-1} appear because of splitting of the peak at 620 cm^{-1} in the Raman spectrum of the $(SiO_2)_{86}$ nanoparticle. A peak close in its position (605 cm^{-1}) is also observed in amorphous SiO_2 [39]. It follows that the positions of peaks in the J spectrum of the $(GaN)_{54}\ (SiO_2)_{50}$ nanoparticle with a SiO_2 core are caused not only by collective vibrations in the GaN subsystem but also by group vibrations in the SiO_2 subsystem.

Analysis of the size dependence of the spectral characteristics in the case of nanoparticle is complicated, as its structure with decreasing particle size changes and eventually becomes non crystal. In addition, the contribution of surface phonons is enhanced with respect to that of the bulk phonons.

6. Structure

The traditional structural characteristic of a physical system is the radial distribution function *g(r)*. This one-dimensional function determines a spherically averaged structure rather than reflects the total pattern of the spatial structure. In a series of cases, a conclusion regarding the physical state of the system can be drawn if the shape of this function is used. The distances between the nearest neighbors and the peak intensities of *g(r)* are important parameters. The functions *g(r)* constructed around the atom nearest to the center of mass of the $(GaN)_{54}\ (SiO_2)_{50}$ nanoparticles are shown in Figure 9 (curves *1*). The function *g(r)* calculated for the atom nearest to the center of mass of the $(GaN)_{129}$ nanoparticle is shown in the inset in Figure 9a, and a similar function for the $(SiO_2)_{86}$ nanoparticle is shown in the inset in Figure 9b. The arrows indicate the peak positions in nanometers. On the one hand, the majority of sharp *g(r)* peaks of the $(GaN)_{54}\ (SiO_2)_{50}$ nanoparticles, including allowed ones, demonstrate that the system state can be regarded as the crystalline one. On the other hand, the majority of peaks are not fully resolved, which is evidence of the fragmentary destruction of the crystalline order. For the nanoparticle with the GaN core, the first peak is localized at 0.163 nm and the second, one at 0.407 nm. The peaks of the function *g(r)* of the $(GaN)_{129}$ nanoparticle at 0.244 and 0.353 nm disappear in this case. For the radial distribution function of the particle with the SiO_2 core, all peaks of the function *g(r)* of the $(SiO_2)_{86}$ nanoparticle are reproduced up to a distance of ~0.9 nm; the first of them is at 0.162 nm. The "tail" parts of the functions *g(r)* (for $r >$ 0.9 nm) for particles with the GaN and SiO_2 cores are characterized by different peaks, which are evidence of the existence of noncrystalline structure at the periphery for the first nanoparticle and of the crystalline type of periphery ordering for the second particle. The periphery of the first particle is mainly represented by SiO_2, and that of the second particle, by GaN. Curves *2* in Figure 9 correspond to the experimental functions *g(r)* for the GaN nanocrystal [20] with an average size of ~10 nm (Figure 9a) and for glass SiO_2 (Figure 9b) [44].

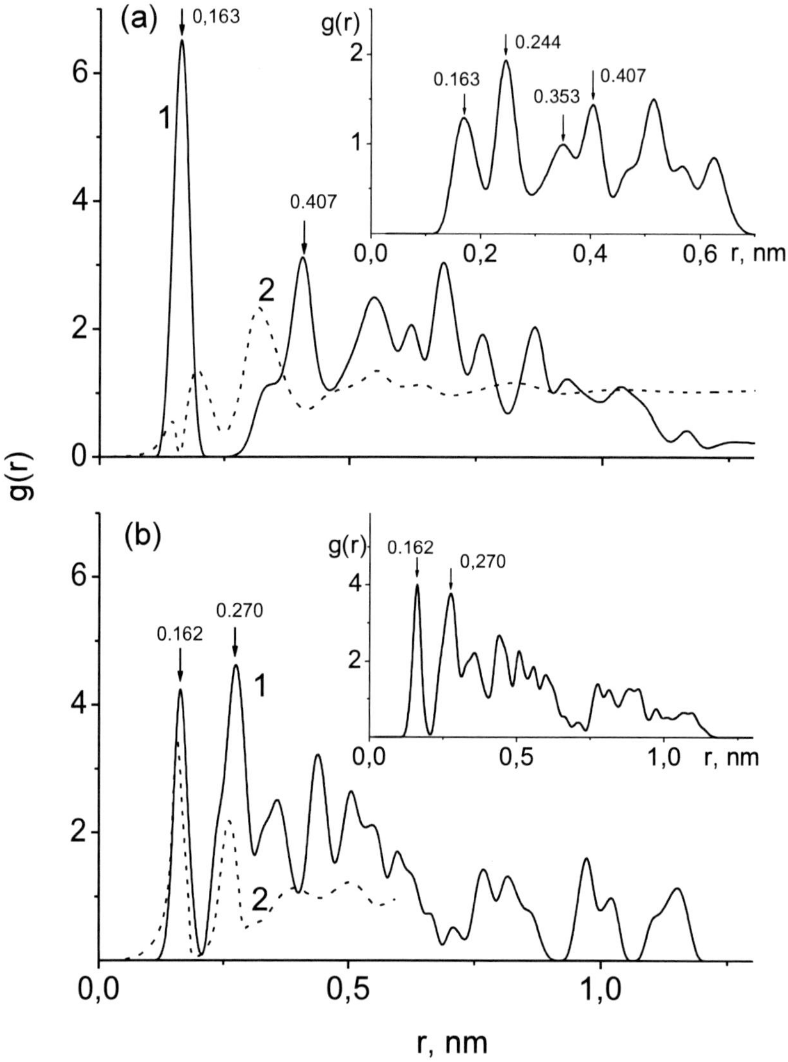

Figure 9. (*1*) Radial distribution functions for the $(GaN)_{54}$ $(SiO_2)_{50}$ nanoparticles with cores formed from (a) GaN and (b) SiO_2; (*2*): (a) the function $g(r)$ of the GaN nanocrystals with an average size of ~10 nm (experiments in [20]) and (b) the function *g*(*r*) for SiO_2 glass (experiments in [44]); the insets shows $g(r)$ of (a) the $(GaN)_{129}$ and (b) the $(SiO_2)_{86}$ nanoparticles.

The radial distribution function for the GaN nanocrystal has many peaks; they disappear at considerably smaller distances than peaks of the function *g(r)* for crystalline GaN.

Consequently, nanocrystalline GaN loses its long-range order, which ordinary crystals of this compound have. The first peak for the nanocrystal is small, and its position corresponds approximately to 0.14 nm. This peak is fully attributed to bound N - N pairs. The second peak is located at 0.194 nm. This value is the average distance between the first Ga - N neighbors. The third peak is located at approximately 0.316 nm. It mainly reflects the Ga - Ga neighborhood and, to a lesser extent, the N - N distances. The GaN nanoparticle under study covered by the SiO_2 layer had an average size of ~2.8 nm. Obviously, the SiO_2 shell had a certain influence on the structure of the GaN core of this particle. This is demonstrated by a significant increase in the first-peak intensity of the function *g(r)* and also by the disappearance of the peaks at 0.244 and 0.353 nm. The positions of the first (0.162 nm) and second (0.265 nm) peak for glass SiO_2 reflect the average nearest Si - O and O - O distances. The two first peaks are located at 0.162 and 0.270 nm for the nanoparticle with the SiO_2 core under study. Based on the form of the function *g(r)*, we can conclude that the nanoparticle-core structure turns out to be very similar to the structure of a free $(SiO_2)_{86}$ nanoparticle. We note that, unlike the experimental function *g(r)*, the calculated radial distribution functions reflect only the local nanoparticle structure.

The method of statistical geometry based on constructing Voronoi polyhedra (VP) determines the spatial structure more precisely. A VP includes a spatial part all points of which are located closer to the center of the polyhedron under study than to other atomic nuclei. To construct a VP, it is necessary that any plane passing through its center divides the space into two parts, so that at least one atom can be located in each half-space. Obviously, a polyhedron cannot be constructed for individual atoms located on the nanoparticle surface. Therefore, the VPs were constructed not for 258 atoms, but for only 200 atoms located closer to the center of the nanoparticle mass; therefore, the statistics represented by the polyhedra reflects the structure of the nanoparticle interior to a larger extent. The VP distribution by the number of faces, obtained by excluding small-scale thermal fluctuations [45], shows the average statistical arrangement of the nearest geometric neighbors. The average number of nearest neighbors is always four in the case of the nanoparticle with the GaN core, and four–six neighbors can be located in the nearest environment of the nanoparticle with the SiO_2 core; in this case, the presence of six neighbors occurs in half of the cases or more (Figure 10). The distributions shown in Figure 10 are constructed based on an analysis of 200000 polyhedrons. The VPs were constructed using the atom coordinates recorded after every 1000 time steps.

To construct the angular distribution of the nearest geometric neighbors, we considered the angles θ with vertices in the nuclei of VP atom-centers and sides formed by lines passing through these vertices and the nuclei of atoms forming the VP faces. The θ distributions for $(GaN)_{54}$ $(SiO_2)_{50}$ nanoparticles shown in Figure 11 have three peaks. In the case of the nanoparticle with the GaN core, these peaks are at 93°, 114°, and 168°, and, in the case of the nanoparticle with the SiO_2 core, the peaks are at 35°, 66°, and 114°. The angle 66° is a supplement to the angle 114°, because the addition of these angles gives the straight angle: 66° + 114° = 180°. The angle 114° corresponds to the maximum stability of the structural elements Si - O - Si with one single bond and one double bond in which the O atom participates [46].

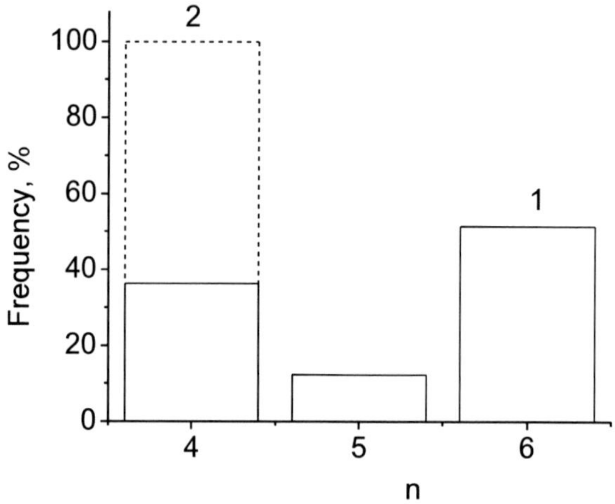

Figure 10. Distribution of the VP faces for $(GaN)_{54}$ $(SiO_2)_{50}$ nanoparticles with cores formed from (*1*) GaN and (*2*) SiO_2.

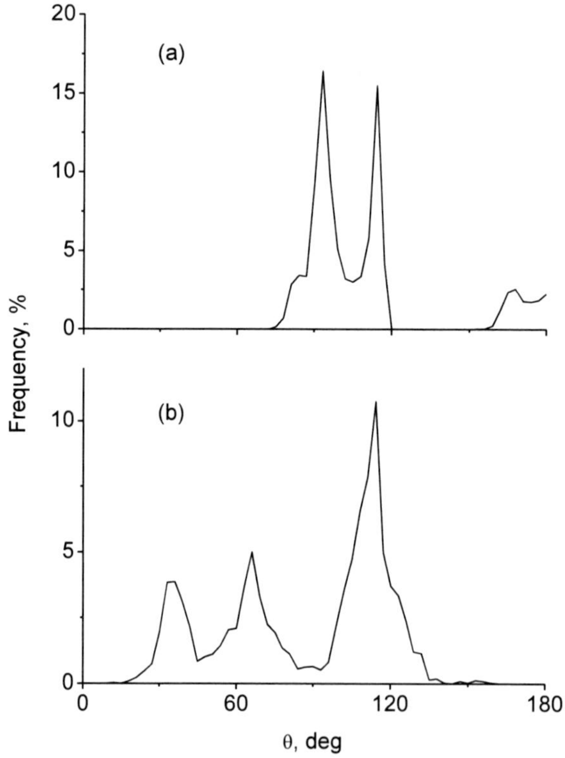

Figure 11. Angular distribution of the nearest geometric neighbors for $(GaN)_{54}$ $(SiO_2)_{50}$ nanoparticles with cores formed from (a) GaN and (b) SiO_2.

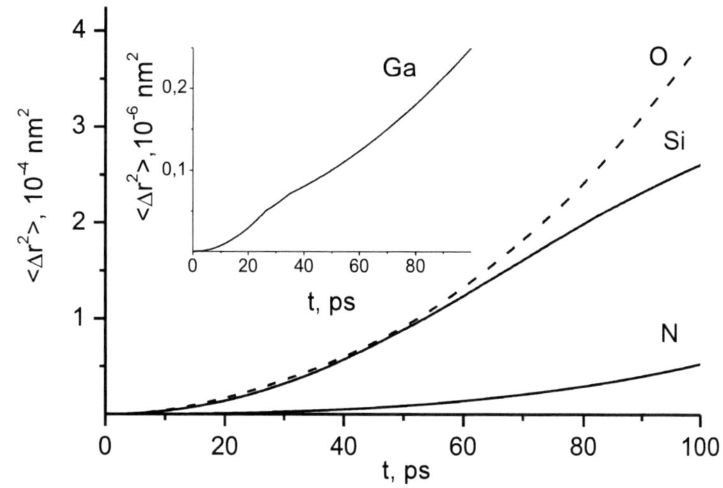

Figure 12. Average squared atomic displacement in the $(GaN)_{54}$ $(SiO_2)_{50}$ nanoparticle with a nucleus formed from SiO_2.

The peak at small angles 35° is indicative of the elongated form of the polyhedra, which follows from the fixing of all interatomic distances without distinguishing between types of atoms. In turn, the presence of elongated VPs indicates the structural inhomogeneity of the obtained nanoparticles.

The character of atomic motion in the nanoparticles was determined calculating the average squared displacement of atoms $\langle(\Delta r)^2\rangle$. Figure 12 shows the dependences $\langle(\Delta r)^2\rangle(t)$ for Ga, N, Si, and O atoms in the nanoparticle with the SiO_2 core. Similar dependences were obtained for the nanoparticle whose core consists of GaN. The value of the atomic displacement in the nanoparticle was determined not only by their mass, but also by the local density, i.e., the character of the nearest environment. An estimate of the GaN density is 6.15 g/cm^3, and the α -quartz density is 2.65 g/cm^3 [22]. Consequently, it can be expected that the atomic displacement in GaN is noticeably smaller than in SiO_2. The mass of the Ga atom is five times larger than that of the N atom. Therefore, the N atoms must have a higher mobility. The mass of the Si atom exceeds that of the O atom by a factor of 1.75. As a consequence, the mobility of O atoms must be noticeably higher. The self-diffusion coefficients *D* in the nanoparticles, which were determined using the average squared displacement, are given in Table 2. It can be seen that the coefficient *D* of N atoms is larger than that for Ga atoms by a factor of 160.7. At the same time, the coefficient *D* of the O atoms is larger than that for the Si atoms by a factor of 2.6. The mobility of the O atoms is higher than that for the N atoms by a factor of 6.6.

Table 2. Self_diffusion coefficients for $(GaN)_{54}$ $(SiO_2)_{50}$ nanoparticles

	Ga	N	Si	O
D, 10^{-12} m^2/s	0.00729	1.172	2.956	7.746

Conclusion

Progress in the synthesis of high-quality nanomaterials makes it possible to investigate one of the most fundamental issues concerning the influence of the size, structure, and surface of nanoparticles on their dynamic properties including the infrared and Raman spectra. Such investigations are necessary to gain better understanding of the basic physics of controlling the parameters of nanoparticles in order to get the required dynamic properties, including the cases of thermal and mechanical load of nanoobjects.

The complex interatomic interactions in $(GaN)_{54}$ $(SiO_2)_{50}$ nanoparticles were described by the Tersoff potential with parameters optimized using density functional theory. N atoms have a significantly higher mobility than Ga atoms. However, the $GaN - SiO_2$ interaction occurs mainly by means of Ga atoms.

The frequency dependence of the real and imaginary parts of the permittivity and the IR absorption spectra of the $(GaN)_{54}$ $(SiO_2)_{50}$ nanoparticles are rather smooth if there is GaN at the particle core and have an oscillatory character in the presence of SiO_2 at the core. The IR absorption spectra of the nanoparticles under study contain a broad intense absorption band in the frequency range $850 \le \omega \le 1350$ cm^{-1}.

As the frequency increases, the reflection coefficient of the IR radiation of the nanoparticle with the GaN core becomes noticeably larger than that of the nanoparticle with the SiO_2 core, starting from a frequency of 750 cm^{-1}.

The calculation results show that, over the larger part of the $0 \le \omega \le 700$ cm^{-1} frequency range, the Raman spectra of nanoparticles consisting of SiO_2 and GaN are continuous. The Raman spectrum of the $(SiO_2)_{86}$ nanoparticle is smooth and contains a wing of Raleigh scattering. Over the frequency range 420–530 cm^{-1}, a break of the spectrum is observed. Over the frequency range covered, the Raman spectrum of the $(SiO_2)_{86}$ nanoparticle is closer to the J spectrum of a SiO_2 amorphous-crystalline film than to the corresponding spectrum of crystalline or microcrystalline SiO_2. Broad Raman spectrum bands of the $(SiO_2)_{86}$ nanoparticle are formed because of overlapping of vibrational bands close in energy. The $(GaN)_{129}$ nanoparticle also has a continuous Raman spectrum. However, as distinct from the J spectrum of the $(SiO_2)_{86}$ nanoparticle, the Raman spectrum of the $(GaN)_{129}$ nanoparticle has a large number of well-defined bands. Over the frequency range $100 \le \omega \le 700$ cm^{-1}, the J spectrum of a GaN film with the wurtzite structure on the whole fits in with the Raman spectrum of the $(GaN)_{129}$ nanoparticle. The form of the Raman spectrum of the $(GaN)_{54}$ $(SiO_2)_{50}$ nanoparticle depends to a substantial extent on the arrangement of components in the volume of this nanoparticle. If GaN forms its core, the J spectrum has a smooth shape, and if SiO_2 occupies its center, the Raman spectrum has a large number of bands. These bands correspond to vibration of groups of atoms of different kinds and atoms of the same kind. The Raman spectrum of the SiO_2 / GaN core-shell nanoparticle has a larger extension, but its "tail" is not a continuous spectrum continuation.

The radial distribution functions of the nanoparticles with the GaN and SiO_2 cores are significantly different. For the first-type nanoparticle, the first peak of the function is resolved and has a high intensity. For the second-type nanoparticle, the second-peak intensity of this function exceeds the intensity of the first peak. In addition, the "tail" parts of the function $g(r)$ of the nanoparticles under study have different shapes, positions, and peak intensities. The tetrahedral structure of the nanoparticle is retained during the entire calculation if its core is SiO_2.

The structure becomes predominantly of the six-coordinate type if the nanoparticle core is GaN. The angular distribution of the nearest geometric neighbors is characterized by two strong peaks at angles of 93° and 114° for the nanoparticle with the GaN core and by one peak with a high intensity localized at 114° for the nanoparticle with the SiO_2 core. The self-diffusion coefficients calculated using the average squared atomic displacements have the following sequence: Ga, N, Si, and O in ascending order of *D*. In this case, Ga is characterized by the a diffusion coefficient that is lower by two orders than those of the other elements in the $(GaN)_{54}(SiO_2)_{50}$ nanoparticle.

References

[1] Lee, D. S.; Steckl, A. J. *Appl. Phys. Lett.* 2003, 83, 2094-2096.

[2] Pan, G.; Kordesch, M. E.; Van Patten, P. G. *Chem. Mater.* 2006, 18, 3915-3917.

[3] Munusamy, P.; Mahalingam, V.; van Veggel, F.C.J.M *Eur. J. Inorg. Chem.* 2008, 23, 3728-3732.

[4] Bockowski, M. *Cryst. Res. Technol.* 2001, 36, 771–787.

[5] Drozdov, Yu. N.; Drozdov, M. N.; Khrykin, O. I.; Shashkin, V. I. *J. Surf. Invest.* 2010, 4, 998–1001.

[6] Cheng, G. S.; Zhang, L. D.; Zhu, Y.; Fei, G. T.; Li, L.; Mo, C. M.; Mao, Y. Q. *Appl. Phys. Lett.* 1999, 75, 2455–2457.

[7] Cao, Y. G.; Chen, X. L.; Lan, Y. C.; Xu, Y. P.; Xu, T.; Liang, J. K. *J. Mater. Res.* 2000, 15, 267–269.

[8] Zhang, X.; Hou, Y. T.; Feng, Z. C.; Chen, J. L. *J. Appl. Phys.* 2001, 89, 6165–6170.

[9] Deguchi, T.; Ichiryu, D.; Toshikawa, K.; Sekiguchi, K.; Sota, T.; Matsuo, R.; Azuhata, T.; Yamaguchi, M.; Yagi, T.; Chichibu, S.; Nakamura, S. *J. Appl. Phys.* 1999, 86, 1860–1866.

[10] Galashev, A. E. *Poverkhnost'*, 2010, No. 8, 103–112.

[11] Galashev, A. E.; Polukhin, V. A.; Izmodenov, I. A.; Galasheva, O. A. *Poverkhnost'*, 2007, No. 10, 60–67.

[12] Galashev, A. E.; Izmodenov, I. A.; Rakhmanova, O. R.; Novruzova, O. A. *Poverkhnost'*, 2007, No. 8, 95–103.
[13] Belyakov, V. A.; Burdov, V. A.; Sidorenko, K. V. *J. Surf. Invest.* 2010, 4, 987–993.
[14] Tersoff, J. *Phys. Rev. B* 1989, 39, 5566–5568.
[15] Tersoff, J. *Phys. Rev. Lett.* 1986, 56, 632–635.
[16] Billeter, S. R.; Curioni, A.; Fischer, D.; Andreoni, W. *Phys. Rev. B* 2006, 73, 155329.
[17] Nishidate, Y.; Nikishkov, G. P. *Comp. Model. Eng. Sci.* 2008, 26, 91–100.
[18] Yasukawa, A. *Jpn. Soc. Mech. Eng.* 1996, 39, 313–320.
[19] Benkabou, F.; Certier, M.; Aourag, H. *Mol. Simul.* 2003, 29, 2001–2009.
[20] Lee, S. H.; Booa, J.-H.; Lee, S. Y.; Yoo, S. H.; Kim, C. G.; Lee, Y. K.; Kim, Y. *J. Vac. Sci. Technol. B* 2004, 22, 2144-2148.
[21] Munetoh, S.; Motooka, T.; Moriguchi, K.; Shintani, A. *Comp. Mater. Sci.* 2007, 39, 334–339.
[22] Le Bail, A. *J. Appl. Crystallogr.* 2005, 38, 389–395.
[23] Bosma, W. B.; Fried, L. E.; Mukamel, S. *J. Chem. Phys.* 1993, 98, 4413–4421.
[24] *Spravochnik khimika* (Chemist's Handbook); Nikol'skii B. P.; Ed.; Khimiya: Leningrad, 1971; Vol. 1.
[25] Dang L. X., Chang, T._M. *J. Chem. Phys.* 1997, 106, 8149–8159.
[26] Bresme, F. *J. Chem. Phys.* 2001, 115, 7564–7574.
[27] Neumann, M. *J. Chem. Phys.* 1985, 82, 5663–5672.
[28] Landau L. D.; Lifshits, E. M. *Course of Theoretical Physics*: *Electrodynamics of Continuous Media*; Pergamon: New York (US), 1984; Vol. 8.
[29] Konenkova, E. V.; Zhilyaev, Yu. V.; Fedirko, V. A.; Zahn, D. R. T. *Appl. Phys. Lett.* 2003, 83, 629–631.
[30] Liang, Y.; Miranda, C. R.; Scandolo, S. *J. Chem. Phys.* 2006, 125, 194524.
[31] West, J. K.; West, D. J.; Hench, L. L. *Physics and Chemistry of Glasses - European Journal of Glass Science and Technology Part B* 1998, 39, 301–304.
[32] Tan, C. Z.; Arndt, J. *J. Phys. Chem. Solids* 2000, 61, 1315–1320.
[33] Tuomisto, F.; Suski, T.; Teisseyre, H.; Krysko, M.; Leszczynski, M.; Lucznik, B.; Grzegory, I.; Porowski, S.; Wasik, D.; Witowski, A.; Gebicki, W.; Hageman, P.; Saarinen K. *Phys. Status Solidi B* 2003, 240, 289–292.
[34] Klein, C.; Hurlbut, C. S. Jr *Manual of Mineralogy*; 20th edn.; Wiley: New York (US), 1985.
[35] Folk, R. L.; Pittman, J. S. *Journal of Sedimentary Petrology* 1971, 41, 1045–1058.
[36] Kingma, K. J.; Hemley, R. J. *Am. Mineral.* 1994, 79, 269–273.
[37] Berezhinsky, L. I.; Maslov, V. P.; Tetyorkin, V. V.; Yukhymchuk, V. A. *Semiconductor Physics Quantum Electronics & Optoelectronics* 2005, 8, 37–40.
[38] Schonbachler, N.; Luthy, W. *Measurements of Raman lines in silica, dimethyl-methylphosphonate and methyl salicylate*; Univ. of Bern Press: Bern (Sw), 2010.
[39] Borowicz, P.; Latek, M.; Rzodkiewicz, W.; Łaszcz, A.; Czerwinski, A.; Ratajczak, J. *Adv. Nat. Sci.: Nanosci. Nanotechnol.* 2012, 3, 045003.
[40] Wang, R-M.; Chen, G.-D.; Lin, J.-Y.; Jiang, H.-X. *Front. Phys. China* 2006, 1, 112–116.
[41] Siegle, H.; Kaschner, A.; Hoffmann, A.; Broser, I.; Thomsen, C.; Einfeldt, S.; Hommel, D. *Phys. Rev. B* 1998, 58, 13619–13626.

[42] Siegle, H.; Loa, I.; Thurian, P.; Eckey, L.; Hoffmann, A.; Broser, I.; Thomsen, C. *Appl. Phys. Lett.* 1997, 70, 909.
[43] McIntosh, C.; Toulouse, J.; Tick, P. *J. Non-Cryst. Solids* 1997, 222, 335–341.
[44] Johnson, P. A. V.; Wright, A. C.; Sinclair, R. N. *J. Non-Cryst. Solids* 1983, 58, 109–130.
[45] Galashev, A. E. *Crystallography Reports* 2002, 47, S169–S176.
[46] Pauling, L. *Am. Mineral.* 1980, 65, 321–323.

In: Gallium Nitride
Editor: Kiera Olivia Peak
ISBN: 978-1-63321-387-6

Chapter 8

AlGaN/GaN-Based Lateral-Type Schottky Barrier Diode with Bonding Pad over Active Structure Prepared on Sapphire Substrate

Jae-Hoon Lee*
Discrete Development Team, System LSI,
Samsung Electronics Co., Ltd, Giheung, Korea

Abstract

In this work, we fabricated a Multi-finger lateral-type AlGaN/GaN Schottky barrier diode (SBD) fabricated on the AlGaN/GaN heterostructure grown on sapphire substrate and also present the electrical properties of the SBD measured by varying temperature. The multi-finger patterned large-area SBD with the bonding pad over active (BPOA) structure was fabricated. The size, the length of the anode perimeter, and the anode-to-cathode distance for both devices are 3000 μm × 3000 μm, 270 mm, and 20 μm, respectively. The fabricated SBD with BPOA structure exhibited excellent device performances, such as forward current of 4.5 A at 1.5 V, leakage current of 6 μA at - 600V, reverse-recovery time (Trr) of 19.4 ns, breakdown voltage of 747 V, and low specific on-resistance (Ron) of 12 mΩ·cm2, resulting in the figure of merit (VBR2/Ron) as high as 47 MW/cm2, which demonstrates that the AlGaN/GaN SBD with BPOA structure has a great potential application to the high power electronics.

1. Introduction

Wide-bandgap III–nitride semiconductors and their ternary solid solutions are very promising as the candidates for both short wavelength optoelectronics and power electronic devices because of the intrinsic advantages of materials such as direct bandgap and high electron peak velocity [1]-[5]. The wide bandgap nature of the AlGaN/GaN heterostructure leads to high device breakdown voltage, which is very important in high power switching applications such as converters and inverters, including solid-state drives for heavy motors,

* E-mail address: jaehoon.lee@samsung.com.

pulsed power for electric vehicles or ships [6]. Also the AlGaN/GaN-based power device exhibits very high switching speed, originated from the high electron velocity of the material, which can simplify the power conversion system and reduce the size of the system. Most of the GaN-based power devices are fabricated on the AlGaN/GaN heterostructure grown on Si due to its large wafer size and cost effectiveness [7,8]. Nevertheless, the growth of GaN on sapphire still has potential advantages in view of the facts that the growth of GaN on sapphire is much easier and the crystalline quality of the grown GaN on sapphire is much better, compared to that on Si. The size of the sapphire substrate steadily enlarged now up to 6 in diameter along with Si-comparable cost. Furthermore, the bad thermal conductivity of the sapphire substrate could be solved by forming backside via-hole contact after thinning the sapphire substrate or by using laser lift-off technique to remove the substrate developed for the vertical-type GaN-based LED [9,10].

It is well known that the usual undoped GaN layer grown on normal growth condition generally exhibits n-type conductivity due to residual donor impurities such as nitrogen vacancies and oxygen atoms which cannot be used as a buffer layer of theAlGaN/GaN-based heterostructure field effect transistor (HFET). Therefore, the semi-insulating GaN layer is required to avoid the parallel conduction and to minimize the off-state leakage current. The semi-insulating GaN layer was successfully obtained by doping compensating acceptors such as C, Fe, or Mg, even though some problems such as redistribution of dopant atoms toward the channel region may exist [11], [12]. These impurities can also exhibit strong memory effect in epitaxial growth thereby precluding abrupt transitions in doping. To avoid these problems, introduction of threading dislocations or other defects during the growth was suggested to obtain the semi-insulating GaN layer. However, the device fabricated on this semi-insulating buffer layer suffers from the trapping/detrapping of electrons due to the degraded crystalline quality which occasionally results in a severe current collapse in AlGaN/GaN HFET [13]. To minimize the trapping effect, therefore, it is necessary to grow a semi-insulating GaN layer while maintaining the crystalline quality as high as possible.

Although remarkable progress has been achieved, an issue related to the surface-states, which significantly deteriorates the device performance as well as the device reliability, still need to be resolved. For AlGaN/GaN-based electronic devices, high Al mole fraction in AlGaN is desirable because it not only allows large forward current flow due to high two dimensional electron gas (2DEG) density, but also leads to small reverse leakage current and high breakdown voltage due to the wide bandgap. However, in reality, Al-GaN/GaN-based electronic devices with high Al content suffer from large reverse leakage currents which are many orders of magnitude larger than theoretically predicted values [14]. The surface-states are believed to be responsible for the leakage current of the AlGaN/GaN heterostructure field effect transistors (HFETs) and occasionally cause a severe current collapse when the HFETs operate under high power and high frequency conditions. To reduce surface leakage current, various methods such as surface passivation, surface treatments with either chemicals or plasma, and post annealing, have been investigated. Dielectric films, such as SiN_x, SiO_2, HfO_2 and Al_2O_3, have been used as surface passivation layers as well as gate insulators for the AlGaN/GaN HFETs [15]-[17].

Various lateral and vertical type GaN-based Schottky barrier diodes (SBDs) have been reported and shown good device performances [18]-[21]. However, most lateral-type SBDs have suffered from low forward current level due to the small size dimension with ring type geometry structure. To increase the on-current, vertical type GaN SBD fabricated on a bulk

GaN wafer has been suggested. The device experiences, however, considerably lower breakdown voltage than expected, because the current flow of the device is parallel to the direction of the dislocation propagation due to its immature growth technique [22], [23]. Recently, it was reported that the use of the p-type GaN barrier controlling layer greatly improves the device characteristics, not only increasing forward current, but also increasing the breakdown voltage with keeping low leakage current [24].

In this work, we suggest a simple method of growing a highly resistive GaN buffer layer grown on sapphire substrate by controlling the size of the nucleation sites. We also proposed a method of reducing the on-resistance of AlGaN/GaN based SBD grown on a highly resistive GaN buffer layer using the bonding pad over active (BPOA) structure and present the electrical properties of the SBD measured by varying temperature. The BPOA allows the placement of bonding pads over active layer without the addition of reinforcing layer [25], which offers a significant reduction in both the overall size of the die and the series resistance of pads.

2. Fabrication of AlGaN/GaN-Based SBD Device with BPOA Structure

The AlGaN/GaN heterostructure for the SBD investigated in this work was grown on 4 inch (0001) c-plane sapphire substrates using metal organic chemical vapor deposition (MOCVD). The heterostructure consists of 30 nm-thick low temperature-grown initial GaN layer, 3 μm-thick highly resistive undoped GaN buffer, and 25 nm-thick AlGaN barrier. A two-step growth technique was proposed to obtain a high sheet resistive GaN. The undoped GaN buffer layer was first grown at relatively low temperature of 950°C for 2 min. and the temperature was elevated to the normal growth temperature of 1100°C to complete the remaining growth for the heterostructure.

The mobility and the carrier concentration of the two dimensional electron gas formed at the AlGaN/GaN heterointerface were 1200 $cm^2/V \cdot s$ and $4\times10^{12}/cm^2$, respectively, estimated by using Hall measurement. For the device fabrication, the active region of the device was defined by inductively coupled plasma (ICP) reactive ion etching using a BCl_3/Cl_2 gas mixture.

After opening the contact hole opening for the cathode using photo-resist mask, Ti/Al/Ni/Au Ohmic metal was deposited, and followed by rapid thermal annealing at 850 °C for 30 s in N_2 ambient, which results in specific contact resistance of 2×10^{-5} Ωcm^{-2}. Next, the deposited 0.8 μm-thick SiO_2 layer for both the passivation and the field plate dielectric was deposited on the AlGaN layer using plasma enhanced CVD (PECVD) at 290℃ with SiH_4/N_2O flow rate of 200/120 sccm. During the deposition, the pressure was kept at 1 mTorr and source RF power was 50 W. To form the Schottky contact, the SiO_2 layer was selectively etched using both photo-resist mask and BOE etchant and then the Ni/Au Schottky metal for the anode was deposited. Another 1.6 μm-thick Si_3N_4 layer was then deposited to form double dielectric passivation structure. To reduce the series resistance of the device, Ti/Al pad metal was deposited to complete a bonding pad over active structure after opening through-holes in the Si_3N_4 layer. The current–voltage (I-V)$(I{-}V)$ characteristics were measured by using an Agilent 4155 parameter analyzer and an STI curve tracer 5000E for low and high currents, respectively. A schematic view of the fabricated SBD with BPOA structure is shown in Figure 1.

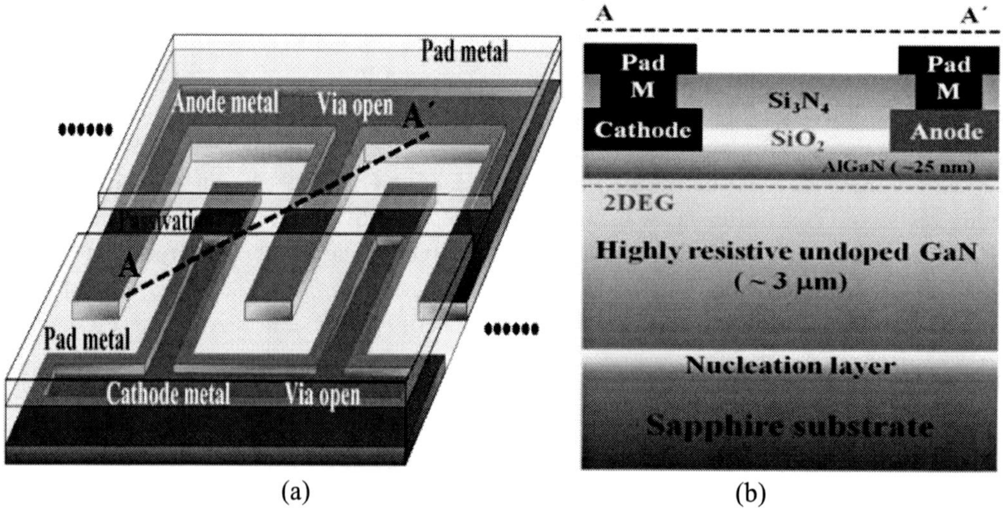

Figure 1. Schematic top (a) and cross-sectional (b) view of the fabricated SBD device with BPOA structure.

3. Growth of Semi-Insulation GaN Buffer Layer

Figure 2 shows a typical electrical property of the undoped GaN buffer layer grown by the proposed sequence.

Extremely high sheet resistance (R_s) of ~ 10^{12} Ω/sq was measured by using a circular type TLM pattern between two ohmic contact pads with inner diameter of 400 μm and inter-electrode distance varied from 5 to 100 μm.

The curve in the figure represents the I-V characteristic between two contacts with inter-electrode distance of 100 μm.

The key feature for the growth of the buffer layer in the initial stage of the growth is that many nucleation sites with smaller grain sizes tend to be generated when the growth proceeds at relatively low temperature of 950 ℃, as shown in Figure 3 (a).

The growth with smaller grain sizes leads to large mismatches in growth orientation and direction which results in generation of many crystal defects such as dislocations and stacking faults in the subsequently grown GaN layer at relatively high temperature of 1100 ℃, as shown in Figure 3 (b).

These defects are generally known to act as deep traps in the GaN layer, which makes the layer very resistive [26]. In contrast, the normal GaN growth without varying temperature results in the growth with larger grain size and less nucleation density, as observed in Figure 3 (c), and hence the density of coalescence boundary decreases as shown in Figure 3(d), which eventually produces improved structural and electrical properties of the layer [27].

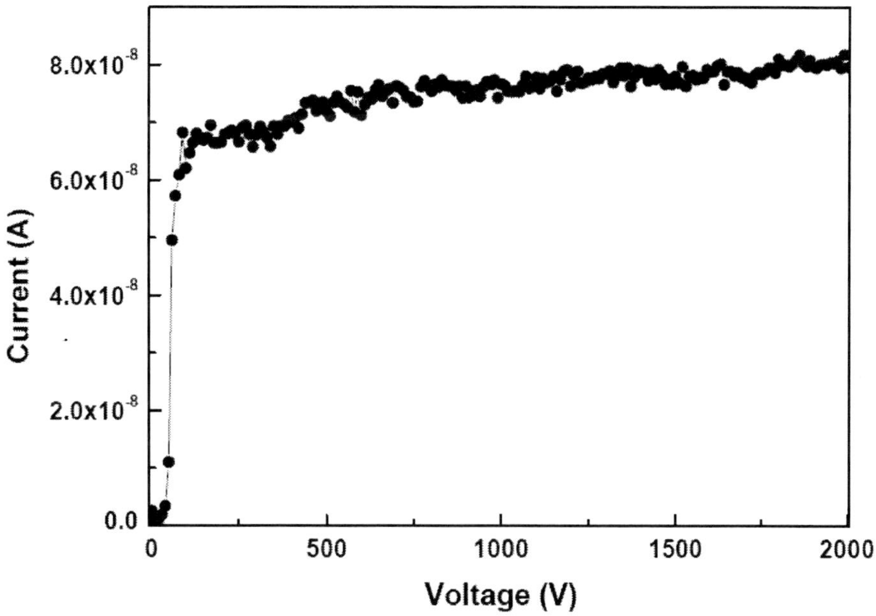

Figure 2. I-V measurement of GaN film with proposed temperature grade growth method.

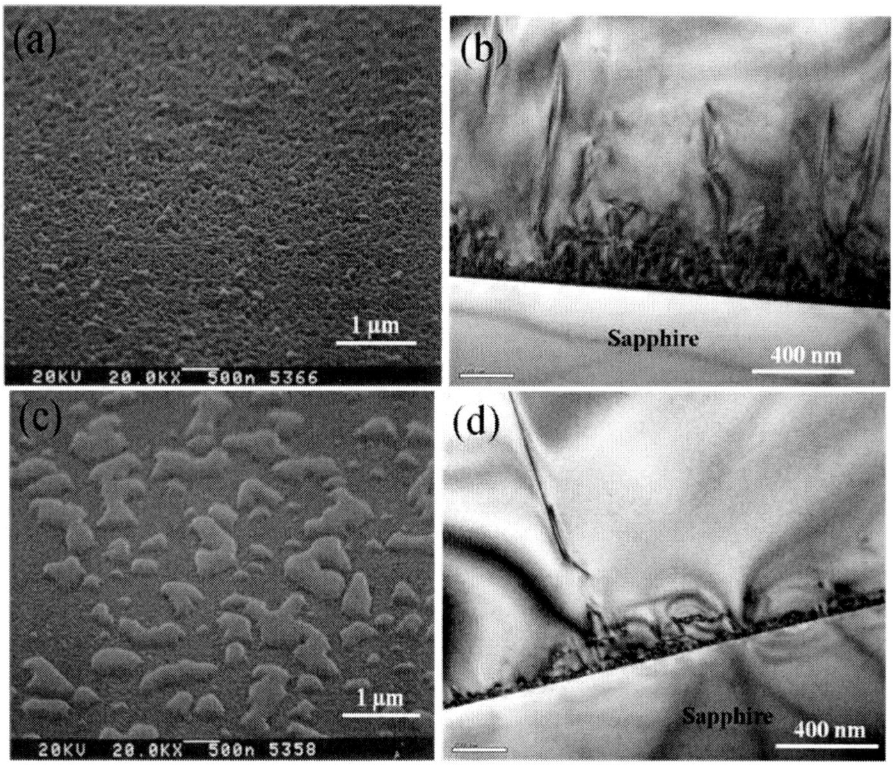

Figure 3. SEM (a), (c) and TEM (b), (d) image of films grown with small grain size and large grain size in the initial growth stage.

The properties for the undoped GaN layers grown by both the normal and the proposed sequence are compared in table I. The full width at half-maximum (FWHM) of asymmetric (102) rocking curve for the sample grown by the proposed sequence is 475 arcsec, which is much narrower than that (5100 arcsec) of previously reported semi-insulating GaN buffer layer [28], even though it is wider than that (240 arcsec) for the undoped GaN layer grown under normal growth condition which usually exhibits n-type conduction. The photoluminescence (PL) spectral intensity of the sample is approximately an order lower in magnitude than that for the sample grown by normal sequence, which indicates that the sample grown by the proposed sequence contains many deep traps as discussed above, mostly known as acceptor-like defects associated with Ga vacancy and edge type dislocation [29]. Lee et al. reported that edge-type dislocations in GaN can act as acceptor-like deep trap center for electrons [30]. The present asymmetry (102) rocking curve results therefore explain that the edge-type dislocations in GaN play a role of compensating n-type background carriers which makes the GaN layer highly resistive as the semi-insulating GaN layer obtained in this work. These results therefore explain that the proposed sequence is very effective in growing a highly resistive GaN buffer layer even without any additional compensation doping.

Table 1. Structural and electrical properties of undopd GaN with small and large grain size in the initial growth state

u-GaN	PL peak Intensity (arb. units)	XRD (arcsec)		Hall	
		(002)	(102)	Rs(Ω/sq)	Mobility (cm^2/Vs)
Normal growth	500	220	280	3×10^3	300
Proposed growth	30	280	475	$\sim 10^{12}$	Not Measurable

4. Characteristics of the Small-Area Algan/GaN SBD According to Electrodes Distance

Figure 4 shows the I-V characteristics of the small-area AlGaN/GaN SBDs without BPOA structure. The diameter of both the outside circular Ohmic contact and the inside Schottky contact is 800 μm and 400 μm, respectively. The distances between anode and cathode are varied from 12 to 22 μm. As increasing the inter-electrode distance, the forward current at 1.5V decreases from 50 to 30 mA as shown in the inset of Figure 4 (a) while the breakdown voltage continually increases from 700 to 1680 V as shown in Figure 4 (b). Especially, the leakage current for inter-electrode distance remains saturated with very low levels until breakdown occurs. In general, the reverse characteristics of the AlGaN/GaN SBDs are less affected by many defects in the GaN buffer layer, mainly caused by dislocations, because the current of the AlGaN/GaN SBDs flows in lateral direction. As mentioned above, the GaN buffer layers of the SBDs were grown by varying the growth temperature to control the size of nucleation sites which results in extremely high buffer resistance (>10^{12} Ω/sq) [31, 32], which ensures very low buffer leakage current. The devices were also well passivated with depositing the SiO_2/Si_3N_4 double dielectric layer which would greatly decreases the surface leakage current. Therefore the breakdown can be determined by the field strength in the buffer layer between two electrodes and hence the breakdown voltage is almost linearly dependent on the inter-electrode distance. The device with inter-electrode distance of 20 μm exhibits a specific on resistance (R_{ON}) of about 2.0 mΩ ·cm^2, estimated from the I-V curves in the figure, and breakdown voltage of 1500V. The resulting figure-of-merit ($V_{BR}{}^2/R_{ON}$) of the device is more than 1.1 GW·cm^{-2}, which is one of the highest values ever reported for the lateral type GaN-based SBDs [33].

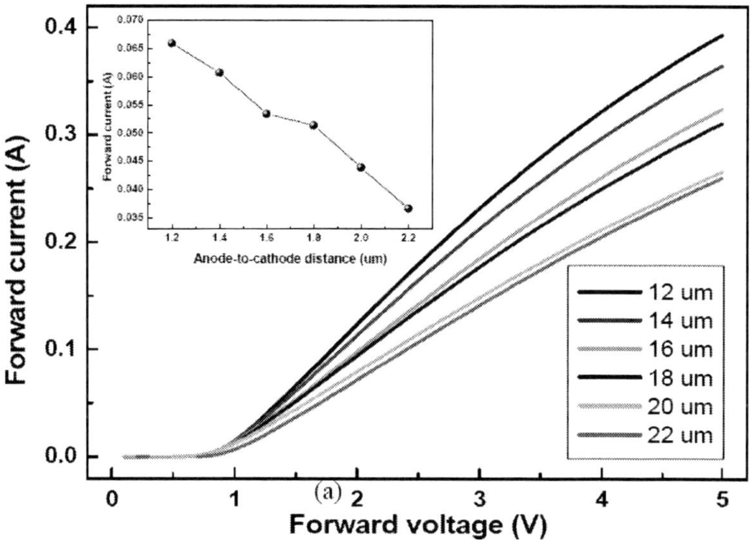

(a)

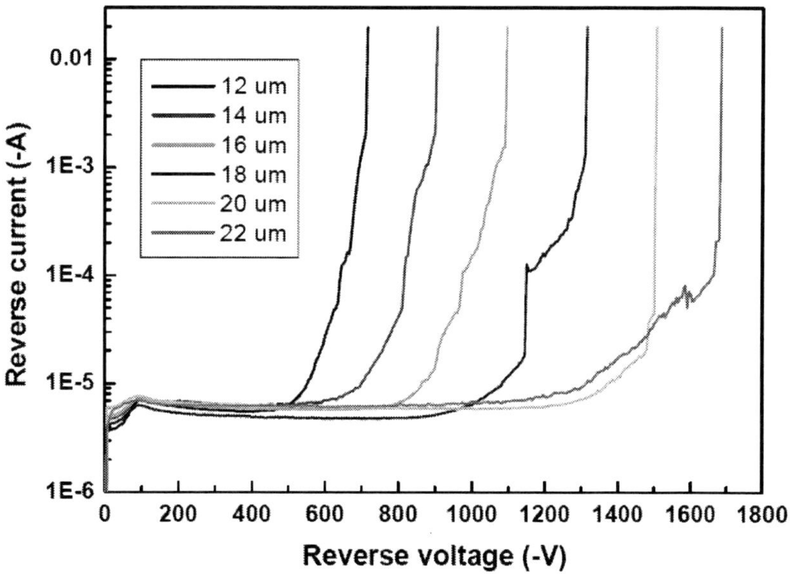

(b)

Figure 4. Forward (a) and reverse (b) I-V characteristics of fabricated AlGaN/GaN-based SBDs with small size according to anode and cathode distance.

5. Characteristics of AlGaN/GaN-Based SBD with Different N_2O Flow Rates

For evaluation, SiO_2 films were deposited on 4 inch Si wafers by varying the flow rate of the N_2O gas flow and the measured film properties are summarized in Table II. Reflective indices of the deposited SiO_2 films are almost same about 1.43, regardless of N_2O flow rates. When N_2O flow rates increases from 120 to 200 SCCM, however, the breakdown field of films at current density of 1mA/cm^2 decreases from 8.2 to 6.8 MV/cm^2 and leakage current at breakdown field of 5 MV/cm increases from 8.8 to 43.5 μA/cm^2, respectively. This explains that the films deposited with higher N_2O flow rates exhibits poor dielectric characteristics related to the defects such as pinholes and weak spots.

Table 2. SiO_2 properties as a function of the N_2O gas flow rates

N_2O flow rate (SCCM)	n	Breakdown field (MV/cm @1mA/cm^2)	Leakage current (μA/cm^2 @5MV/cm)	BOE etch rate(Å/s)	Uniformity (%)
120	1.43	8.2	8.8	48.9	±6.6
160	1.43	7.7	11.3	48.1	±6.4
180	1.44	7.3	19.6	47.4	±6.2
200	1.44	6.8	43.5	45.5	±5.4

Figure 5 shows the I-V characteristics of the small-area AlGaN/GaN SBDs. The devices were passivated with SiO_2 layer which was deposited by varying the N_2O flow rate from 120 to 200 sccm. The diameter of the circular Schottky contact is 400 μm and the distance between anode and cathode is 20 μm. The forward turn-on voltages of the SBDs are all the same about 0.7 V, regardless of N_2O flow rates. The experimental I-V characteristics were analyzed using the following equations [34]:

$$I = I_o[\exp(\frac{qV}{nkT}) - 1], \qquad I_o = AA^*T^2\exp(\frac{-q\Phi_{BO}}{kT}) \tag{1}$$

where, I_o is the reverse saturation current, q is the electric charge, n is the ideality factor, T is the absolute temperature, k is the Boltzmann constant, Φ_{BO} is the Schottky barrier height, A is the contact area, and A^* is the effective Richardson constant. The extracted ideality factor of the SBD with N_2O flow rate of 120 sccm were about 1.3. The typical barrier heights of SBD at room temperature, extracted from the intercept of the y-axis of the Figure 5 (a), were 0.82 eV. In contrast to the forward characteristics, however, the N_2O flow rate drastically influences the reverse characteristic of the SBD as shown in Figure 5 (b). When the N_2O flow rates was 120 sccm, the lowest reverse leakage current of - 0.2 nA at a reverse voltage of -50V was measured which is remarkably lower than the corresponding value of -100 μA for the non-passivated reference-SBD (Ref-SBD), as shown in Figure 5 (b). This explains that the high quality of the SiO_2 layer, prepared at the N_2O flow rate of 120 sccm, well passivates the surface of the SBD. When the N_2O flow rates increased to 200 sccm, however, the reverse characteristic became even worse.

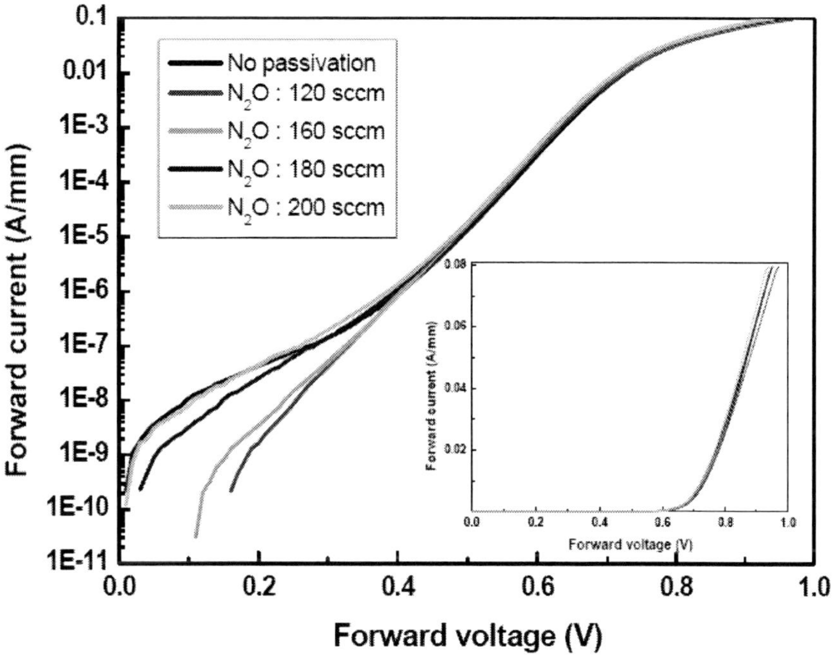

(a)

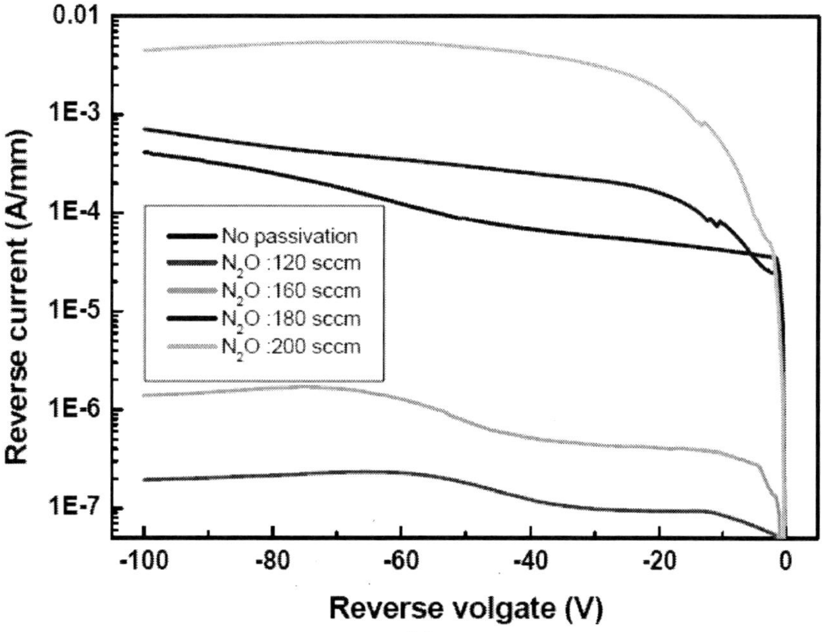

(b)

Figure 5. The semilogarithmic scale I-V forward (a) and reverse (b) characteristics of fabricated circular type AlGaN/GaN-based SBDs with different N_2O flow rates. The insets in Figure 5 (a) shown the leaner scale forward I-V characteristics of SBDs.

The leakage current of the SBD measured at a reverse voltage of -50V was -4 mA, one order higher in magnitude compared to the value of the Ref-SBD. It is believed that this unexpected degradation in electrical characteristic is because high plasma density generated under high N_2O flow rate may severely cause surface damage which is responsible for increasing surface defects (such as unstable formation of surface oxide) or surface traps [35]-[37].

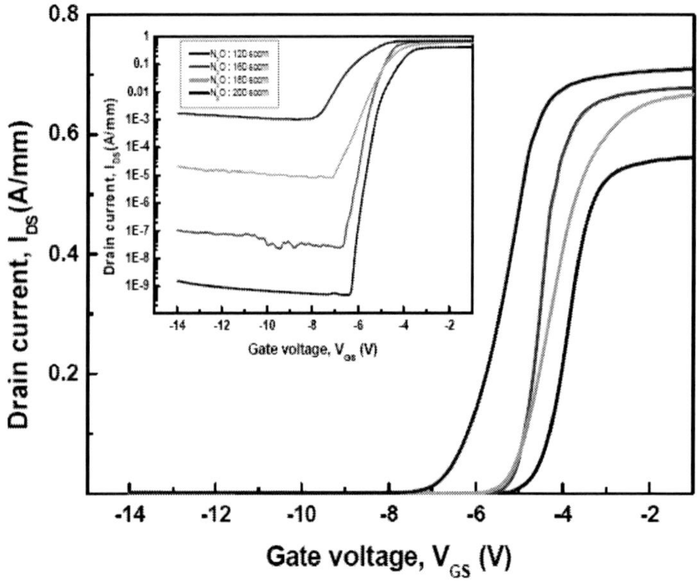

(a)

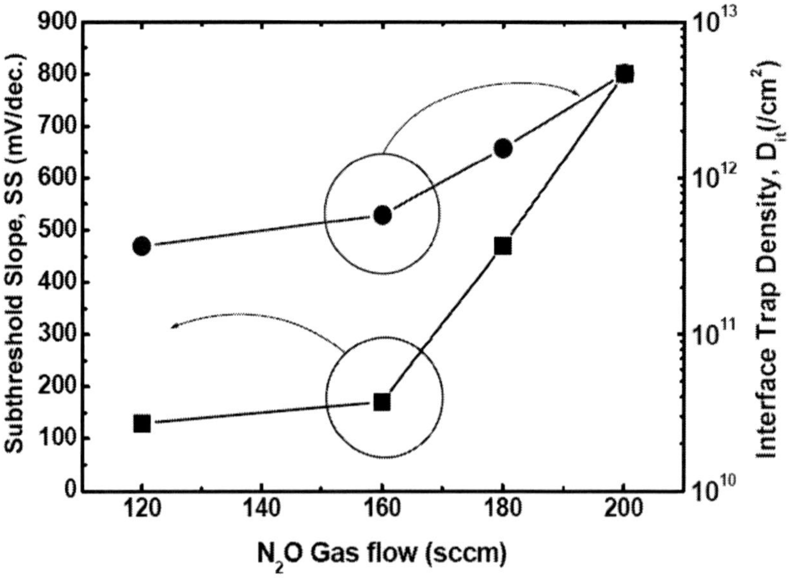

(b)

Figure 6. (a) The transfer characteristics *(I_{DS} - V_{GS})* and (b) the substreshold slope and interface trap density of the MOSHFETs with different N_2O flow rates.

To evaluate the trap density on AlGaN surface, small-sized normally-on MOSHFETs with gate oxide thickness of 25 nm were fabricated. The gate oxide was deposited on the AlGaN surface with the same condition as the AlGaN/GaN SBDs. The gate length, the gate width, and the gate-to-drain distance of the device were 2, 140, and 20 μm, respectively. The transfer characteristics (I_{DS} - V_{GS}) of the MOSHFETs at a fixed V_{DS} of 8 V are shown in Figure 6.

The 2DEG concentration in AlGaN/GaN heterostructure is mainly due to the polarization charge, which can be increased or decreased according to the charge state at the AlGaN surface. If the surface state is acceptor-like, then the 2DEG concentration decreases to meet the charge neutrality (opposite if the surface state is donor-like). With surface passivation, the acceptor-like surface state becomes neutralized and then the 2DEG concentration can be increased. As the N_2O flow rate increases from 120 to 200 SCCM, the threshold voltage is negatively increased from -4.5 to -6.4 V. This is because positive surface charge created at AlGaN surface (such as Si-rich SiO_2, Al-rich Al_xO_y and/or Ga-rich Ga_xO_y) increases the 2DEG density to shift the threshold voltage to more negative [38]-[40], as the flow rate of N_2O increases, which would also increases the maximum drain current from 560 to 710 mA/mm as shown in Figure 6 (a). The figure also shows that the on/off current ratio (I_{on}/I_{off}) and the subthreshold slope (SS) of the device significantly degraded as the N_2O flow rate increases. It is well known that SS and I_{on}/I_{off} are closely related to the interface trap density (D_{it}) between gate dielectric and semiconductor [41]. The interface trap density can be estimated by the following equation,

$$D_{it} = \left(\frac{\log(e)S}{\frac{kT}{q}} - 1\right) \cdot \frac{C_{ox}}{q} \tag{2}$$

assuming the traps are isotropic, where C_{ox} is the capacitance of gate dielectric per unit area. The estimated D_{it} for the MOSHFET with N_2O flow rates of 120 SCCM was 7.1×10^{11} cm^{-2} which is lower than that (5.3×10^{12} cm^{-2}) for the MOSHFET with N_2O flow rates of 200 sccm.

6. Characteristics of the Large-Area AlGaN/GaN-SBDs Measured by Varying Temperature

Based on the optimized SiO_2 deposition process, the multi-finger patterned large-area SBD with BPOA structure (GaN-SBD) was fabricated. The size, the length of the anode perimeter, and the anode-to-cathode distance of GaN-SBD device are 3000 μm × 3000 μm, 270 mm, and 20 μm, respectively. The width of Schottky contact, Ohmic contact, and oxide field plate for GaN-SBD is 10, 10, and 3 μm, respectively. The GaN-SBDs have been packaged using a standard TO-220 package. Figure 7 (a) shows forward I-V characteristics of the packaged GaN-SBD, measured by varying temperature. For comparison, the those of a commercial SiC Schottky diode (SiC-SBD) and a hyperfast Si diode (Si-diode) are also shown in Figure 7 (b) and (c) [42], [43]. The turn-on voltage and forward current of the GaN-SBD at 1.5V is about 0.8 V and 4.5 A at room temperature, respectively. The corresponding values of the SiC-SBD and the Si diode are 0.9 V and 6 A and 0.7 V and 6A, respectively.

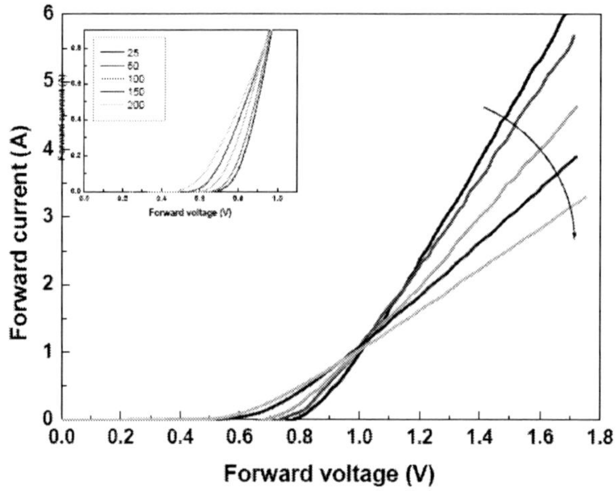

(a)

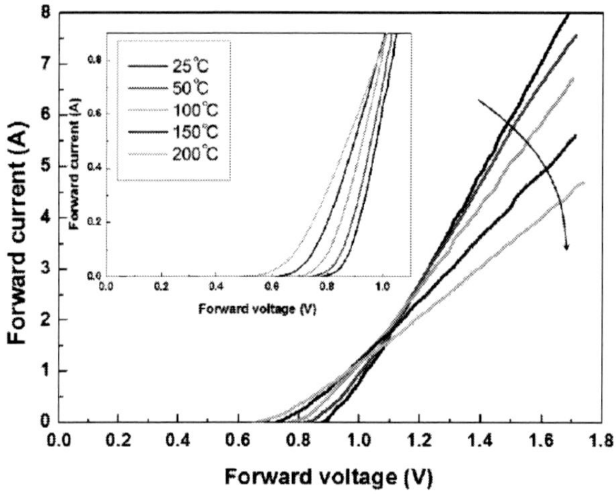

(b)

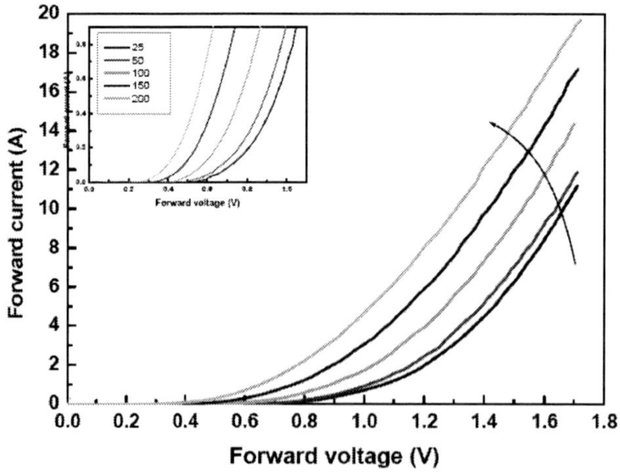

(c)

Figure 7. Forward I-V characteristics for GaN-SBD (a), SiC-SBD (b), and Si-diode (c) by varying temperature.

From the I-V curve, the estimated specific on-resistance (R_{ON}) of for the GaN-SBD with dimension of 3000 × 3000 μm^2 is 12 $m\Omega \cdot cm^2$. As increasing temperature, the turn-on voltage of the GaN-SBD continually decreases, as shown in set of Figure 7 (a), due to the reduction of the band gap energy which is inversely proportional to the temperature. However, the slope of the current decreases as temperature increases in spite of the decreased turn-on voltage. This results from the fact that the electron mobility for a majority carrier device decreases as temperature increases, which was also observed in SiC-SBD [44], [45]. It is believed that the increase in 2DEG density is not significant in this temperature range. The forward current of the GaN-SBD and the SiC-SBD at 1.5V is about 2.5 A and 3.5 A at 200°C, which is 42~45% lower in magnitude compared to the value at room temperature. On the other hand, corresponding values of Si-diode increased to 13 A, almost two times higher in magnitude due to thermal carrier generation compared to the value at room temperature.

Figure 8 shows a characteristic for typical forward voltage versus temperature at a sensing current of 1 mA. The sensing current must be small enough so that the heat generated at this current becomes negligible. From the figure, we can obtain the so called "K factor," which is the thermal calibration factor equal to the reciprocal of the voltage-temperature coefficient for the device under test at a fixed sense current [46]. The factor is 1.1, 1.3, and 1.8 mV/℃ for GaN-SBD, SiC-SBD, and Si-diode. The measured parameter for the GaN-SBD is superior to those obtained from the Si-diode.

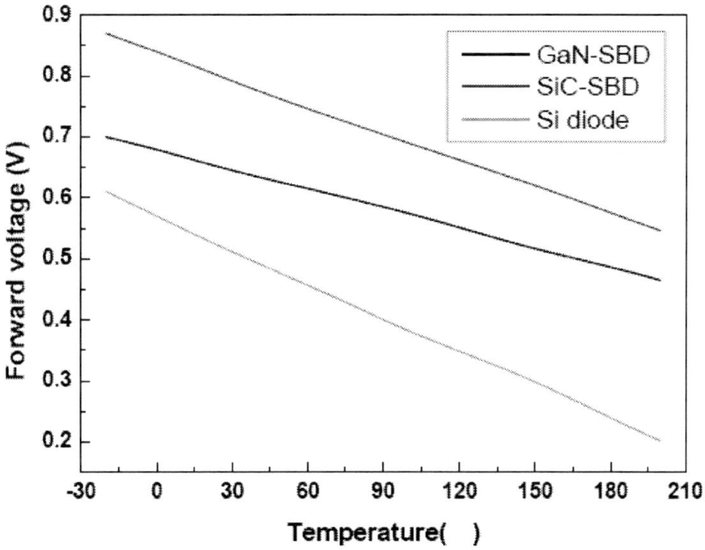

Figure 8. K factor of GaN-SBD, SiC-SBD, and Si-diode at 1 mA current.

Figure 9 shows the reverse characteristics of GaN-SBD, SiC-SBD, and Si diode at room temperature and 175 ℃. The leakage current of the GaN-SBD at a reverse voltage of - 600 V was 6 and 9.8 μA at room temperature and 175 ℃ as shown in inset of Figure 9 (a), respectively. The corresponding values of the SiC-SBD and the Si-diode are (0.1 and 3 μA) and (0.09 and 17 μA) as shown in inset of Figure 9 (b) and (c), respectively. Though the reverse leakage current of the GaN-SBD at room temperature is higher than that of the Si-diode, but the leakage current of the GaN-SBD at high temperature is much smaller than that of the Si-diode.

The soft breakdown voltage of the GaN-SBD at leakage current of 250 μA was 747 and 767 V at room temperature and 175 ℃, respectively. The corresponding values of the SiC-SBD and the Si-diode are (970 and 1060 V) and (680 and 717 V), respectively. In our previous work [47], we reported that when temperature increases, the reverse breakdown voltage of the GaN based SBD decreases due to the defect-assisted breakdown. The higher breakdown voltage at high temperature obtained in this work explains that the optimized two-step growth mentioned above improves the crystal quality of semi-insulating GaN buffer layer.

The figure of merit V_{BR}^2/R_{ON} of the GaN-SBD is 47 $MW \cdot cm^{-2}$, which is one of the highest value ever reported for the lateral type GaN- based SBD considering the forward current is as high as 4 A [48].

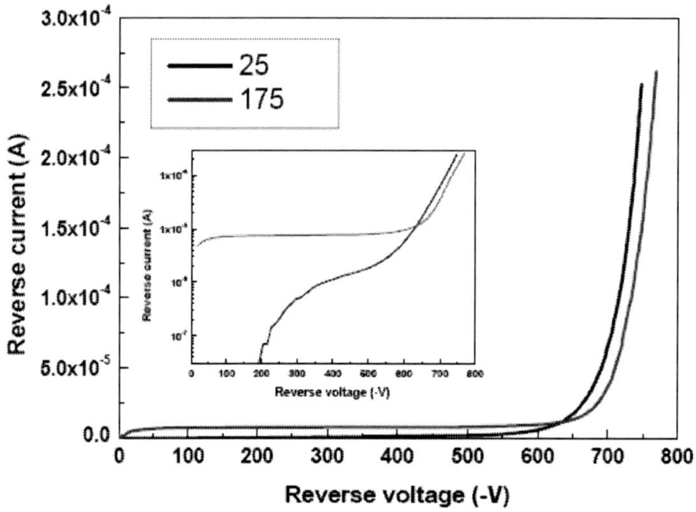

(a)

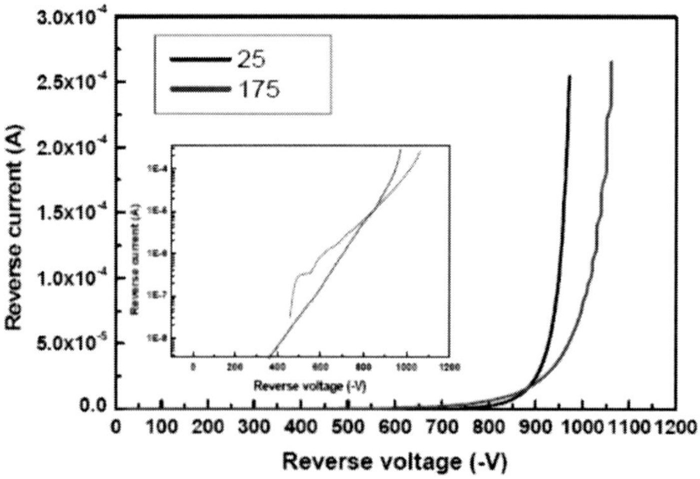

(b)

Figure 9. Continued on next page.

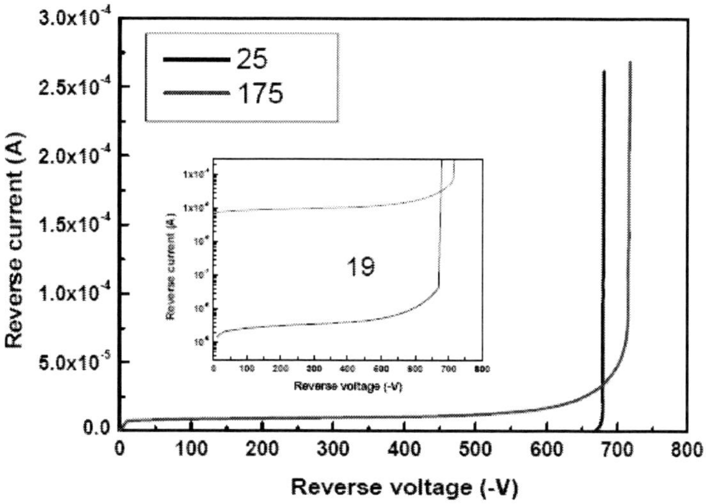

(c)

Figure 9. Reverse I-V characteristics for GaN-SBD (a), SiC-SBD (b), and Si-diode (c) at room temperature and 175 ℃. The inset shows the log scale I-V characteristics.

Figure 10 shows the C-V measurement for diodes. The capacitances measured at reverse voltage of 100V are 160, 76, and 20 pF for GaN-, SiC-, and Si diode, respectively. Compared to that of SiC-diode, a higher capacitance of GaN-diode is due to large chip size (9 mm^2). The capacitance of the GaN-diode was reduced to 60 pF with smaller chip size of 4.5 mm^2 (not shown).

A typical reverse recovery wave form of the GaN-SBD is shown in Figure 11 and compared to that of the SiC-SBD and Si diode [49]. The switching conditions were set with the forward current of I_F = 4 A, the supply voltage of 400 V, and the di/dt of 200 A/μs at room temperature and 175 ℃.

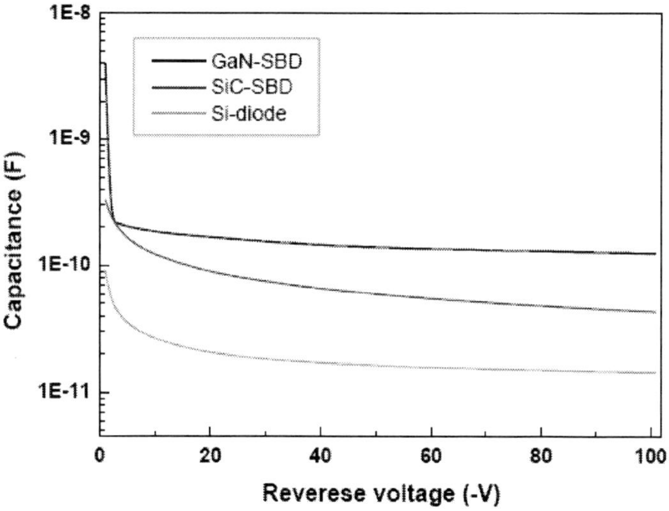

Figure 10. C-V measured data for GaN-diode, SiC-diode, and Si diode at reverse voltage of 100V.

Table 3. Reverse recovery parameter measurement data for the GaN-SBD, SiC-SBD, and the Si-diode at room temperature and 175 ℃

Parameter	Temp.(℃)	I_{rr} (A)	t_{rr} (ns)	Q_{rr} (nC)	Test condition
GaN-SBD	25	1.88	19.4	19.7	V_R= 400 V di/dt =200 A/μs I_F = 4 A
	175	1.80	19.6	19.9	
SiC-SBD	25	1.08	11.8	7.4	
	175	1.12	13.2	8.2	
Si-diode	25	3.58	26.6	51.7	
	175	8.14	116	403	

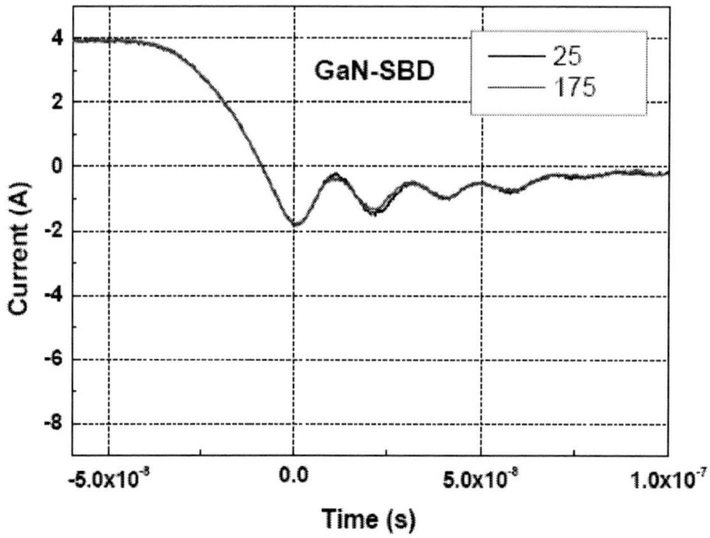

(a)

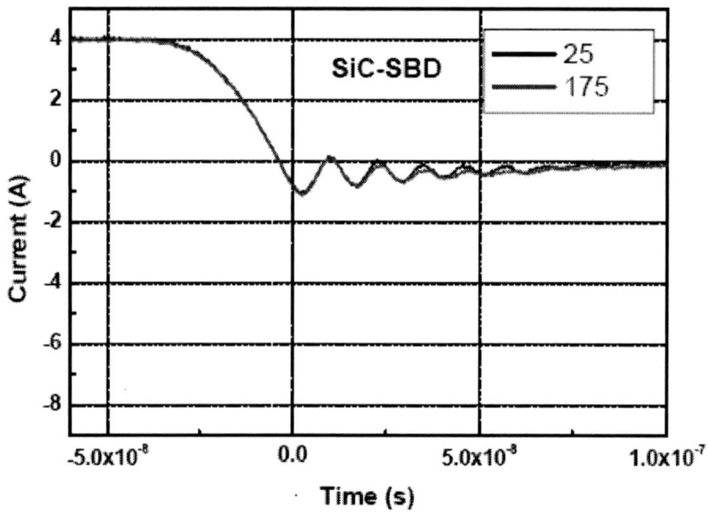

(b)

Figure 11. Continued on next page.

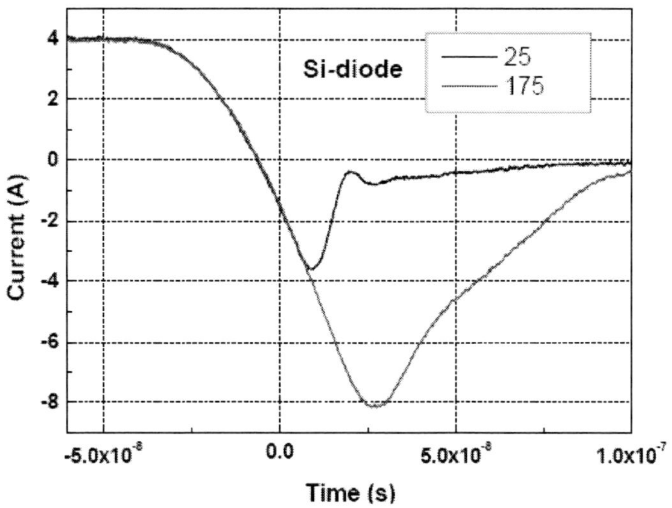

(c)

Figure 11. Reverse-recovery waveforms comparing GaN-SBD (a), SiC-SBD (b), and Si diode (c) at room temperature and 175 ℃.

The measured parameters, such as recovery current (I_{rr}), time (t_{rr}), and charge (Q_{rr}), are summarized in table III. In spite of larger dimension of 9 mm^2, all the measured data for the GaN-SBD are superior to those obtained from the commercial Si diode. The t_{rr} of the GaN-SBD is 19.4 and 19.6 ns at room temperature and 175℃, much lower than the corresponding value of 26.6 and 116 ns for the Si diode due to the high electron velocity of the originated material, which is also comparable to those obtained for small-area GaN-based SBDs reported elsewhere [22], [48].

The temperature variations of GaN-SBD for the reverse recovery characteristics are much better than those of Si-diode. The Q_{rr} of Si-diode increases from 51.7 nC at 25℃ to 403 nC at 175 ℃, while that of GaN-SBD increases from 19.7 nC at 25℃ to 19.9 nC at 175 ℃. The recovery characteristics of GaN diode seem to be worse than those of SiC diode. This is believed to be due to large chip size of the GaN SBD. When the chip size is reduced from 9 to 4.5 mm^2, the recovery characteristics of GaN decreased to 1.02 A of I_{rr}, 11.2 ns of t_{rr}, and 6.9 nC of Q_{rr}, respectively, which is comparable to those obtained from the SiC-SBD.

Conclusion

Multi-finger lateral type AlGaN/GaN Schottky barrier diode (SBD) with bonding pad over active (BPOA) structure was fabricated and exhibited excellent device performances, such as forward current of 4.5 A at 1.5 V, leakage current of 6 μA at - 600V, reverse-recovery time (t_{rr}) of 19.4 ns, breakdown voltage of 747 V, and low specific on-resistance (R_{on}) of 12 $m\Omega \cdot cm^2$, resulting in the figure of merit (V_{BR}^2/R_{on}) as high as 47 MW/cm^2, which demonstrates that the AlGaN/GaN SBD with BPOA structure has a great potential application to the high power electronics.

References

[1] Y.-F. Wu, D. Kapolnek, J. P. Ibbetson, P. Parikh, B. P. Keller, and U. K. Mishra, "Very-high power density AlGaN/GaN HEMT's," *IEEE Electron Device Lett.*, vol. 48, pp.586 - 590, 2001.

[2] S. Nakamura, M. Senoh, S. Nagahama, N. Iwasa, T. Yamada, T. Mat-sushita, Y. Sugimoto, and H. Kiyoku, "High-Power, Long-Lifetime InGaN Multi-Quantum-Well-Structure Laser Diodes," *Jpn. J. Appl. Phys.*, vol. 36, pp. L1059-L1061, 1997.

[3] Y. Uemoto, M. Hikita, H. Ueno, H. Matsuo, H. Ishida, M. Yanagihara, T. Ueda, T. Tanaka and D. Ueda, "Gate injection transistor (GIT)-A normally-off AlGaN/GaN power transistor using conductivity modulation," *IEEE Trans. Electron Devices*, vol. 54, no.12, pp.3393-3399, 2007.

[4] J. H. Lee, D. Y. Lee, B. W. Oh, and J. H. Lee, "Comparison of InGaN-based LEDs grown on conventional sapphire and cone-shape-patterned sapphire substrate," *IEEE Trans. Electron Devices*, vol. 57, no. 1, pp. 157-163, 2010.

[5] O. Ambacher , J. Smart , J. R. Shealy , N. G. Weimann , K. Chu , M. Murphy , W. J. Schaff , L. F. Eastman , R. Dimitrov , L. Wittmer , M. Stutzmann , W. Rieger and J. Hilsenbeck, "Two-dimensional electron gases induced by spontaneous and piezoelectric polarization charges in N and Ga-face AlGaN/GaN heterostructures," *J. Appl. Phys.*, vol. 85, no. 6, pp.3222 - 3233 , 1999.

[6] S. Iwakami, O. Machia, I. Yoshimichi, B. Ryohei, Y. Masataka, E. Toshihiro, K. Nobuo, G. Hirokazu, and I. Akio, "Evaluation of AlGaN/GaN Heterostructure Field-Effect Transistors on Si Substrate in Power Factor Correction Circuit," *Jap. J. Appl. Phys.*, Vol. 46, No. 29, pp. L721-L723, 2007.

[7] F. Semond, P. Lorenzini, N. Grandjean, and J. Massies, "High electron mobility AlGaN/GaN heterostructures grown on Si(111) by molecular beam epitaxy," *Appl. Phys. Lett.*, vol. 82, pp.335-337, 2001.

[8] Kai Cheng, M. Leys, S. Degroote, B. Van Daele, S. Boeykens, J. Derluyn, M. Germain, G. Van Tendeloo, J. Engelen, and G. Borghs, "Flat GaN Epitaxial Layers Grown on Si(111) by Metalorganic Vapor Phase Epitaxy Using Step-Graded AlGaN Intermediate Layers", *J. Electron. Mater.*, Vol. 35, No. 4, 2006.

[9] J. H. Lee, J. T. Oh, S. B. Choi, Y. C. Kim, H. I. Cho, and J. H. Lee, "Enhancement of InGaN-based vertical LED with concavely patterned surface using patterned sapphire substrate," *IEEE Photon. Technol. Lett.*,vol. 20, no. 5, pp.345-347, 2008.

[10] J. H. Lee and K. S. Kim, "Gallium nitride based semiconductor devices and methods of manufacturing the same," U.S. Patent 20120061727A1, Mar. 15, 2012.

[11] Y. Ohba and A. Hatano, "A study on strong memory effects for Mg doping in GaN metalorganic chemical vapor deposition," *J. Cryst. Growth*, vol. 145, pp. 214, 1994.

[12] S. Heikman, S. Keller, S. P. DenBaars, and U. K. Mishra, "Growth of Fe doped semi-insulating GaN by metalorganic chemical vapor deposition," *Appl. Phys. Lett.*, vol. 81, no. 3, pp.439, 2002.

[13] S. C. Binari *et al.*, "Trapping effects and microwave power performance in AlGaN/GaN HEMTs," *IEEE Trans. Electron Devices*, vol. 48, pp.465, 2001.

[14] H. Hasegawa and S. Oyama, "Mechanism of anomalous current transport in n-type GaN Schottky contacts," *J. Vac. Sci. Technol. B,* vol. 20, no. 4, pp. 1647–1655, Jul. 2002.

[15] J. Derluyn, S. Boeykens, K. Cheng, R. Vandersmissen, J. Das, W. Ruythooren, S. Degroote, M. R. Leys, M. Germain, and G. Borghs, "Improvement of AlGaN/GaN high electron mobility transistor structures by in situ deposition of a Si_3N_4 surface layer," *J. Appl. Phys.*, vol. 98, no. 5, pp. 054 501-1–054 501-5, Sep. 2005.

[16] N. Onojima, M. Higashiwaki, J. Suda, T. Kimoto, T. Mimura, and T. Matsui, "Reduction in potential barrier height of AlGaN/GaN heterostructures by SiN passivation," *J. Appl. Phys.*, vol. 101, no. 4,pp. 043 703-1–043 703-6, Feb. 2007.

[17] C.Wang, N. Maeda, M. Hiroki, H. Yokoyama, N. Watanabe, T. Makimoto, T. Enoki, and T. Kobayashi, "Mechanism of superior suppression effect on gate current leakage in ultrathin Al_2O_3/Si_3N_4 bilayer-based AlGaN/GaN insulated gate heterostructure field-effect transistors," *Jpn. J. Appl. Phys.*, vol. 45, no. 1A, pp. 40–42, Jan. 2006.

[18] G. T. Dang, A. P. Zhang, F. Ren, X. A. Cao, S. J. Pearton, H. Cho, J. Han, J. I. Chyi, C.-M. Lee, C.-C. Chuo, S. N. George Chu,and R. G. Wilson, "High voltage GaN Schottky rectifiers," *IEEE Trans. on Electron Devices*, vol. 47, no. 4, pp. 692-696, 2000.

[19] P. Zhang, J. W. Johnson, and F. Ren, "Lateral $Al_xGa_{1-x}N$ power rectifiers with 9.7kV reverse breakdown voltage," *Appl.Phys. Lett.*, Vol. 78, No. 6, pp. 823-825, 2001.

[20] M. Miyoshi, Y. Kuraoka, K. Asai, T. Shibata, M. Tanaka, and T. Egawa, "Improved reverse blocking characteristics in AlGaN/GaN Schottky barrier diodes based on AlN template," *Electron. Lett.*, vol. 43, no. 17, pp. 953–954, 2007.

[21] J. H. Lee, J. H. Jeong, and J.H. Lee, "Enhanced Electrical Characteristics of AlGaN-Based SBD With In Situ Deposited Silicon Carbon Nitride Cap Layer," *IEEE Electron Device Lett.*, vol. 33, pp. 492-493, 2012.

[22] J. W. Johnson, A. P. Zhang, W. B. Luo, F. Ren, S. J. Pearton, S. S. Park, Y. J. Park, and J. I. Chyi, " Breakdown voltage and reverse recovery characteristics of free-standing GaN Schottky rectifiers", *IEEE Trans. on Electron Devices*, vol. 49, no. 1, pp.32–36, 2002.

[23] S-C. Lee, J-C Her, S-S Kim, M-W Ha, K-S Seo, Y-I Choi and M-K Han, " A new vertical GaN Schottky barrier diode with floating metal ring for high breakdown voltage", *Proc. ISPSD'04.* pp 319–322, 2004.

[24] D. Shibata, K. Kaibara, T. Murata, Y. Yamada, T. Morita, Y. Anda, M. Ishida, H. Ishida, T. Ueda, T. Tanaka, and D. Ueda, "GaN-based Multi-Junction Diode with Low Reverse Leakage Current Using P-type Barrier Controlling Layer," in *IEDM Tech. Dig.*, pp. 26.2.1 - 26.2.4, 2011.

[25] Y. Liu, S. Irving, and T. Luk, "Thermosonic wire bonding process simulation and bond pad over active stress analysis," *IEEE Trans. Electron. Packag. Manuf.,* vol. 31, no. 1, pp. 61–71, 2008.

[26] M. Sumiya et al., "Systematic analysis and control of low-temperature GaN buffer layers on sapphire substrates," *J. Appl. Phys.*, vol. 93, pp.1311, 2003.

[27] L. Sugiura, K. Itaya, J. Nishio, H. Fujimoto, and Y. Kokubun, "Effects of thermal treatment of low-temperature GaN buffer layers on the quality of subsequent GaN layers," *J. Appl. Phys.*, vol. 82, pp.4877 , 1997.

[28] Geng-Yen Lee, Hsueh-Hsing Liu, and Jen-Inn Chyi, "High-Performance AlGaN/GaN Schottky Diodes With an AlGaN/AlN Buffer Layer," *IEEE Electron Device Lett.*, vol. 32, no. 11, pp. 1519, 2011.

[29] J. Elsner et al., "Deep acceptors trapped at threading-edge dislocations in GaN," *Phys. Rev. B*, Condens. Matter, vol. 58, no. 19, pp. 12 571, 1998.

[30] S. M. Lee, M. A. Belkhir, X. Y. Zha, and Y. H. Lee, "Electronic structures of GaN edge dislocations," *Phys. Rev. B*, vol. 61, pp. 16033, 2000.

[31] J. H. Lee, J. H. Jeong, and Jung-Hee Lee, "Normally-off GaN power MOSFET grown on sapphire substrate with highly resistive undoped buffer layer," *IEEE Electron Device Lett.*, vol. 33, no. 10, pp. 1429-1431, 2012.

[32] P. B. Klein, S. C. Binari, K. lkossi, A. E. Wichenden, D. D. Koleske, and R. L. Henry, " Current collapse and the role o carbon in AlGaN/GaN high electron mobility transistors grown by metalorganic vapor-phase epitacy, " *Appl. Phys. Lett.*, vol. 79, no. 21.
pp. 3527-3529, 2001.

[33] Geng-Yen Lee, Hsueh-Hsing Liu, and Jen-Inn Chyi, "High-Performance AlGaN/GaN Schottky Diodes With an AlGaN/AlN Buffer Layer," *IEEE Electron Device Lett.*, vol. 32, no. 11, pp. 1519-1521, 2011.

[34] S. K. Cheung and N.W. Cheung, "Extraction of Schottky diode parameters from forward current–voltage characteristics," *Appl. Phys. Lett.*, vol. 49, no. 2, pp. 85–87, Jul. 1986.

[35] P. D. Ye, B. Yang, K. K. Ng, J. Bude, G. D. Wilk, S. Halder, and J. C. M. Hwang, "GaN metal oxide-semiconductor high electronmobility transistor with atomic layer deposited Al2O3 as gate dielectric," *Appl. Phys. Lett.*, vol. 86, no. 6, pp. 063 501-1–063 501-3, Feb. 2005.

[36] H. -M. Kim, C. -Huh, and S.-J Park, "Reduction in leakage current of InGaN-based light-emitting diodes by N20 plasma passivation," *Proc of SPIE*, vol. 4776, pp. 247-254, 2002.

[37] H. C. Chiu, J. H. Wu, C. W. Yang, and F. H. Huang, "Low-Frequency Noise in Enhancement-Mode GaN MOS-HEMTs by Using Stacked Al2O3/Ga2O3/Gd2O3 gate dielectric, " *IEEE Electron Device Lett.*, vol.33, pp. 958-960, 2012.

[38] Y. Cai, Y. Zhou, K. M. Lau, and K. J. Chen, "Control of threshold voltage of AlGaN/GaN HEMTs by fluoride-based plasma treatment: From depletionmode to enhancement mode," *IEEE Trans. Electron Devices*, vol. 53, no. 9, pp. 2207–2215, Sep. 2006.

[39] M. A. Khan, X. Hu, G. Simin, J. Yang, R. Gaska, and M. S. Shur, "AlGaN/GaN Metal-oxide-semiconductor heterostructure field effect transistors on SiC substrates," *Appl. Phys. Lett.*, vol. 77, pp. 1339–1341, Aug. 2000.

[40] L. M. Lin, Y. Luo, P. T. Lai, and K. M. Lau, "Influence of oxidation and annealing temperatures on quality of Ga_2O_3 film grown on GaN," *Thin Solid Films*, vol. 515, no. 4, pp. 2111–2115, Dec. 2006.

[41] Neugroschel, C. T. Sah, K. M. Han, M. S. Carroll, T. Nishida, J. T. Kavalieros, and Y. Lu, "Direct-current measurement of oxide and interface traps on oxidized silicon," *IEEE Trans. Electron Devices*, vol. 42, pp. 1657–1662, 1995.

[42] Commercial SiC diode (Z-Rec Schottky)_Part number : C3D0660A, Manufacturer: Cree Co., Ltd, and voltage and current ratings : 600V/6A@1.5V.

[43] Commercial Si diode (Hyperfast II diode)_Part number: FFP08H60, Manufacturer: Fairchild semiconductor Co., Ltd, and voltage and current ratings : 600V/8A@1.8V.

[44] R. Singh, D. C. Capell, A. R. Hefner, J. Lai, and J. W. Palmour, "High- Power 4H-SiC JBS rectifiers," *IEEE Trans. Electron Devices*, vol. 49, no. 11, pp. 2054–2063, Nov. 2002.

[45] Bahat-Treidel, E. Hilt, O. Zhytnytska, R. Wentzel, A. Meliani, C. Wurfl, and J. Trankle, G., "Fast-Switching GaN-Based Lateral Power Schottky Barrier Diodes With Low Onset Voltage and Strong Reverse Blocking," *IEEE Electron Device Lett.*,, vol.33, no.3, pp.357,359, March 2012.

[46] L. Kim and M.W. Shin, "Thermal resistance measurement of LED package with multichips," *IEEE Trans. Compon. Packag., Manuf. Technol.*, vol. 30, no. 4, pp. 632–636, Dec. 2007.

[47] W. T. Lim, J. H. Jeong, J.H. Lee, S. B. Hur, J. K. Ryu, K. S. Kim, T. H. Kim, S. Y. Song, J. I. Yang, and S. J. Pearton, "Temperature dependence of current-voltage characteristics of Ni-AlGaN/GaN Schottky diodes," *Appl. Phys. Lett.*, vol. 97, pp. 242103, Dec. 2010.

[48] Geng-Yen Lee, Hsueh-Hsing Liu, and Jen-Inn Chyi, "High-Performance AlGaN/GaN Schottky Diodes With an AlGaN/AlN Buffer Layer," *IEEE Electron Device Lett.*, vol. 32, no. 11, pp. 1519-1521, 2011.

[49] T. McNutt, A. Hefner, A. Mantooth, J. Duliere, D. Berning, R. Singh, "Silicon-Carbide PiN and Merged PiN- Schottky Power Diode Models Implemented in the Saber Circuit Simulator," *Trans. on Power Electronics*, vol. 19, no. 3, May 2004.

In: Gallium Nitride
Editor: Kiera Olivia Peak
ISBN: 978-1-63321-387-6

Chapter 9

IMPROVEMENT IN HEAT DISSIPATION CAPABILITY OF GaN-BASED LIGHT-EMITTING DIODES

Ray-Hua Horng[1,2,*]***, Sin-Liang Ou***[3] ***and Dong-Sing Wuu***[3,4]
[1]Graduate Institute of Precision Engineering,
National Chung Hsing University, Taichung, Taiwan
[2]Department of Photonics & Advanced Optoelectronic Technology Center,
National Cheng Kung University, Tainan, Taiwan
[3]Department of Materials Science and Engineering,
National Chung Hsing University, Taichung, Taiwan
[4]Department of Electrical Engineering, Da-Yeh University, Changhua, Taiwan

Abstract

Recently, high brightness GaN-based light emitting diodes (LEDs) have attracted a lot of interest owing to their excellent device characteristics. The GaN-based LEDs are widely used for a variety of applications such as general lighting, liquid crystal display and traffic signal. In order to achieve high performance of these devices, it is urgently required to enhance the optoelectronic properties of LEDs. It is well-known that the poor heat dissipation of sapphire substrate is the main limitation to fabricate the high-efficiency LEDs. For a conventional LED, the heat is easily accumulated in the device, resulting in the performance deterioration. In this chapter, the improvement heat dissipation of GaN-based LEDs is proposed and discussed.

In this chapter, we have proposed several techniques including the employment of cup-shaped copper-diamond heat spreader, the combination of beveled sapphire and cup-shaped copper sheet, and the addition of diamond like carbon layer into LED structure. To develop these techniques, many methods consisting of copper electroplating, chemical mechanical polishing, metal mirror coating, wafer bonding, laser lift-off, and so on. After fabricating the LEDs by using these techniques, the surface temperature, thermal resistance, and light output power of LED devices are investigated in detail. With the aids of above-mentioned techniques, the results reveal that not only the heat dissipation but also the light extraction of LEDs can be improved efficiently.

* E-mail address: huahorng@dragon.nchu.edu.tw; Address: Graduate Institute of Precision Engineering, National Chung Hsing University, No.250, Kuo Kuang Road, Taichung 40227, Taiwan, Department of Photonics & Advanced Optoelectronic Technology Center, National Cheng Kung University, No.1, University Road, Tainan 701, Taiwan (Corresponding author)

Keywords: light-emitting diodes, heat dissipation, cup-shaped heat spreader, beveled sapphire, diamond like carbon layer

1. Introduction

Recently, the high brightness light emitting diodes (LEDs) have been developed quickly [1,2]. Because of their high performances in optical and electrical characteristics, LEDs have been used in various applications such as traffic signals, large full-color displays, solid-state lighting, and backlighting for liquid crystal displays [3]. Moreover, blue and UV emission LEDs can be obtained by exciting some phosphors for implementing white LEDs, which possess a great potential to substitute the mercury vapor lamp in general lighting devices. As a result, improving the optical and electrical properties of LED devices is very important. At present, there exist remarkable enhancements in the light extraction and internal quantum efficiency of GaN-based light emitting diodes for high-efficiency and high-brightness applications. However, most researches were focused on the epitaxial structures and thin GaN LED processes. The development in sapphire-based, high-power InGaN LEDs is mainly impeded by their poor thermal dissipation [4]. Additionally, the sapphire-based devices using LED packaging were usually mounted on the heat sink by silver paste. For these LEDs, the Joule heat created from the device can be dissipated from the sapphire substrate to the silver paste and then to the heat sink subsequently. However, since the sapphire substrate and silver paste both have high levels of thermal resistivity; the sapphire-based LEDs would suffer from the power saturation at high injection currents, which is attributed to the elevated junction temperature [5]. It is well-known that the copper metal is highly helpful to the heat dissipation of optoelectronic devices because of its high thermal conductivity. In our previous study, the direct contact of sapphire substrate with a cap-shaped copper sheet has been proposed to improve the thermal dissipation of sapphire-based InGaN LEDs [6]. On the other hand, the diamond material also possesses high thermal conductivity (2300 W/mK). It indicates that the diamond-based solder pastes also possess high thermal conductivity [7]. Based on previous studies, the issues of thermal dissipation for LEDs were always focused on either the resistance minimization of device or package designs for thermal management [6-9]. In this chapter, the employment of a diamond-added copper (DAC) heat spreader fabricating with a cap-shaped sheet was proposed and applied for the heat dissipation of LED chip. The sapphire-based InGaN LED on the DAC substrate was fabricated by using a composite electroplating process. The thermal conductivity of the DAC substrate was measured. Moreover, the thermal and optical characteristics of the LED fabricated on the DAC substrate were also analyzed in detail.

Several techniques consisting of roughness surface, high reflection mirror and flip-chip package were developed to improve the light extraction and output power of LEDs [10]. Besides, the thin GaN LED process also can be employed to enhance the thermal characteristics of LEDs [11]. Nevertheless, there exist some disadvantages in these techniques, such as the poor heat dissipation in flip-chip package and the difficult fabricating processes in thin GaN LED. In this chapter, we also proposed the method of beveled sapphire substrate to improve the light extraction of LEDs. By using this method, the light generated from quantum well layer can be reflected efficiently when the optimal beveled angle of substrate is obtained. In addition, the sapphire substrate would be mounted on a cup-shaped copper heat spreader to enhance the thermal dissipation of LED devices. The light extraction was also improved by the metal mirror coating on cup-shaped copper heat spreader. Consequently, both light extraction and thermal dissipation of LEDs can be enhanced.

As mentioned above, the flip-chip LED structure depends on a sapphire-free thermal path, indicating that the heat created from the device could be dissipated through the solder to a submount with high thermal conductivity. However, the enhancement in thermal dissipation is not enough for LED devices by using flip-chip LED process due to the poor thermal resistance between the submount and solder. On the contrary, the fabrication of LED chips using the thin-film method consists of two processes. One is the removal of the sapphire via the laser lift-off method, and the other is the transfer of the remaining structure to a substrate with high thermal conductivity by applying the wafer-bonding process. Although the thermal dissipation was improved by using the thin-film process, there was still much room for enhancement. Diamond material possesses a high thermal conductivity exceeding 2000 W/mK. Nevertheless, it is unsuitable for LED fabrication owing to its electrical insulation and expensive price. On the other hand, diamond like carbon (DLC) is another material with good thermal conductivity, which is approximately 475 W/mK. DLC is a member of a carbon-based family, and includes a mixed carbon phase with sp^2 (graphitic) and sp^3 (diamond) bonds [12]. Due to its good thermal and electrical conductivities, DLC becomes a potential material for thermal dissipation of LED devices. Commonly, manufacturers coat DLC material on the metal-core printed circuit board (MCPCB) to improve the heat dissipation of commercial LED devices. However, there is no research investigated the heat dissipation effects of growing DLC film on the mirror of LED device. Therefore, in this chapter, the LED device was also fabricated on a Si substrate by wafer bonding and laser lift-off techniques. This approach can ensure the transfer of the LED device from sapphire to Si substrate enhances the heat dissipation. Moreover, a DLC film prepared by sputtering on the mirror of a LED device can be also employed to improve heat dissipation. This work will investigate the thermal resistances, surface temperatures, and output powers of LEDs embedded with and without a DLC layer.

2. Experimental

2.1. Fabrication of LED with Cup-Shaped Copper-Diamond Heat Spreader

For this fabrication, blue InGaN LEDs (λ_p = 455 nm) were prepared on the sapphire substrates by using low-pressure metalorganic chemical vapor deposition (MOCVD). The LED epitaxial structure consisted of a low-temperature GaN buffer layer (30 nm), an undoped GaN layer (1.5 μm), a Si-doped n-type GaN layer (3 μm), an InGaN/GaN multiple-quantum well stack, and an Mg-doped p-type GaN layer (0.3 μm). After the device process (the LED chip with the size of 1 mm^2 area), the back side of the sapphire was subsequently lapped, polished, and coated with an Ag reflector. Composite electroplating and self-aligned lithography were then used to prepare the sapphire-based InGaN LEDs with the cup-shaped heat spreaders. For this composite electroplating process, the copper plate with 5 N purity was employed as the anode. The electrolyte solution was composed of copper sulfamate (200 g/L, 0.5 M), hydrochloric acid (1 mL/L), sulfuric acid (65 g/L, 0.38 M), sodium dodecyl sulfate (1 mL/L), poly (ammonium acrylate) (1 mL/L), and thallium sulfate (1 g/L). The commercial diamond powder (5 g/L) with a particle diameter of 0.5 μm was mixed in the solution for composite electroplating process. Then the electrolyte solution was circulated using the pump in this process. The amount of diamond powders was able to control through the electroplating current density. The electrolyte was kept at room temperature by a control module. The DAC heat spreader was formed with a thickness of 250 μm and a base area of 9 mm^2. A sapphire-based LED with a cup-shaped heat spreader using the pure Cu metal was also fabricated to compare their characteristics. Eventually, these devices were attached to the MCPCBs (with the diameter of 2 cm and thickness of 1.5 mm) by using SnAgCu solders. The current–voltage characteristics (I-V curve) of these LEDs were measured via an Agilent 4155B semiconductor parameter analyzer. The output power was analyzed by using an integration sphere detector (CAS 140B, Instrument Systems). The surface thermal property of the LED device was estimated by thermal infrared imaging at thermal equilibrium at an injection current of 700 mA under a close ambient. The total thermal resistance (R_{th}) of LED was evaluated by a T3Ster Master system based on thermal transients. Thermal diffusivity of the DAC substrate was measured by the laser-flash method.

2.2. Fabrication of LED with Beveled Sapphire and Cup-Shaped Copper Sheet

Before fabricating this LED device, the optimal beveled angle of shaped sapphire mounted on the cup-shaped copper was estimated by Trace-Pro software (commercial light-simulation software based on geometrical optics). In the simulation, the length of cathetus was maintained at 50 μm and the beveled angles were changed from 15° to 80°.

The epitaxial structures of InGaN LED (λ_p = 455 nm) on (0001) sapphire substrate by MOCVD was the same with that described above. After the depositing process of the LED chip (with 1 mm^2 area), the backside of sapphire was lapped and polished by chemical mechanical polishing (CMP) to generate a beveled sapphire. Then the sapphire-based LED was mounted on the cup-shaped copper heat spreader (with a thickness of 250 μm and a base area of 9 mm^2). Subsequently, the LED device was attached on a MCPCB using the diamond-added AgSnCu solders. Furthermore, the conventional LED and the LED with original sapphire embedded in cup-shaped copper heat spreader were both prepared on MCPCBs as the contrasted samples. The analyses of angular emission were characterized by placing LED chips on a rotating stage, which was connected to a fiber optic-coupled spectrometer.

2.3. Fabrication of LED Including Diamond Like Carbon Layer

This work proposed the fabrication of vertical-type LEDs (V-LEDs) on Si substrates. The InGaN LEDs (λ_p = 455 nm) were deposited by low-pressure MOCVD on c-plane sapphire substrates. The LED structures consisted of a low-temperature GaN buffer layer (50 nm), a undoped GaN layer (1 μm), a n-type GaN:Si layer (3 μm), six pairs of InGaN/GaN multiple quantum well active layers, and a p-type GaN:Mg layer (0.4 μm). In the device process, a Ni/Ag/Ni/Pt metal structure grown by e-beam evaporation on the p-GaN layer and annealed at 550 °C under a N_2 atmosphere was served as both the ohmic contact and reflective mirror. A DLC layer with a thickness of 2 μm was then prepared by sputtering, and an Au layer deposited by electroless plating was prepared sequentially on the mirror. The DLC films were deposited by RF sputtering of a graphite target with a diameter of 10 cm and 99.999% carbon content. The distance between the substrate and target was kept at 10 cm. When the base pressure was decreased to 5×10^{-7} Torr, Ar gas was introduced into the sputtering chamber under a flow rate of 15 sccm. During the DLC deposition process, the working pressure was fixed at 2×10^{-3} Torr. In addition, the RF input power and growth temperature were maintained at 200 W and 120 °C, respectively. Then the Au layer grown on the DLC and the In/Au layer deposited on the Si wafer were diffusion-bonded at 200 °C for 120 min to generate Au/In/Au. After the wafer bonding process, a laser lift-off technique was applied for the sapphire removal. Figs. 1(a) and 1(b) display the sample structures after the wafer bonding and laser lift-off processes, respectively. The remaining undoped GaN was immediately removed by using inductively coupled plasma (ICP) etching. The mesa pattern of the LED chip was defined as a size of 1 mm^2. The Cr/Au metal was grown by thermal evaporation to create the n- and p-pad electrodes. The working pressure and growth temperature of Cr/Au were kept at 7×10^{-7} Torr and 50 °C, respectively. After fabricating the n- and p-pad electrodes, the optimal ohmic contacts were carried out by annealing at 350 °C under a N_2 ambient for 15 min. The R_{th}, the temperature distribution of the sample surface, the I-V characteristics and the output powers of LED devices were analyzed in detail.

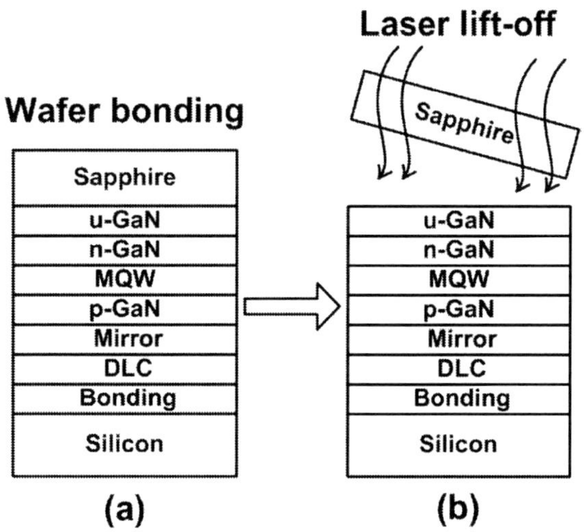

Figure 1. Schematic of the LED structure embedded with a DLC layer after (a) wafer bonding and (b) sapphire removing by laser lift-off technique [13].

3. Analysis of Experimental Data

3.1. Analysis of LED with Cup-Shaped Copper-Diamond Heat Spreader

Fig. 2(a) exhibits the photo of the sapphire-based LED with the DAC substrate mounted on the MCPCB. Fig. 2(b) shows the Ag reflector on the DAC substrate. In this figure, the image of panel shown in Fig. 2(a) was magnified. Fig. 2(c) shows a cross-sectional image of the abovementioned package taken by scanning electron microscopy (SEM). It can be found that the DAC substrate was contacted to the sapphire closely without any voids. Moreover, the cup shape of the reflector can be well controlled by self-aligned lithography. In order to analyze the diamond in the DAC heat spreader, the cross-sectional region of substrate was observed by SEM and characterized by elemental analysis, as shown in Fig. 3. It can be observed the diamond powders (particle size <5 μm) were uniformly distributed on the copper substrate. As shown in Fig. 3(b), the elemental analysis reveals that diamond indeed existed in the DAC heat spreader.

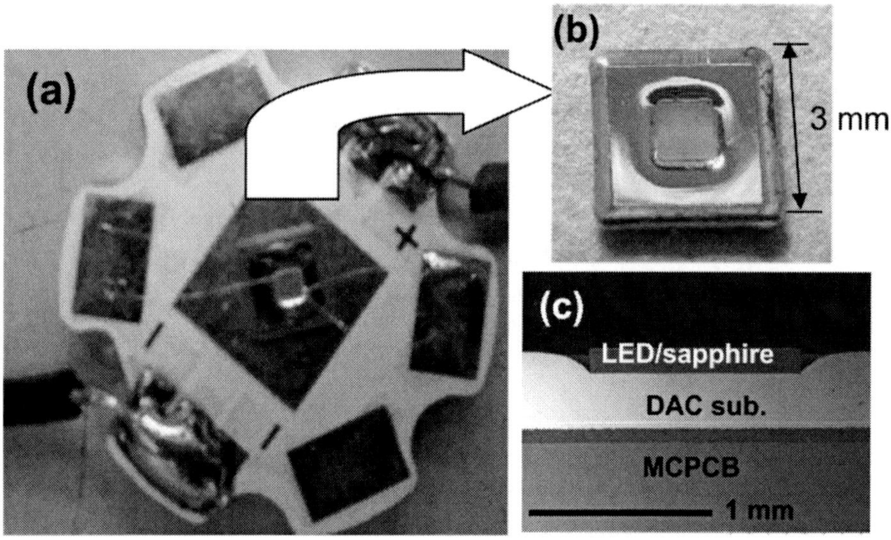

Figure 2. (a) Packaged LED sample, (b) the magnified image of the sapphire-based LED with the DAC heat spreader, and (c) cross-sectional SEM image [14].

It is important for the evaluation of the thermal behaviors of LED devices with and without cup-shaped DAC and copper heat spreader. For this purpose, the most straightforward method is the measurement of thermal resistance. Fig. 4 shows the thermal capacitance as a function of the total thermal resistance for these various LEDs. It was found that the total thermal resistance of the LED surrounded with the DAC substrate was measured to be 5.8 K/W, which was obviously lower than that with pure copper (7.7 K/W) and with the sapphire substrate (10.5 K/W). The differences of total thermal resistance among these three samples depended on the presence of DAC or Cu, which can contribute to the decrease in the thermal resistance. Especially, the total thermal resistance can be greatly reduced as the LED with the DAC heat spreader was used, indicating that the DAC was highly helpful to the reduction in the LED thermal resistance.

Fig. 5 shows the surface temperature distributions (@700 mA, 3 W) of these LED chips with the DAC heat spreader, the pure copper heat spreader, and the sapphire substrate. The surface temperature of the LED with the DAC heat spreader was determined to be 49.0 °C, which was lower than that with the pure copper heat spreader (52.75 °C) and with the sapphire substrate (59.16 °C). Based on the results, the LEDs with DAC and copper heat spreaders possessed more rapid temperature grading than that with sapphire substrate. The surrounding temperatures were measured to be 65, 45, and 42 °C for the sapphire-based LED, the LEDs with the copper and DAC heat spreaders, respectively. This suggests that the junction temperature of LED device can be effectively dissipated by Cu or DAC heat spreader.

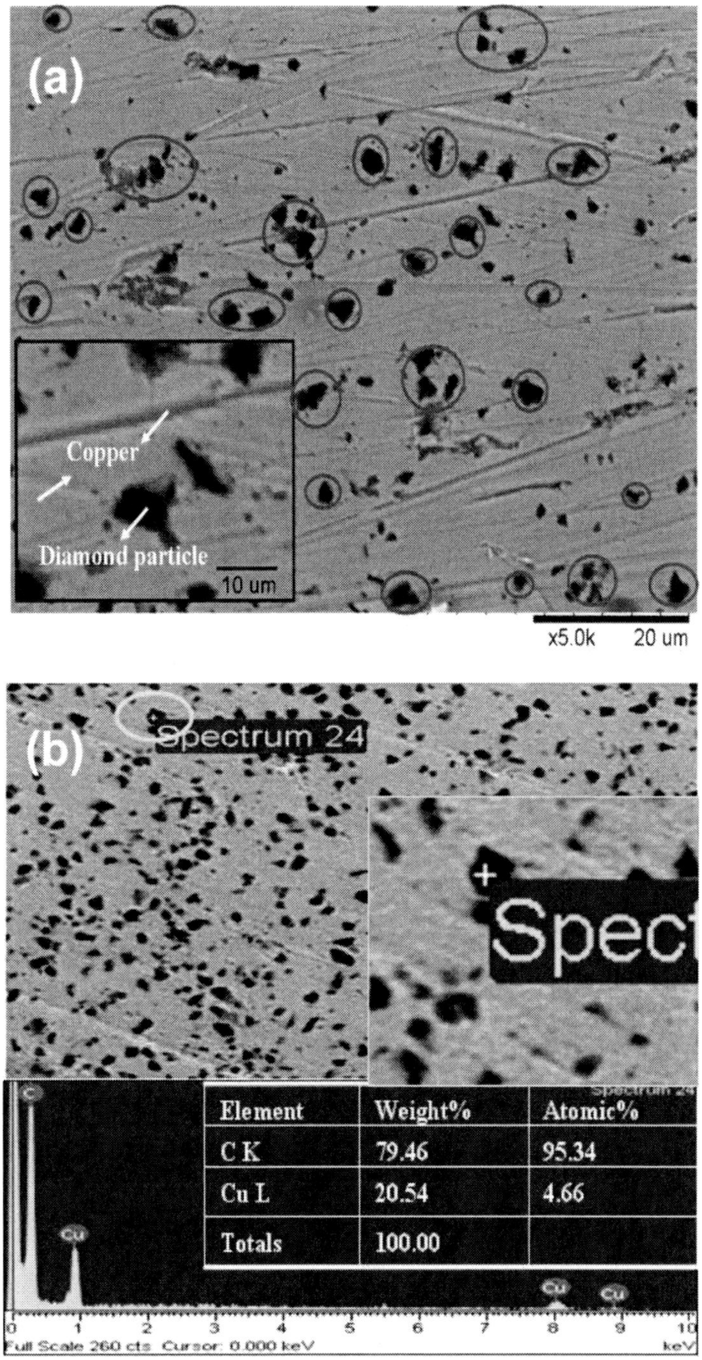

Element	Weight%	Atomic%
C K	79.46	95.34
Cu L	20.54	4.66
Totals	100.00	

Figure 3. (a) Cross-sectional SEM image taken on the substrate and (b) elemental analysis for the LED with a DAC heat spreader [14].

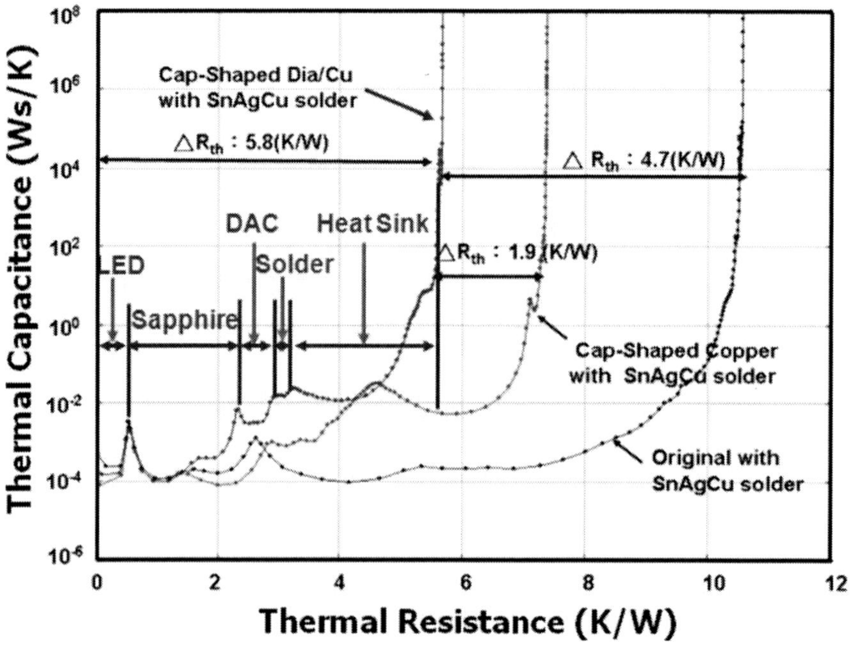

Figure 4. Thermal capacitance as a function of total thermal resistance for the LEDs with DAC, pure copper, and sapphire mounted on MCPCBs by SnAgCu solders [14].

The thermal dissipation is dependent on the thermal conductivity (k) of materials. As a result, the thermal conductivity of the DAC substrate should be estimated. Additionally, it is well known that the thermal conductivity and thermal diffusivity are both the most important thermophysical parameters to describe the heat transport characteristics of materials. For the measurement of thermal diffusivity, the laser flash method was confirmed as a fast and precise technique. In this technique, a very short pulse (<1 ms) of heat was employed to the front face of the sample by using a laser flash, and then the temperature change of the rear face can be measured with an infrared (IR) detector [15]. Therefore, the thermal diffusivity of the sample can be achieved. The thermal property was obtained from the following equation [16].

$$k = D \times q \times c \tag{1}$$

where D, q, and c were denoted the thermal diffusivity, mass density, and specific heat, respectively. The thermal diffusivity of the DAC substrate was measured to be 0.7179 cm^2/s, which was much higher than that of pure copper (0.117 cm^2/s). Furthermore, the thermal conductivity of the DAC substrate was evaluated to be 492.4 W/mK from equation (1), which was also higher than that of pure copper (457.4 W/mK) [16]. This proves that the DAC substrate can provide better heat dissipation than that of the pure copper and sapphire. This result agrees well with those of the thermal resistance and surface temperature of LED chips, as shown in Figs. 4 and 5.

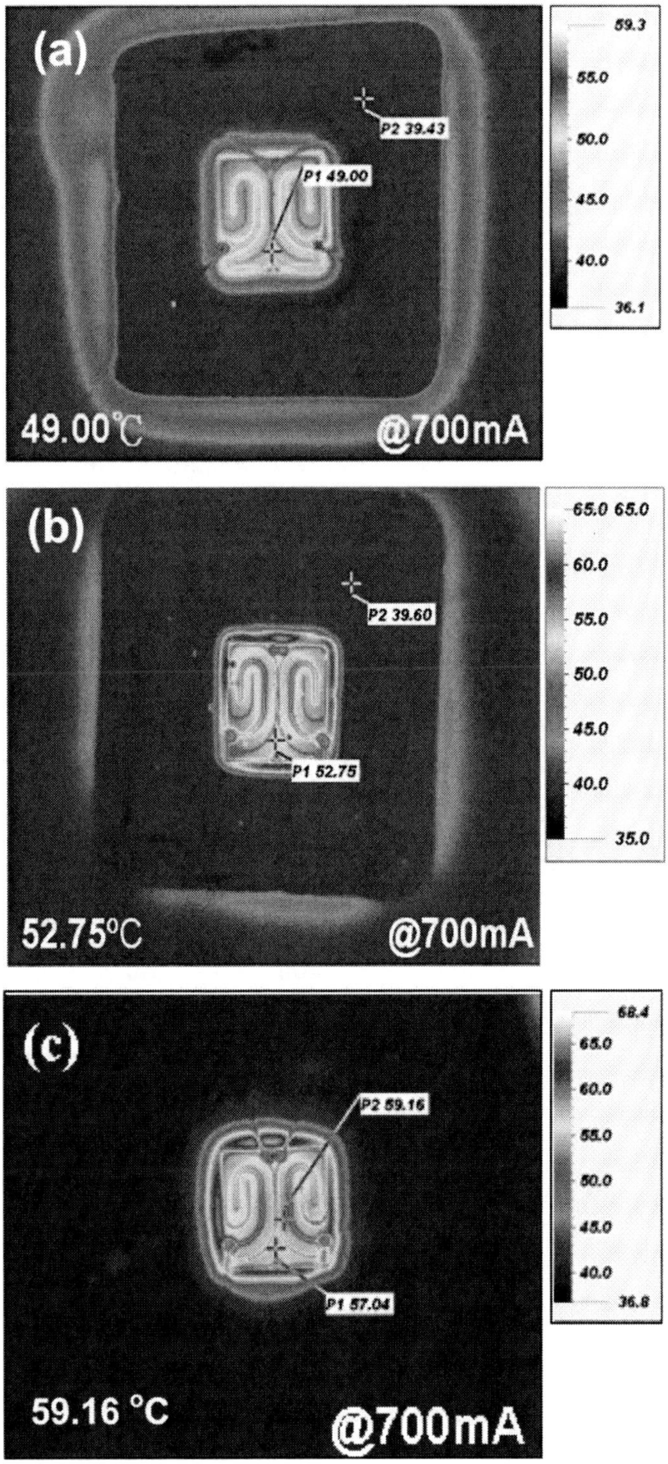

Figure 5. Surface temperature distributions (@700 mA, 3 W) of various LEDs with (a) DAC, (b) pure copper, and (c) sapphire substrates [14].

The measurement of the LED device performance is another method to analyze the heat dissipation characteristics. Fig. 6 exhibits the output powers of these three LED samples as a function of the driving current from 0 to 1000 mA. The LED fabricated with the DAC substrate possessed superior performance in the output power due to its preformed reflector and high heat dissipation. At an injection current of 350 mA, the output powers were determined to be 313, 287, and 253 mW for these LEDs with the DAC heat spreader, the copper heat spreader, and sapphire substrate, respectively. Among these three devices, the output power of LED with DAC substrate was highest with increasing the injection current to 1000 mA. When all the LEDs were deposited with the same epitaxial structure and prepared with equal chip size, the improvement in the output power could be primarily ascribed to the relatively high capacity of the DAC substrate, which can dissipate the heat efficiently.

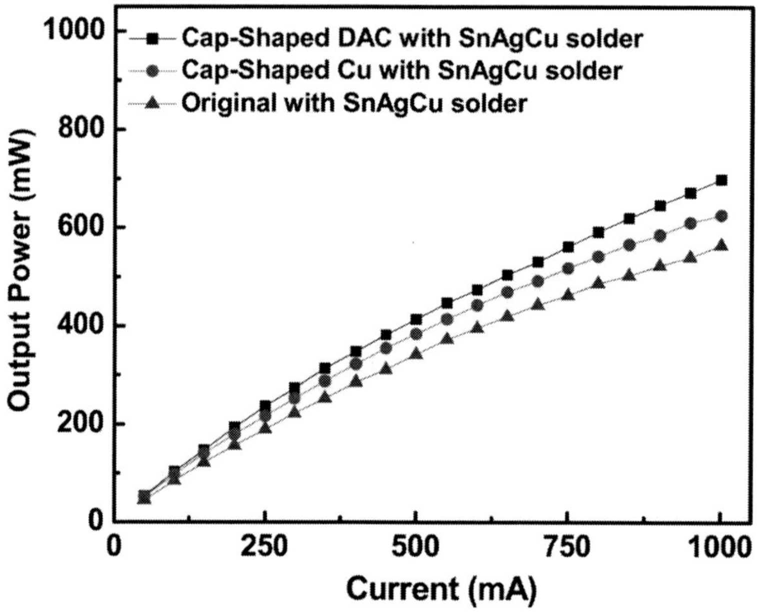

Figure 6. Output powers of these LEDs with DAC, pure copper, and sapphire substrate mounted on the MCPCBs by using SnAgCu solders [14].

3.2. Analysis of LED with Beveled Sapphire and Cup-Shaped Copper Sheet

Fig. 7 displays the schematic diagram of LED structure and light extraction efficiencies with various beveled angles of shaped sapphire (15°-80°). According to our simulation, the highest light extraction efficiency of 37.16 % can be achieved at a beveled angle of 75°. Light beam trace images of the LEDs with original and 75° beveled sapphires are shown in Fig. 8. In this simulation, 500 light rays were assumed to emit randomly from the multiple quantum well. It is found that the density of light beam reflected from the beveled sapphire is more than that from the original sapphire. This implies the larger included angle is resulted from the shaped sapphire structure and expected to favor the enhancement in the light extraction. In contrast, the smaller included angle in the original sapphire could decrease the light extraction and accumulate the heat created from LED chip. Consequently, the LED prepared with 75°

beveled sapphire and embedded in cup-shaped copper heat spreader could be the optimal structure in our research.

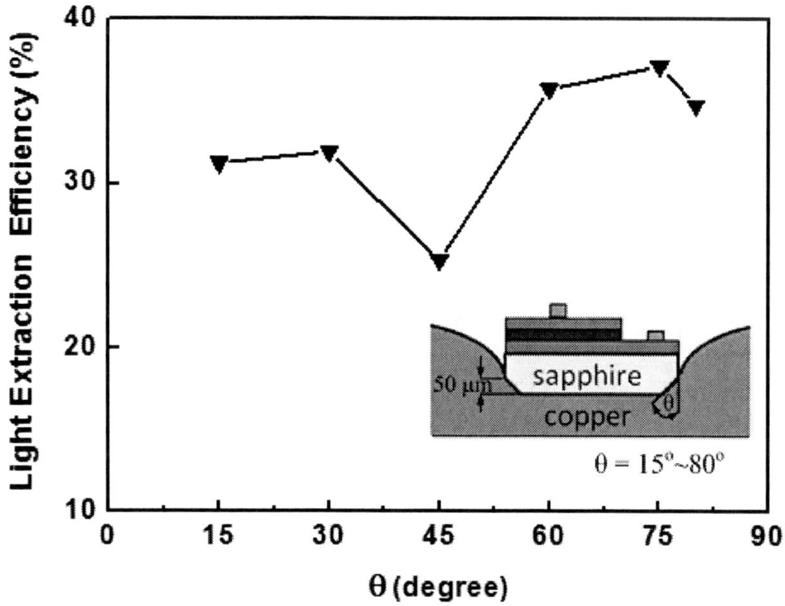

Figure 7. Simulated light extraction efficiencies of LEDs with beveled sapphires and embedded in cup-shaped copper sheets as a function of beveled angle (15°-80°) [17].

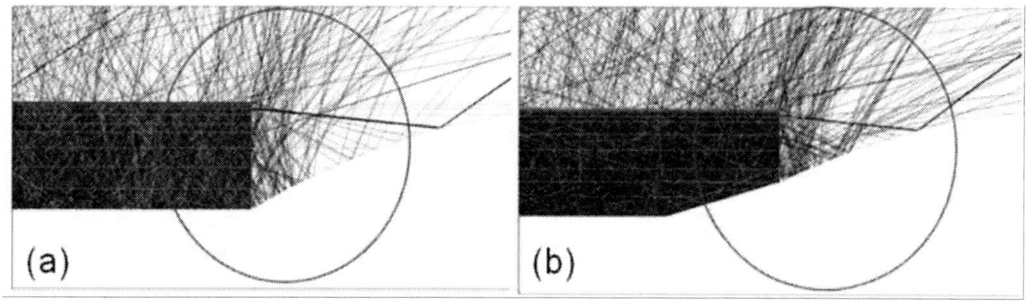

Figure 8. Ray trace simulation taken from the LED chips with (a) original and (b) beveled sapphires both embedded in cup-shaped copper sheets [17].

In order to realize the effect of geometrical shape on the optical performance, the LED devices were characterized by the angular emission measurement. The angular emission analysis was carried out under a driving current of 20 mA, and the step size of detection angle was fixed at 3°. The experimental angular emission patterns of LEDs with original and beveled sapphires both embedded in the cup-shaped copper sheets are shown in Fig. 9. The patterns should be normalized to the intensities at normal. We can observe that the LED with beveled sapphire presents a broader pattern and higher light emission intensity at the normal direction than that with original sapphire. Based on our analysis, the light extraction angles of LEDs with original and beveled sapphires were determined to be 120° and 130°, respectively. It can be seen that the LED with beveled sapphire not only possesses larger extraction angle, but also higher extraction intensity at the normal direction. The result could be ascribed to the redirection of emitted photon, employing the beveled sapphire in the LED structure to couple light out efficiently.

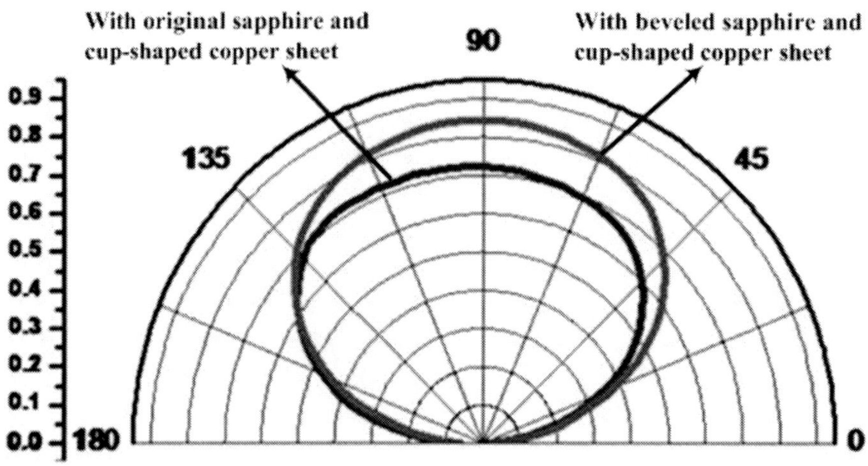

Figure 9. Angular emission patterns (normalized to the intensities at normal) of LEDs with original and beveled sapphires both embedded in cup-shaped copper sheets [17].

In order to characterize the device performances, the LED samples which we employed in this research are labeled as samples A, B and C. The conventional LED with an original sapphire and without the cup-shaped copper sheet is labeled as sample A. The LED with an original sapphire embedded in the cup-shaped copper sheet is labeled as sample B. Moreover, the LED with a beveled sapphire embedded in the cup-shaped copper sheet is labeled as sample C. Fig. 10(a) shows the schematic diagrams of these three samples. Fig. 10(b) exhibits the light output powers as a function of injection current for the samples A, B and C. It is obvious that the output powers of the samples B and C are higher than that of sample A in the whole measured currents, revealing the output power can be enhanced by using the cup-shaped copper sheet. The output powers were measured to be 325.4, 372.6, and 395.6 mW for the samples A, B and C, respectively, with an injection current of 350 mA. As the copper heat spreader was fabricated, the silver was employed as a high reflective mirror and coated on the cup-shaped copper sheet, which leaded to improvement in the light extraction. In addition, when the injection current was increased, the power saturation behavior formed in the sample A was not observed in the samples B and C. With increasing the injection current to 1000 mA, the output powers of samples A, B and C were determined to be 639.5, 830.6, and 877.1 mW, respectively. Because the heat created from LED chip can be dissipated easily by using the copper sheet, the junction temperature is restricted efficiently. Thus, the carriers are confined successfully within the quantum wells, resulting in a steady increase in output power with as the injection current is increased [18]. Furthermore, high reflection mirror fabricated in the cup-shaped copper sheet also possesses the usability of light extraction. It was also found that the sample C had the highest output power because of the adoption of beveled sapphire, which was in good agreement with the simulated result shown in Fig. 9. This indicates the lateral emitted light can be reflected efficiently via an employment of the shaped sapphire substrate.

Fig. 10(c) presents the power efficiencies as a function of injection current for samples A, B and C. Apparently, sample C possessed higher power efficiency than that of samples A and B. The power efficiencies of samples A, B and C under an operating current of 350 mA were determined to be 27.6%, 31.5% and 33.3%, respectively. As discussed above, the enhancement in the power efficiency of sample C can be assigned to the copper heat spreader with high thermal conductivity and efficient light reflection resulted from the beveled sapphire. There were several studies investigated on ameliorating one of the LEDs characteristics, consisting of thermal dissipation, light extraction and electrical property [19,20]. However, these characteristics were related to each other closely. In this research, the heat dissipation and light extraction were both enhanced by applying the shaped sapphire and cup-shaped copper heat spreader. The increment in light extraction efficiency could result in the enhancement in the light output power. The cup-shaped copper heat spreader can improve the heat dissipation performance to increase the life time of the LED device [21]. Further increasing the driving current to 1000 mA, the power efficiencies of samples A, B and C were evaluated to be 15.5%, 20% and 21.8%, respectively.

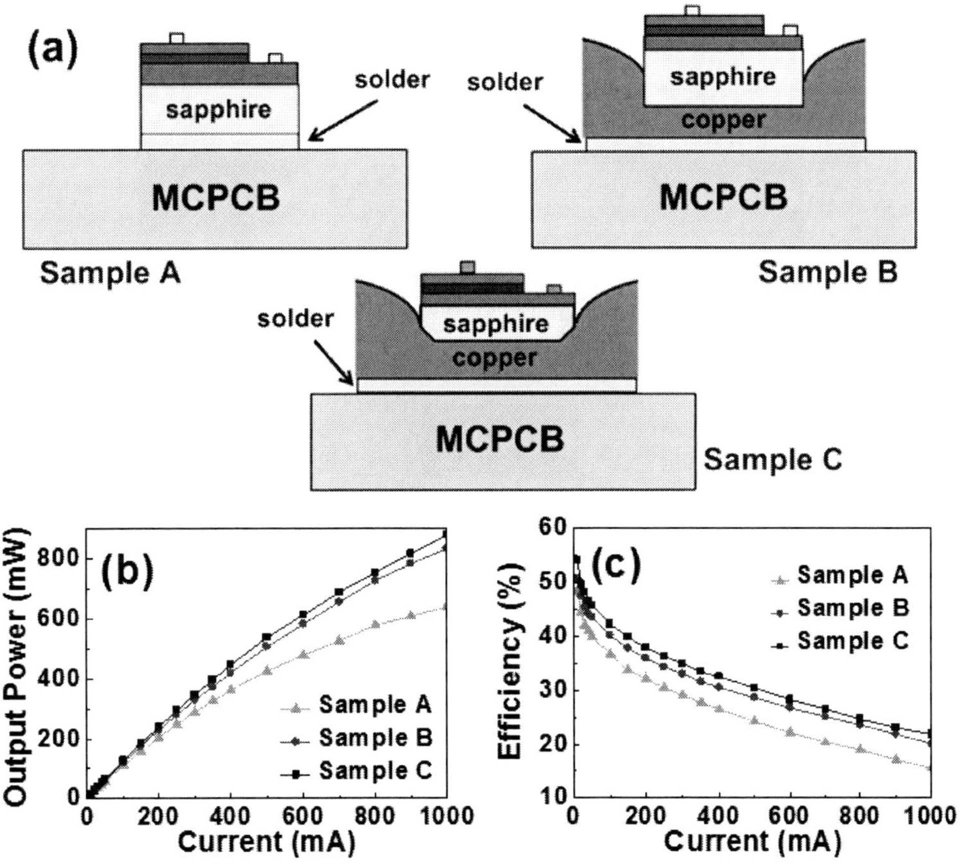

Figure 10. (a) Schematic diagrams of samples A, B and C. (b) Output powers as a function of driving current for the samples A, B and C. (c) Power efficiencies as a function of driving current for the samples A, B and C [17].

3.3. Analysis of LED Including Diamond Like Carbon Layer

The measured cumulative structure functions for the LEDs with and without an embedded DLC layer are shown in Fig. 11. By adding the DLC layer into the device, the total thermal resistance was measured to be 13.2 K/W, which was lower than that of the LED without the DLC layer (16.5 K/W). The thermal transfer in DLC layer both dominated by the electrons and phonons (lattice vibrations) was faster than that in metals [22]. A lot of sp^3 bonds in the DLC layer are beneficial for the heat diffusion, leading to uniform thermal dissipation from the LED device to the substrate and then to the heat sink. Furthermore, the reduction in total thermal resistance of the sample with a DLC layer can be clarified through the good thermal diffusivity of DLC. According to previous works, the thermal diffusivities of Au (0.2-0.7 cm^2/s) [23] and DLC (5.2 cm^2/s) [24] films were estimated. Due to the current crowding, the proposed LED device created considerable heat near the n-pad electrode. Before embedding the DLC layer into the structure, there were few effective thermal paths in the LED sample because of low thermal diffusivity, even if the heat was conducted into the Au bonding layer. On the contrary, the addition of a DLC layer with higher thermal diffusivity into the LED sample would induce rapid heat transfer in the lateral direction, generating the more effective thermal paths. This caused a reduction in the total thermal resistance by adding the DLC layer.

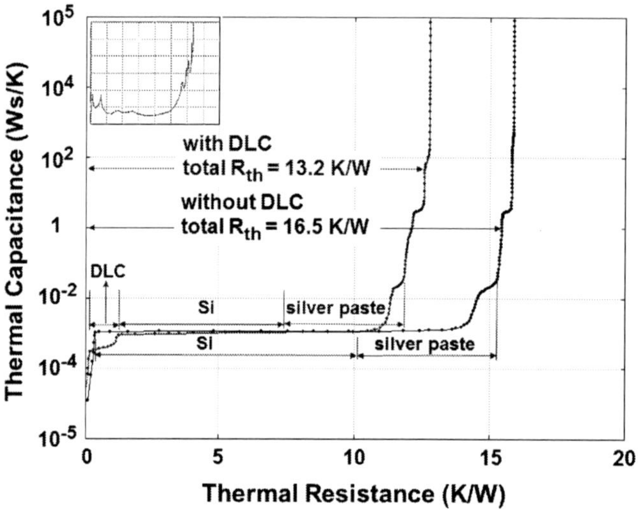

Figure 11. Cumulative structure functions of the LEDs samples embedded with and without a DLC layer [13].

In order to confirm the heat dissipation performance, the samples were characterized by thermal infrared imaging. Fig. 12 exhibits the surface temperature distributions of the samples with and without a DLC layer at the injection currents of 350 and 700 mA, respectively. Under an injection current of 350 mA, the surface temperatures of LED chips with and without adding a DLC layer are measured to be 42-43.67 °C and 44.5-46.93 °C, whereas the temperature differences (ΔT) are 1.67 °C and 2.43 °C, respectively. This reveals the device with a DLC layer possesses a lower surface temperature and uniform temperature distribution. With increasing the injection current to 700 mA, the LED without embedding a

DLC layer has a higher surface temperature of 82.45-101.1 °C and an un-uniform ΔT of 18.65 °C. Through the addition of a DLC layer into LED sample, the surface temperature and ΔT can be decreased efficiently to 55.51-65.34 °C and 9.83 °C, respectively. The insertion of a DLC layer into the LED sample decreases the surface temperature and uniformity of temperature difference, especially under higher injection currents. In comparison to the LED sample embedded with a DLC layer, the electrodes of the device without adding a DLC layer presented greater heat accumulation. The heat accumulation results in the electrode deterioration, which can decrease the device reliability and lifetime.

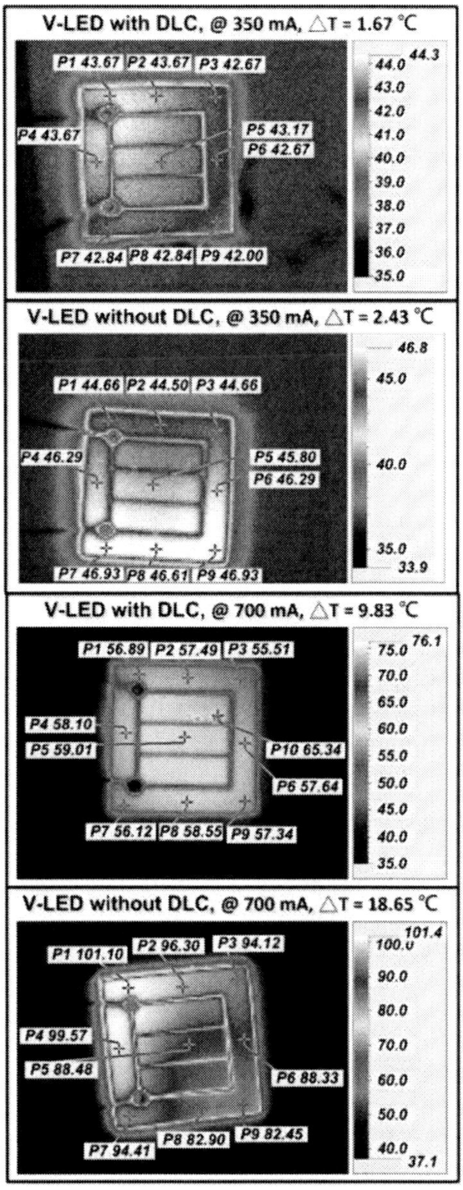

Figure 12. Surface temperature distributions for the LEDs embedded with and without a DLC layer under the driving currents of 350 and 700 mA [13].

I-V characteristics of the LEDs with and without embedding a DLC layer are shown in Fig. 13. The inset of this figure displays the LED structure embedded with a DLC layer after performing the mesa and electrode processes. We can observe that the turn-on voltages (V_f) at various injection currents for these samples are almost the same, implying that the insertion of a DLC layer into the LED sample does not influence the electrical properties. The V_f values at various injection currents of 20, 350, and 700 mA were measured to be 2.6, 3.1, and 3.3 V, respectively.

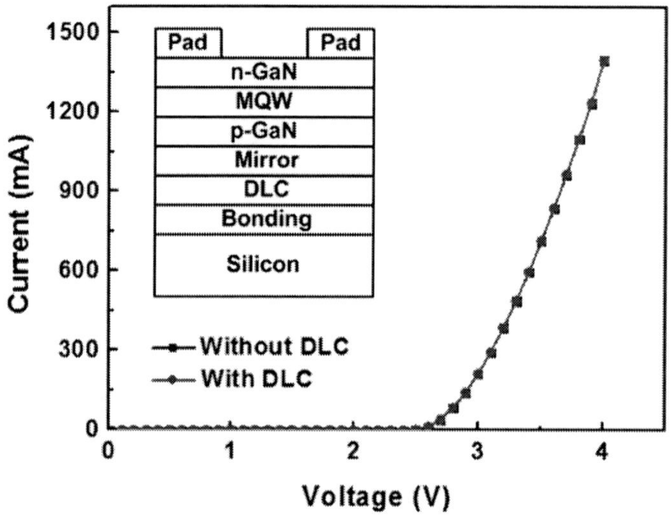

Figure 13. I-V characteristics of the LEDs samples embedded with and without a DLC layer. The inset is the LED structure embedded with a DLC layer [13].

The output powers and wall plug efficiencies as a function of injection current for LEDs embedded with and without a DLC layer are shown in Fig. 14. The output powers of the LED sample embedded with a DLC layer are higher than that without inserting a DLC layer in all measured currents (0-1400 mA). Moreover, the results of wall plug efficiency presented the similar tendency to the output power. Under a driving current of 700 mA, the output powers of LEDs embedded with and without a DLC layer were measured to be 604 and 500 mW, respectively. The wall plug efficiency increased from 21.4% to 25.2% as a DLC layer was inserted into the sample. The output power and wall plug efficiency (@ 700 mA) for the sample with a DLC layer can be enhanced by 20.8% and 17.7%, respectively, in comparison to that without inserting a DLC layer. The degradation in output power of LED device is typically ascribed to inefficient carrier confinement in the quantum wells with increasing the temperature [25]. As mentioned above, the total thermal resistance of the LED sample with a DLC layer can be reduced, restricting the junction temperature and enhancing the output power.

The total thermal resistance of the LED chip also can be estimated by using the following equation [26].

$$R_{th} = \Delta T/(P_e - P_{op}) \qquad (2)$$

where R_{th} is the total thermal resistance, ΔT is the difference between the room temperature (approximately 27 °C) and junction temperature, P_e is the input power, and P_{op} is the output power. In our research, the ΔT values of the samples can be achieved by applying R_{th}, P_e, and P_{op}. As discussed in Fig. 11, the R_{th} values were measured to be 13.2 and 16.5 K/W, respectively, for LEDs samples with and without inserting a DLC layer. Besides, as displayed in Fig. 13, the LEDs samples with and without a DLC layer possess the same value of V_f (3.3 V, @700 mA). Therefore, the P_e values (@700 mA) of these two samples were evaluated to be 2.31 W by using $P_e = 0.7 \times V_f$. The P_{op} values (@700 mA) of these two samples (Fig. 14) were analyzed to be 0.604 and 0.5 W, respectively. Consequently, the ΔT values can be calculated to 22.52 and 29.86 °C for LEDs samples embedded with and without a DLC layer, and the junction temperatures were evaluated to be 49.52 and 56.86 °C, respectively. As discussed in Fig. 12, the measured temperatures of these two LED samples (55.51-65.34 °C and 82.45-101.1 °C) were both higher than the theoretical results. In the measurement of total thermal resistance using a T3Ster cooling system, the heat created from the LED can be dissipated during microseconds. Nevertheless, the measurement of surface temperature is based on the thermal equilibrium, which can lead to the thermal accumulation easily. It can be explained the reason why the surface temperatures obtained by the thermal infrared image are higher than the junction temperatures estimated by R_{th} calculation. Furthermore, the temperature difference between the R_{th} calculation and surface temperature measurement for the LED samples embedded with a DLC layer was obviously smaller than that without a DLC layer. The result proves the DLC layer provides the rapid thermal transfer in both vertical and horizontal directions, inducing a smaller temperature difference.

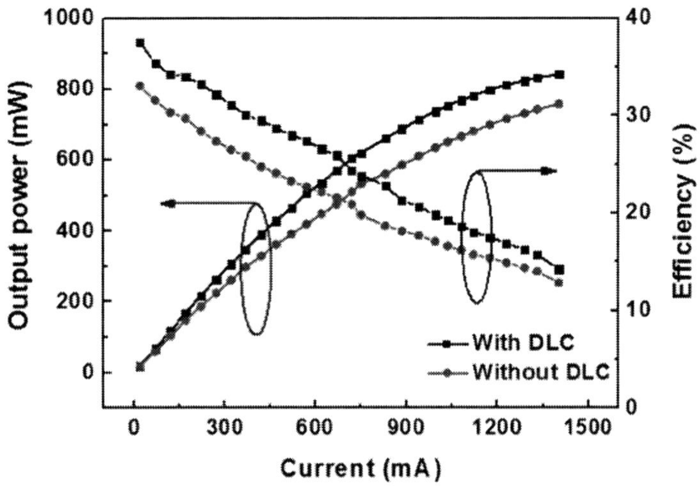

Figure 14. Output powers and wall plug efficiencies as a function of operating current for LEDs samples with and without inserting a DLC layer [13].

Conclusion

Three techniques consisting of the cup-shaped diamond-added copper (DAC) heat spreader, the combination of beveled sapphire and cup-shaped copper sheet, and the insertion of diamond like carbon layer into LED sample have been proposed in this chapter. The composite electroplating method was used to fabricate a sapphire-based LED with a DAC composite on a cup-shaped copper sheet. From our results, the thermal dissipation performance of the device can be enhanced greatly by the DAC heat spreader because of its high thermal conductivity. The employment of DAC heat spreader can not only decrease the LED thermal resistance, but also improve the light output power. In addition, as the cup-shaped copper sheet was combined with the beveled sapphire technique, the device performance can be further improved. Based on the angular emission measurement, the LED fabricated with beveled sapphire possesses higher light extraction than that with original sapphire. It confirms that the combination of these two techniques is indeed helpful to the light extraction and thermal characteristic of LEDs. On the other hand, due to the high thermal conductivity and good thermal diffusivity of DLC, the heat can be dissipated in both vertical and horizontal directions. Thus, the LED with inserting a DLC layer has a lower thermal resistance and surface temperature than those without a DLC layer. This reveals the design of the LED structure embedded with a DLC layer can improve the heat dissipation and optoelectronic performance efficiently.

References

[1] Amano, H.; Sawaki, N.; Akasaki, I.; Toyoda, Y. *Appl. Phys. Lett.* 1986, 48, 353-355.

[2] Chen, K. C.; Chuang, R. W.; Su, Y. K.; Lin, C. L.; Hsu, H. C.; Huang, J. Q.; Yang, K. F. *57th IEEE Electron. Compon. Technol. Conf.*, Reno, NV, 2007, 734-736.

[3] Chang, S. J.; Chang, C. S.; Su, Y. K.; Lee, C. T.; Chen, W. S.; Shen, C. F.; Hsu, Y. P.; Shei, S. C.; Lo, H. M. *IEEE Trans. Adv. Packaging* 2005, 28, 273-277.

[4] Choi, Y. S.; Kim, S. J. *Solid-State Electron.* 2006, 50, 1522 (2006).

[5] Fred Schubert, E. *Light-Emitting Diodes*, Cambridge University Press, New York, 2003.

[6] Horng, R. H.; Chiang, C. C.; Hsiao, H. Y.; Zheng, X.; Wuu, D. S.; Lin, H. I. *Appl. Phys. Lett.* 2008, 93, 111907-1~111907-3.

[7] Horng, R. H.; Hong, J. S.; Tsai, Y. L.; Wuu, D. S.; Chen, C. M.; Chen, C. J. *IEEE Trans. Electron. Device* 2010, 57, 2203-2207.

[8] Fujii, T.; Gao, Y.; Sharma, R.; Hu, E. L.; DenBaars, S. P.; Nakamura, S. *Appl. Phys. Lett.* 2004, 84, 855-857.

[9] Horng, R. H.; Hsiao, H. Y.; Chiang, C. C.; Wuu, D. S.; Tsai, Y. L.; Lin, H. I. *IEEE J. Sel. Top. Quantum Electron.* 2009, 15, 1281-1286.

[10] Lee, C. E.; Lee, Y. C.; Kuo, H. C.; Tsai, M. R.; Lu, T. C.; Wang, S. C. *Semicond. Sci. Technol.* 2008, 23, 025015-025025.

[11] Chang, C. L.; Chuang, Y. C.; Liu, C. Y. *Electrochem. Solid-State Lett.* 2007, 10, H344-H346.

[12] Robertson, J. *Mater. Sci. Eng. R* 2002, 37, 129-281.

[13] Horng, R. H.; Kao, W. C.; Ou, S. L.; Wuu, D. S. *Appl. Phys. Lett.* 2012, 101, 171102-1~171102-4.

[14] Horng, R. H.; Hu, H. L.; Lin, R. C.; Peng, K. C.; Chiang, Y. C. *Electrochem. Solid-State Lett.* 2011, 14, H215-H217.

[15] Graham, S.; McDowell, D. L.; Curzio, E. L.; Dinwiddie, R. B.; Wang, H.; Porter, W. *J. Compos. Mater.* 2003, 37, 73-87.

[16] Yoshida, K.; Morigami, H. *Microelectron. Reliab.* 2004, 44, 303-308.

[17] Horng, R. H.; Hu, H. L.; Ou, S. L.; Lin, R. C.; Hsu, C. P. *IEEE Photonics Technol. Lett.* 2012, 24, 1421-1423.

[18] Cao, X. A.; Arthur, S. D. *Appl. Phys. Lett.* 2004, 85, 3971–3973.

[19] Akasaka, T.; Gotoh, H.; Saito, T.; Makimoto, T. *Appl. Phys. Lett.* 2004, 85, 3089-3091.

[20] Schubert, E. F.; Kim, J. K. *Science* 2005, 308, 1274-1278.

[21] Xi, Y.; Xi, J. Q.; Gessmann, T.; Shah, J. M.; Kim, J. K.; Schubert, E. F.; Fischer, A. J.; Crawford, M. H.; Bogart, K. H. A.; Allerman, A. A. *Appl. Phys. Lett.* 2005, 86, 031907-1~031907-3.

[22] Wei, Q.; Sankar, J.; Sharma, A. K.; Oktyabrsky, S.; Narayan, J.; Narayan, R. J. *J. Mater. Res.* 2000, 15, 633-641.

[23] Käding, O. W.; Skurk, H.; Matthias, E. *J. Phys. IV France* 1994, 4, C7-619~C7-622.

[24] Fujii, K.; Yokota, S. H.; Shōhata, N. *MRS proceedings* 1986, 75, 765.

[25] Nakamura, S.; Senoh, M.; Iwasa, N.; Nagahama, S. *Appl. Phys. Lett.* 1995, 67, 1868-1870.

[26] Faekas, G.; Vader, Q. V. V.; Poppe, A.; Bognar, G. *Proceedings of 9th International Workshop on Thermal Investigations of ICs and Systems* 2003, 213-218.

In: Gallium Nitride
Editor: Kiera Olivia Peak

ISBN: 978-1-63321-387-6

Chapter 10

GALLIUM NITRIDE BULK AND NANOSTRUCTURES FOR SPINTRONIC APPLICATION

E. Salmani[2], O. Mounkachi[1,*], H. Ez-Zahraouy[2], H. El Moussaoui[1], R. Masrour[3], M. Hamedoun[1] and A. Benyoussef[1,2]

[1]Institute for Nanomaterials and Nanotechnologies, MAScIR, Rabat, Morocco
[2]Laboratory of Materials, Processes, Environment and Quality, LMPHE (URAC 12), Faculté des Sciences, Université Mohammed V-Agdal, Rabat, Morocco
[3]Cady Ayyed University, National School of Applied Sciences, Sidi Bouzid, Safi, Morocco

Abstract

In the present chapter we review the recent achievements of experimental and theoretical developments in understanding the magnetic, optical and electronic properties of transition-metal-doped Gallium Nitride semiconductors, which show ferromagnetic order at relatively high temperature and good structural compatibility with existing III–V devices. Electronic structure, magnetic and optical properties of GaN:TM (TM: Cr, V, Mn, Fe, Co, Ni), based dilute magnetic semiconductors (DMS), are investigated using first-principles calculations for bulk and nanofilms. We show that density functional theory is a very powerful tool for: 1- studying the effects of local doping, defects effect, size and disorder effect on the magnetic and optical properties of these materials, and predicting properties of new materials.

Keywords: Gallium Nitride, Spintronic, Dilute magnetic semiconductors, III–V devices, Density functional theory

[*] E-mail address: o.mounkachi@mascir.com (Corresponding author)

Introduction

Electrical and magnetic properties of materials are used in the modern information technology. The information processing is performed in solid state circuits where the logical states "1" and "0" are associated to the electric potential. The electron spin is not used during the processing. However to store a large volume of information we use the magnetic record, i.e. electron spin in ferromagnetic metals. The aim of the spin electronics (spintronics) is to create new electronic components which would use both the electrical and spin dependent phenomena for the information processing. Today, the technologies converge towards devices capable of storing and processing information, without the necessity of transferring the electric current. The usual semiconductors which are used in the microelectronics (Si,GaAs) are not ferromagnetic. On the other hand it is quite difficult to use ferromagnetic metals (Fe,Ni) together with conventional semiconductors, because of complicate properties of metal semiconductor interfaces. The concentrated magnetic semiconductors ($CuCr_2S_4$, Sr_2FeMoO_6, Sr_2CrReO_6) are ferromagnetic at high temperature. However the crystal structure of these materials does not allow incorporating them in modern microelectronic devices. In order to use semiconductors in microelectronics we change their electric properties. Just as in the case of electric properties, the magnetic properties of semiconductors can be changed by doping with magnetic impurities. The functionalization of semiconductor materials through doping with magnetic elements is an exciting and a controversial area of research that has seen broad scientific discussions. Starting with Semimagnetic materials to dilute magnetic semiconductors (Diluted Magnetic Semiconductors, DMS) and ferromagnetic nano-objects in semiconductors, important discoveries have been made in the understanding of the physical mechanisms involved in the ferromagnetism of these materials, but the problem is still open [1-10]. One promising approach is to unify the fundamental aspects of growth theory and characterization, let them interact with each other in a self-consistency cycle. For a long time much attention has being paid to II-VI semiconductors, where the valence of the cation coincide to the valence of well studied magnetic ions (Mn, Cr). However antiferromagnetic interactions are very important in the semiconductors and obtained values of the Curie temperature is very low (a few K in ZnMnTe). In the last decade the scientific community had much interest towards diluted magnetic semiconductors since they combine conventional semiconductor properties with magnetic properties [1-10]. This is the most obvious way to achieve polarization and the injection of a polarized current in a semiconductor because this solution, DMS / semiconductor overcomes the fundamental problem of disagreement conductivity between a junction ferromagnetic metal / semiconductor [11, 12]. The observation of ferromagnetism in conventional diluted magnetic semiconductors (DMSs) (III-V and II-VI) has opened a new field of research focused on the use of spin in semiconductors, motivated by promising applications in devices based semiconductor spintronics. An important current problem towards practical use of the spin-transport in semiconductor devices is the design of the materials that make possible the injection of spin-polarized electrons into a semiconductor at room temperature. This will make new features unavailable in conventional devices of semiconductors. More accurately, the DMSs are known to have a long length spin diffusion compared to transition metals. The diluted magnetic semiconductor with ferromagnetism above room temperature, with formation of a 100% spin polarization a broad band with an impurity could pave the way for performing the polarization and injection spin with high efficiency needed to achieve spintronics technology, such as the spin valve transistor, spin light emitting diodes, magnetic memories based magnetic semiconductor (MRAM), the sensors and ultrafast optical switches, these spintronics devices will be smaller, more flexible and more robust than those currently manufactured by Si chips and circuit elements. The search for developing dilute magnetic semiconductors (DMSs) with high Curie temperature (T_C) and half metalicity proprieties has been the subject of interest for a new generation of spintronics like giant magnetoresistance and magnetoresistance sensors [13,14]. One of the major technical barriers that must be overcome to realize the practical implementation of spintronic devices is the development of suitable spin-polarized materials that will effectively allow spin-polarized carriers to be injected, transported, and manipulated. Most of the past attention was focused on the (Ga,Mn)As system. TM doped GaN as promising workbench for future applications in spintronics [15] only little is known about the (Ga,TM)N materials system for nanostructure. Gallium nitride (GaN) nanostructures have triggered extensive interest in the past two decades due to their potential applications in blue and ultraviolet light-emitting diodes and high-temperature/ high power electronic and spintronic devices. Recently, films of (Ga,Fe)N fabricated by means of molecular beam epitaxy (MBE) were reported with Curie temperature above room temperature (RT) [16-19]. However, the origin of the exhibited ferromagnetic is still under the investigation. This is probably due to many experimental parameters: preparation methods, measurement techniques, substrate choose, thickness of the thin layer, unexpectedly carriers introduced during the synthesis. In this respect, the effects of parameters such as TM defects, thickness, and magnetic stability are investigated in this paper in terms of electronic structure and optical proprieties of TM doped GaN bulk and thin films. In this chapter, the electronic structure and magnetic proprieties of (Ga,TM)N where TM (V,Cr,Mn,Fe,Co and Ni) bulk and nanostructure with and without defect is investigated based on the first principles calculations and a material of new TM doped GaN-based DMS is proposed for spintronic application. Our calculations are based on KKR-CPA method within the density functional theory (DFT). Mechanism of exchange interaction between magnetic ions in (Ga,TM)N with acceptor defects is investigated. In TM doped GaN, we show that layer of thickness and acceptor defect can switch the magnetic ordering from disorder local moment (DLM) to ferromagnetic (FM) order , the optical absorption spectra obtained by ab initio calculations confirm the ferromagnetic stability based on the charge state of magnetic impurities.

Magnetic and Optical Properties of TM Doped GaN Bulk

To solve the Density Functional Theory (DFT) one-particle equations one uses multiple-scattering theory, i.e. the KKR Green's function (KKR-GF) method and the KKR-CPA for low impurity concentration. The coherent potential approximation (CPA) was employed to describe the random disordered distribution of impurities. In this work, we use the KKR-CPA-SIC-LDA package, in which SIC is implemented into MACHIKANEYAMA2009, [20,21,24]. LDA-SIC approximation can be viewed as an extension of LDA in the sense that the self-interaction correction the Kohn-Sham wave function are projected onto a set of localized basis orbital's. For LDA approximation we use a Vosko, Wilk and Nusair (VWN) [22].

Here, the relativistic effect has been considered using the scalar relativistic approximation. Moreover, the potential form is approximately given by the muffin-tin model.

In such a model, the wave functions in the muffin-tin spheres were expanded with the real harmonics up to $l = 2$, where l is the angular momentum defined at different sites. For this aim, we use higher K-points up to 500 in the irreducible part of the first Brillouin zone. In the present work, our GaN calculations will be performed on the wurtzite structure being. This crystalline structure has hexagonal unit cell with two lattice parameters ("a" and "c"). The optimized structures are obtained by calculating the total energy of GaN as a function of the ratio c/a. To perform this optimization the ratio c/a has been changed.

The total energy of the GaN supercell was calculated for different values of the ratio c/a. And the minimum of the function E(c/a) corresponds to the optimal value of c/a. It is found that a minimum value for the energy was obtained for c/a ≈1.63, here we specify the lattice parameter of GaN "a" and take the value a=3.170Å and c=5.166Å. In fact, these parameter values are comparable with those obtained experimentally [23]. In order to achieve a good packing, we should add initially "empty" spheres (ES) with (Z = 0) representing atomic inter-sites. TM impurities are introduced randomly into sites of the Ga in GaN semiconductor with ferromagnetic interaction. TM are substituted randomly and of parallel moment between atoms $Ga_{0.95}(TM_{0.05},up)N$, then the system has a finite magnetization. The DLM (disorder local moment) state is described as $Ga_{0.95}(TM(_{0.025},up),TM(_{0.025},down))N$, the magnetic moment of TM atoms point randomly with respect to each other and the system has no magnetization.

The difference energy, $\Delta E = E_{DLM} - E_{FM}$, between the ferromagnetic and the spin glass states, in the case of TM (V, Cr, Mn, Co, Ni and Fe) doped , with LDA approximation (see Table 1), was computed in order to confirm the stability of the FM or DLM states. This quantity determines the stabilization of the magnetic phases in the DMS. $\Delta E > 0$ corresponds to the fact that the ferromagnetic phase is more stable than spin glass phase, and vice versa. The positive larger ΔE is the more stabilizated ferromagnetism becomes. For TM_1 (Fe, Co and Ni)-doped GaN the spin-glass state is stable (Table 1) because the minority *d* (TM) band is unoccupied Figure 1, no half metallic character was predicted, so there is no carrier to mediate the ferromagnetic state. The calculated magnetic energy differences for TM (Fe, Co and Ni)-doped GaN are in agreement with those already calculated theoretically using the local density approximation [25,26,27].

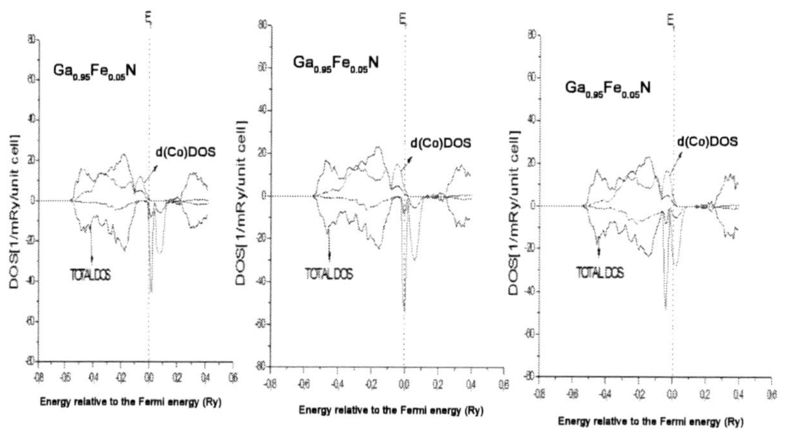

Figure 1. Total and partial DOS for (Ga,TM)N (TM:Fe,Co and Ni).

For TM_2 (V,Cr and Mn) doped GaN the ferromagnetic stability (see Table 1) and half metallic character (see Figure 2) are observed so there is *p*-type carrier to mediate the ferromagnetic state. By electronic structure calculations we can see a large delocalization of Cr-*d* levels situated in GaN gap.

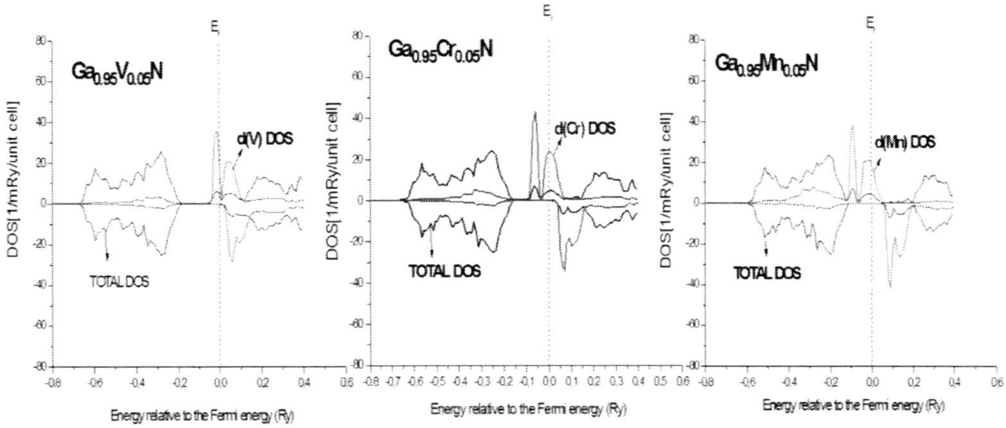

Figure 2. Total and partial DOS for (Ga,TM)N (TM:V,Cr and Mn).

Table 1. The magnetic energy difference $\Delta E = E_{DLM} - E_{FM}$ between the FM state and DLM state for $Ga_{0.925}TM_{0.05}N$ bulk

Bulk	**ΔE(meV)**
$Ga_{0.95}V_{0.05}N$	1.53
$Ga_{0.95}Cr_{0.05}N$	4.55
$Ga_{0.95}Mn_{0.05}N$	2.26
$Ga_{0.95}Fe_{0.05}N$	-8. 3
$Ga_{0.95}Co_{0.05}N$	-6.08
$Ga_{0.95}Ni_{0.05}N$	-1.63

The occupied majority *d* bands are in the host valance band maximum (VBM), the unoccupied minority *d* bands is above the conduction band maximum (CBM), TM_2 is partially filled so the Fermi level locate on majority *t* states and half metallic character was predicted (see Figure 2). As was stated earlier, theory predicts ferromagnetism to occur in TM-doped GaN, only if additional holes carriers are introduced [25-28].

To reveal the relations between electronic structure and optical properties in GaN based DMS the optical absorption spectra modeling is computed for Fe doped GaN (for DLM configuration) and Cr doped GaN (for FM configuration) and compared to find the valence state of TM (Cr,Fe) atoms in (Ga,TM)N. When a TM atom is in a substitutional position it is surrounded by four Ga atoms located on the vertices of a tetrahedron. The tetrahedral crystal field of Ga ions splits 3*d*-states of TM, which are mainly located in the band gap, into e_g and t_{2g}-states. Due to the tetrahedral arrangement of the ligands the t_{2g}-states of the Fe atom hybridize with TM 4*p*-states (see Figure 3). For $Ga_{0.95}Fe_{0.05}N$ which the unoccupied minority *d* bands is above the valance band, we have showed that the states near the Fermi energy are majority dominated by Fe^{3+} valence ground state configuration.

This is in good agreement with optical absorption spectra modeling. Indeed, the calculated X-ray absorption spectrum at the K-edge of $Ga_{0.95}Fe_{0.05}N$ corresponding to the transfer of 1*s* electrons to valence 3*d*-states mediated by 1*s*– 4*p* dipole transition and 4*p*–3*d* hybridization (see Figures. 4). Besides, we observed merging of one absorption lines corresponding to 1s–t_{2g} electron transfers into one line A_1 (0.85eV) (see Figure 5).

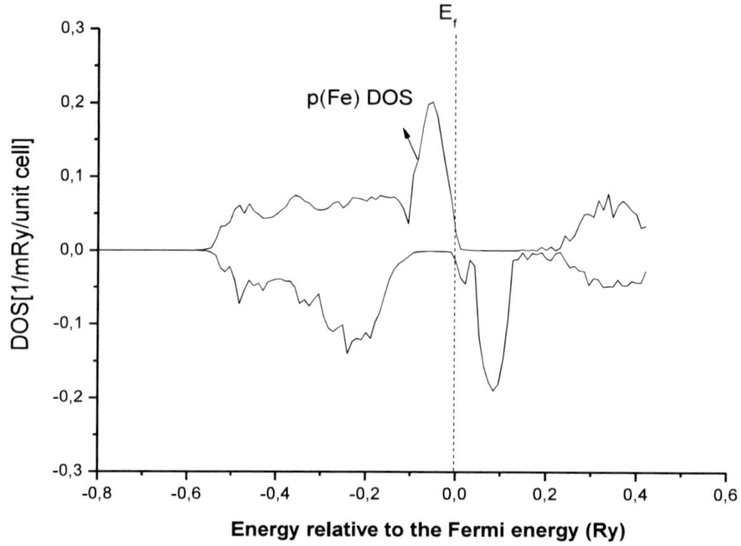

Figure 3. Partial DOS (*p*-Fe) for (Ga,Fe)N.

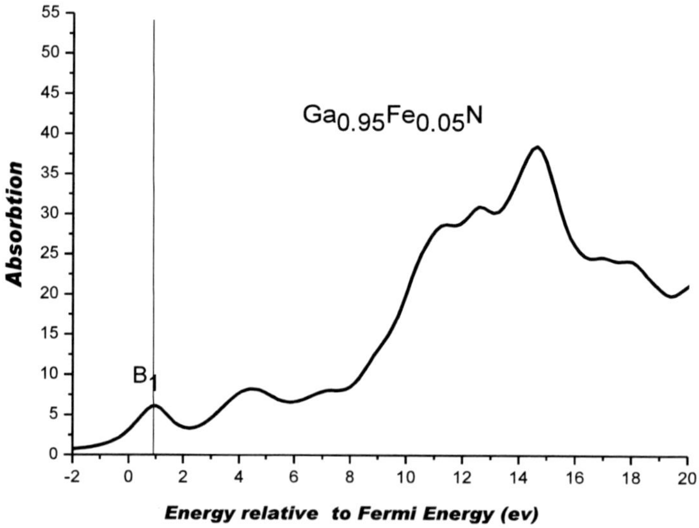

Figure 4. Ray absorption spectrum at the K-edge of $Ga_{0.95}Fe_{0.05}N$.

For $Ga_{0.95}Cr_{0.05}N$, the Fermi energy passes through the spin-up density of states and caused a delocalization of Cr *d*-electrons. It is found that the global energy minimum is obtained in the Cr^{3+} state, because the 3*d-up* state of Cr is partially filled. Indeed, the calculated X-ray absorption spectrum at the K-edge of $Ga_{0.95}Cr_{0.05}N$ corresponding to the transfer of 1*s* electrons to valence 3*d*-states mediated by 1*s*–4*p* dipole transition and 4*p*–3*d* hybridization (see Fi.6).

Besides, there exist two bands of empty *d* states in the gap for down configuration: t_{2g}(up) for spin up A_1(0.22 eV), t_{2g}(down) for spin down A_2 (1.48eV) (Figure 6), consequently they are merging of two absorption lines corresponding to 1s–t2g (A1) and 1s-t2g (A2).

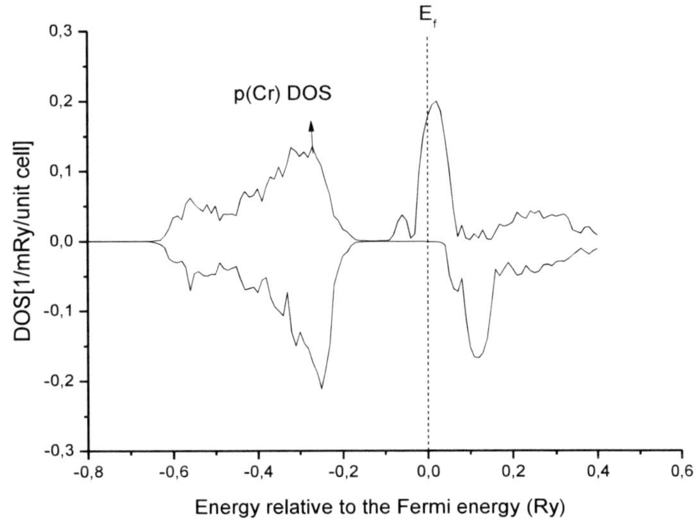

Figure 5. Partial DOS (*p*-Fe) for (Ga,Cr)N.

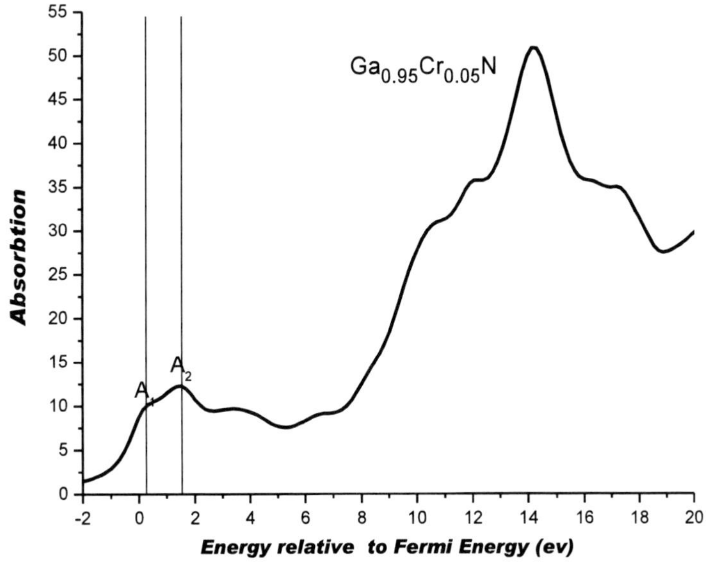

Figure 6. X-ray absorption spectrum at the K-edge of $Ga_{0.95}Cr_{0.05}N$.

Magnetic and Optical Properties of TM Doped GaN Nanofilms

The electronic structures are calculated by using the Korringa–Kohn–Rostoker (KKR) method combined with coherent potential approximation (CPA). Such method is also extended to study the electronic structures of Mn-doped ultrathin ZnO nanofilms [29].

Table 2. The magnetic energy difference $\Delta E = E_{DLM} - E_{FM}$ between the FM state and DLM state for $Ga_{0.925}TM_{0.05}N$ 2 layers

2 layers	ΔE(meV)for
$Ga_{0.95}V_{0.05}N$	2.19
$Ga_{0.95}Cr_{0.05}N$	3.25
$Ga_{0.95}Mn_{0.05}N$	3.01
$Ga_{0.95}Fe_{0.05}N$	0.63
$Ga_{0.95}Co_{0.05}N$	0.78
$Ga_{0.95}Ni_{0.05}N$	0.19

All studied GaN layers, of various thicknesses (2 layers and bulk), are constructed from 3D wurtzite structure. In order to simulate an infinite plane, we consider a periodic boundary conditions, in the layer plane (x and y directions), the film surface is flat. The electronic structure and magnetic properties of TM-doped GaN thin film with and without acceptor defect in totally doping for which the doping is performed homogeneously in the whole layers of the film. Gallium nitride (GaN) nanostructures have triggered extensive interest in the past two decades due to their potential applications in blue and ultraviolet light-emitting diodes and high-temperature/ high power electronic and spintronic devices. Recently, films of (Ga,Fe)N fabricated by means of molecular beam epitaxy (MBE) were reported with Curie temperature above room temperature (RT), Fe doped GaN films show the RT ferromagnetic behavior [23,24]. However, the origin of the exhibited ferromagnetic is still under investigation. This is probably due to many experimental parameters: preparation methods, measurement techniques, substrate choose, thickness of the thin layer, unexpectedly carriers introduced during the synthesis. In this respect, the effects of parameters such as TM defects, thickness, and magnetic stability are investigated in this paper in terms of electronic structure and optical proprieties of TM doped GaN thin films.

Table 2 shows our calculated magnetic energy differences, $\Delta E = E_{AFM}-E_{FM}$, between the ferromagnetic and the spin glass state (AFM under disorder) for TM doped 2 layer. This quantity determines the stabilization of the magnetic phases in the DMS. For all TM-doped GaN 2 layer, the ferromagnetic is stable. The calculated magnetic stability for TM1 (V,Cr and Mn) doped GaN 2 layers agree with those already calculated theoretically for (V,Cr and Mn) doped GaN bulk system using the local density approximation. While the calculated magnetic stability for TM_2 (Fe,Co and Ni) doped GaN 2 layers was changed from already calculated theoretically for TM_2 (Fe,Co and Ni) doped GaN bulk from spin glass state to ferromagnetic state. Taking the case of For $Ga_{0.95}Fe_{0.05}N$ and $Ga_{0.95}Mn_{0.05}N$ 2 layers, the ferromagnetic phases are more stable compared to the spin glass state. This stability can be explained from computed density of state analysis (see Figure 7). Fe and Mn atoms introduce *d* levels in the gap of semiconductor; the splitting of the five-time degenerate atomic minority-spin 3*d* level into the *e* and *t* subgroups is caused by the influence of the crystal environment, the *d* major spin-band of Fe and Mn is partially filled due to large *p–d* exchange interaction valance band and a gap was formed between spin up and spin down *d* states of Fe, Mn atoms, while For $Ga_{0.95}Fe_{0.05}N$ bulk the Fe impurities provide the localized moment, but without acceptor or donor aspect, there are no carriers to mediate ferromagnetic interaction resulting in a spin glass state.

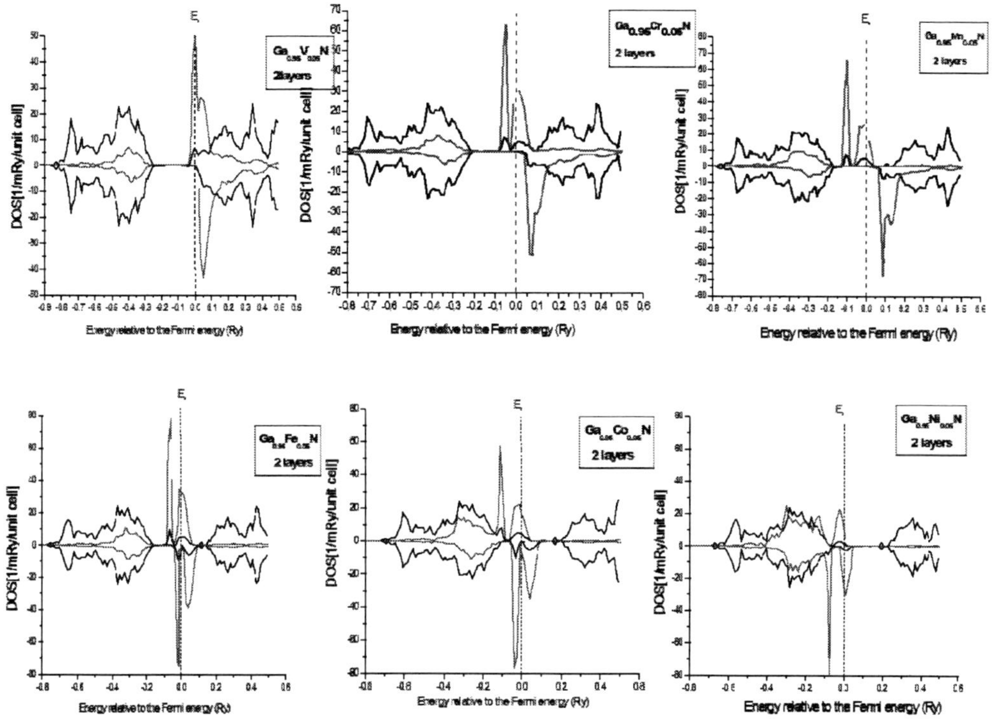

Figure 7. Total and partial DOS for (Ga,TM)N (TM:V,Cr,Co,Ni and Mn) 2 layer.

In the case of $Ga_{0.95}Fe_{0.05}N$ bulk the Fermi energy lies in a region of vanishing electron density and this is due to the fact that the Fe^{3+} valence ions does not introduce any carriers, the states near the Fermi energy are dominated by Fe^{3+} valence ground state configuration, these impurity states show a large exchange splitting, leading to the high-spin configuration of *d* electrons and this exchange splitting is consistently larger than the crystal-field splitting (see Figure 1). For $Ga_{0.95}Fe_{0.05}N$ 2 layers (Figure 8) we observed an inverse comportment, the crystal-field splitting larger than exchange splitting and nearly half metal character was observed. For $Ga_{0.95}TM_{0.05}N$ 2 layers, the Fermi energy passes through the spin-up density of states leading to a delocalization of TM *d*-electrons; we found that the global energy minimum is now obtained in the TM^{3+} state because the 3*d* state of TM is partially filled. As was stated earlier, theory predicts ferromagnetism to occur in Mn-doped GaN, only if additional holes carriers are introduced [25-28].

Defect Effect in Magnetic Properties of TM Doped GaN Bulk and Nanofilms

Point defects have an important effect on the electronic and magnetic properties of semiconducting materials. In particular, the identification and the quantification of such points help to understand the microscopic processes leading to the specific properties of semi-conductors. Defects and impurities in semiconducting materials may be native or embedded during the manufacturing process with a major contribution from defects induced by implantation.

Table 3. The magnetic energy difference $\Delta E = E_{DLM} - E_{FM}$ between the FM state and DLM state for $Ga_{0.925}TM_{0.05}Vc_{0.025}N$ bulk

Bulk	ΔE(meV)
$Ga_{0.925}V_{0.05}Vc_{0.025}N$	2.10
$Ga_{0.925}Cr_{0.05}Vc_{0.025}N$	1.25
$Ga_{0.925}Mn_{0.05}Vc_{0.025}N$	2.91
$Ga_{0.925}Fe_{0.05}Vc_{0.025}N$	2.45
$Ga_{0.925}Co_{0.05}Vc_{0.025}N$	1.28
$Ga_{0.925}Ni_{0.05}Vc_{0.025}N$	-0.7

Table 4. The magnetic energy difference $\Delta E = E_{DLM} - E_{FM}$ between the FM state and DLM state for $Ga_{0.925}TM_{0.05}Vc_{0.025}N$ 2 layer

2 layers	ΔE(meV)
$Ga_{0.925}V_{0.05}Vc_{0.025}N$	1.83
$Ga_{0.925}Cr_{0.05}Vc_{0.025}N$	3.42
$Ga_{0.925}Mn_{0.05}Vc_{0.025}N$	3.19
$Ga_{0.925}Fe_{0.05}Vc_{0.025}N$	0.38
$Ga_{0.925}Co_{0.05}Vc_{0.025}N$	0.89
$Ga_{0.925}Ni_{0.05}Vc_{0.025}N$	-0.01

The latter is a fundamental step in the elaboration of doping materials. However, ion implantation damage in materials and their evolution are still very limited and deserve deeper investigations [30–39]. The properties of defects created (their nature, charge state, and their evolution with thermal treatments) are not yet sufficiently developed, except for some molecular dynamics simulations. Among many candidates, GaN and its derived doped materials have been recently in the focus of an intense and ongoing topic research in the condensed matter physics and related topics. A huge number of experimental works have been prepared for the TM doped GaN system. However, despite of possibility refinements in the synthetic method and of treatment improvements to control the various defects concentrations, the GaN doped system still remains far from a clear picture. In particular, the fundamental nature of the magnetic interactions between transition metal impurities, and the origin of the magnetism in TM doped GaN are still under development. Experimental results and theoretical calculations proposed that ferromagnetism in such a doped material has been due to the existence of intrinsic point defects. This suggestion has triggered extensive studies and investigations. The main objective of this work is to investigate the role of *p*-type defects on magnetic properties of GaN material doped with some TM concentrations in bulk and nanofilms. More precisely, we discuss the magnetic transition between DLM and FM states in such a system. This can be done by calculating the magnetic energy difference $\mathbf{\Delta E} = E_{DLM} - E_{FM}$ between the FM state and DLM state (or spin glass state). This quantity can determine the stabilization of the magnetic phase in the DMS using the KKR-CPA simulation method. In particular, the negative value of this quantity corresponds to the fact that the spin glass phase is more stable than FM state and vice versa. In the case of TM-doped GaN with concentrations up to 5 % of TM with acceptor defect, it has been shown that the system has a ferromagnetic behavior for bulk and nanofilms except for the case of Ni doped GaN (see table 3 and 4) acceptor defect. In this doped (Ga, Ni)N system, the spin glass state is stable due to the absence of itinerant carrier.

Our performed calculations, presented in Table 3 and 4, on magnetic energy difference for the same concentration of TM-doped in GaN with Ga vacancies defect, recover the theoretical results obtained without Ga vacancies defect (table 1 and 2). Moreover, certain native point defects play a crucial role in stabilizing the FM state in a doped GaN system. In connection with this result, it is shown in table 2 that the origin of the ferromagnetic is intrinsic.

However, the sample with defects is DLM, while those sharing defects may become ferromagnetic. It follows from these results that we can convert FM states to DLM ones by introducing holes into the system. For (Ga,Fe)N and (Ga,Co)N bulk systeme the magnetic ground state change from DLM to ferromagnetic by acceptor defect (Ga vacancies). Indeed, (Figure 9), show from elecronic structure calculations, the delocalization of holes carriers from Vc (vacancy site) to *p*-state of N, proved a large delocalization of (Fe,Co)-*d* levels. By effect of acceptor defect in (Ga,Fe)N the 3*d* electrons of Fe shift more to the valence band Figure 9. and increases the exchange splitting.

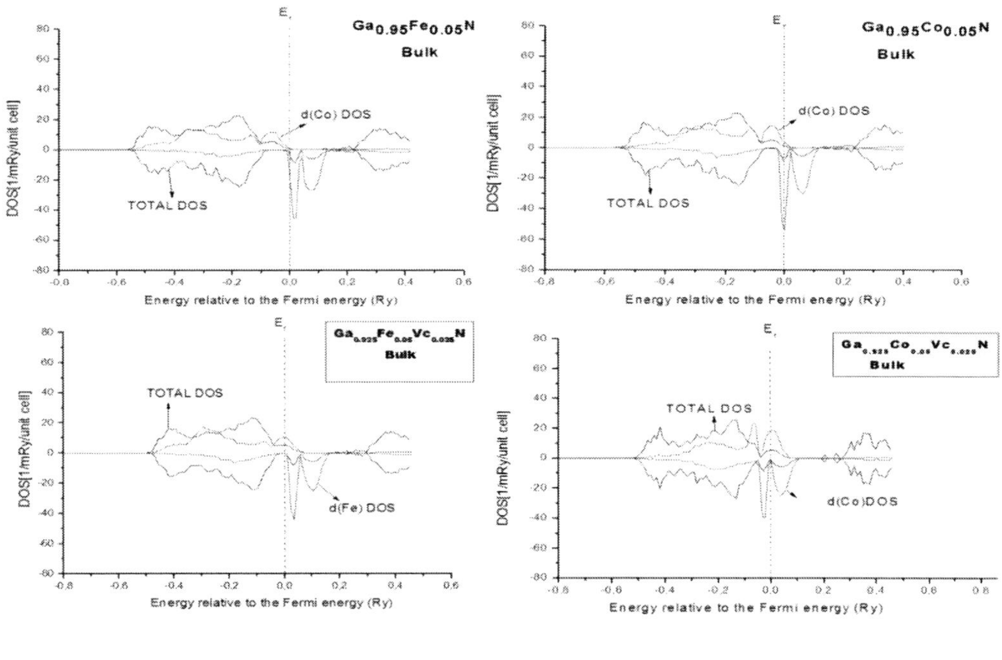

Figure 8. Total and partial DOS for (Ga,TM)N (TM:Co and Fe) 2 layer and bulk with and without defect.

As shown in (Figure 8) impurity-*d* states show substantial hybridization with valence-*p* states. Considering the local symmetry of the Ga-substitutional site, it is concluded that $3d_\alpha$ orbitals, which have the same symmetry as the functions xy, yz and zx, hybridize well with *p* orbitals of valence bands. Consequently, they make bonding states (t_b), and their anti-bonding counterparts (t_a). On the other hand, $3d_\beta$ orbital's, which have the same symmetry as the functions $3z^2$-r^2 and x^2-y^2; extend to the interstitial region to make nonbonding states (e). It was already proposed that the origin of the ferromagnetism in DMSs was the double exchange mechanism.

From the viewpoint of that mechanism, it is a key point to stabilize the ferromagnetic state if there are mobile carriers with d character or not. For p-type (Ga,Fe)N ($Ga_{0.925}Fe_{0.005}Vc_{0.025}N$) the ferromagnetic state is stabilized because there are carriers in itinerant t_a states. These itinerant electrons in t_a states participate in the double exchange mechanism to lower the kinetic energy of the ferromagnetic state. On the other hand, if the *d*-orbitale is completely occupied then this reduction of energy via hopping is no possible and the system energy gets lowered by antiferromagnetic spin alinment of neighboring Fe atoms by super exchange.

Size Effect in Magnetic Properties of TM Doped GaN

For all TM doped GaN 2 layers, the ferromagnetic phases are more stable compared to the spin glass state, whereas for TM doped GaN bulk only TM (Fe,Co and Ni) doped GaN are ferromagnetic, this stability can be explained from computed density of state analysis. For TM doped GaN 2 layers, TM atoms introduce *d* levels in the gap of semiconductor; the splitting of the five-time degenerate atomic minority-spin 3*d* level into the e and t subgroups is caused by the influence of the crystal environment. The magnetism in the GaN-based DMS under size effect (in the case of Fe and Mn doped GaN) is investigated for with 2 and bulk see Figure 10. For Mn doped GaN bulk and nanofilm according to electronic structure calculations, the *d* major spin-band of Mn is partially filled due to large *p–d* exchange interaction valance band and a gap was formed between spin up and spin down d states of Mn atoms. For Fe doped GaN bulk the Fe impurities provide a localized moment, but without acceptor or donor aspect, there are no carriers to mediate ferromagnetic interaction resulting in a spin glass state. The Fermi energy lies in a region of vanishing electron density and this is due to the fact that the Fe^{3+} valence ions does not introduce any carriers, the states near the Fermi energy are dominated by Fe^{3+} valence ground state configuration, these impurity states show a large exchange splitting, leading to the high-spin configuration of *d* electrons and this exchange splitting is consistently larger than the crystal-field splitting (see Figure 10). For Fe doped GaN 2 layers (Figure 9) we observed an inverse comportment, the crystal-field splitting larger than exchange splitting and nearly half metal character was observed, the Fermi energy passes through the spin-up density of states leading to a delocalization of Fe *d*-electrons; we found that the global energy minimum is now obtained in the Fe^{4+} state because the 3*d* state of Fe is partially filled. As was stated earlier, theory predicts ferromagnetism to occur in Fe-doped GaN, only if additional holes carriers are introduced [25-28].

Table 5. The magnetic energy difference ΔE = EDLM − EFM between the FM state and DLM state for $Ga_{0.95}TM_{0.05}N$ for 2 layers and bulk

	ΔE(meV)for 2 layers	ΔE(meV)for Bulk
$Ga_{0.95}V_{0.05}N$	2.19	1.53
$Ga_{0.95}Cr_{0.05}N$	3.25	4.55
$Ga_{0.95}Mn_{0.05}N$	3.01	2.26
$Ga_{0.95}Fe_{0.05}N$	0.63	-8. 3
$Ga_{0.95}Co_{0.05}N$	0.78	-6.08
$Ga_{0.95}Ni_{0.05}N$	0.19	-1.63

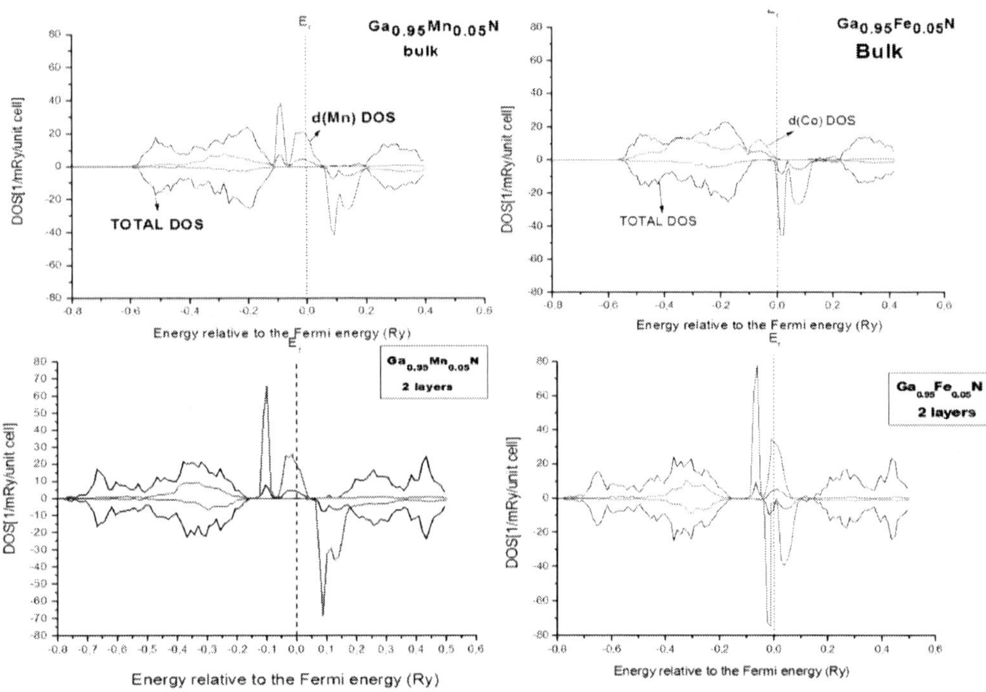

Figure 9. Total and partial DOS for (Ga,TM)N (TM:Mn and Fe) 2 layers and bulk.

Conclusion

In summary, we have showed that the doping TM atoms have an attractive interaction in GaN nanofilm or bulk with and without p-type defect, we showed that layer of thickness and acceptor defect can switch the magnetic ordering from DLM to FM. In the FM ordering, the material is half-metallic and therefore ideal for spin injection, the optical absorption spectra was computed by ab initio calculations and confirm the ferromagnetic stability based on the charge state of magnetic impurities. We show that density functional theory is a very powerful tool for: 1- studying the effects of local doping, defects effect, size and disorder effect on the magnetic and optical properties of these materials, and predicting properties of new materials.

References

[1] M. Zaja¸ C. R. Doradzin´ski, J. Gosk, J. Szczytko, M. Lefeld-Sosnowska, M. Kamin´ska, and A. Twardowski, *Appl. Phys. Lett.,* 78, 1276 (2001).

[2] T. Sasaki, S. Sonoda, Y. Yamamoto, K. Suga, S. Shimizu, K. Kindo, and H. Hori, *J. Appl. Phys.,* 91, 7911 (2002).

[3] G. T. Thaler, M. E. Overberg, B. Gila, R. Frazier, C. R. Abernathy, S. J. Pearton, J. S. Lee, S. Y. Lee, Y. D. Park, Z. G. Khim, et al., *Appl. Phys. Lett.,* 80, 3964 (2002).

[4] M. L. Reed, N. A. El-Masry, H. H. Stadelmaier, M. K. Ritums, M. J. Reed, C. A. Parker, J. C. Roberts, and S. M. Bedair, *Appl. Phys. Lett.,* 79, 3473 (2001).

[5] R. Y. Korotkov, J. M. Gregie, and B. W. Wessels, Appl. Phys. Lett. 80, 1731 (2002).

[6] S. Kuwabara, T. Kondo, T. Chikyow, P. Ahmet, and H. Munekata, Jpn. *J. Appl. Phys.,* 40, L724 (2001).

[7] S. J. Pearton, C. R. Abernathy, M. E. Ovenberg, G. T. Thaler, D. P. Norton, N. Theodoropoulou, A. F. Hebard, Y. D. Park, F. Ren, J. Kim, et al., *J. Appl. Phys.,* 93, 1 (2003).

[8] A. Y. Polyakov, A. V. Govorkov, N. B. Smirnov, N. Y. Pashkova, G. T. Thaler, M. E. Overberg, R. Frazier, C. R. Abernathy, S. J. Pearton, J. Kim, et al., *J. Appl. Phys.,* 92, 4989 (2002).

[9] J. Kim, F. Ren, G. T. Thaler, R. Frazier, C. R. Abernathy, S. J. Pearton, J. M. Zavada, and R. G. Wilson, *Appl. Phys. Lett.,* 82, 1565 (2003).

[10] S. Dhar, O. Brandt, A. Trampert, K. J. Friedland, Y. J. Sun, and K. H. Ploog, *Phys. Rev.,* B 67, 165205 (2003).

[11] H. Toyosaki, T. Fukumura, K. Ueno, M. Nakano, and M. Kawasaki, *Jpn. J. Appl. Phys.,* Part 2 44, L896 (2005).

[12] W. G. Wang, C. Ni, T. Moriyama, J. Wan, E. Nowak, and John Q. Xiao, *Appl. Phys. Lett.,* 88, 202501 (2006).

[13] S. Yuasa, T. Nagahama, A. Fukushima, Y. Suzuki, and K. Ando, *Nat. Mater.*, vol. 3, p. 868, 2004.

[14] A. FERT: Nobel lecture: development and future of spintronics. *Rev. Mod. Phys.,* 80, 1517, 2008.

[15] H. Ohno, *Science,* 281, 951 (1998).

[16] T. Dietl, H. Ohno, and F. Matsukura, *Phys. Rev.,* B 63, 195205 (2001).

[17] T. Jungwirth, J. K_nig, J. Sinova, J. Kucˇera, and A. H. MacDonald, *Phys. Rev.,* B 66, 012402 (2002).

[18] H. Ohno, D. Chiba, F. Matsukura, T. Omiya, E. Abe, T. Dietl, Y. Ohno, and K. Ohtani, *Nature,* 408, 944(2000).

[19] H. Boukari, P. Kossacki, M. Bertolini, D. Ferrand, J. Cibert, S. Tatarenko, A. Wasiela, J. A. Gaj, and T. Dietl, *Phys. Rev. Lett.,* 88, 207204 (2002).

[20] H. Akai, P.H. Dederichs, *Phys. Rev.,* B47 (1993) 8739.

[21] H. Akai, *Phys. Rev. Lett.,* 81 (1998) 3002.

[22] S.H. Vosko, L. Wilk, M. Nusair, *Can. J. Phys.,* 58 (1980) 1200.

[23] T. Fukumura, Zhengwu Jin, A. Ohtomo, H. Koinuma, M. Kawasaki, *Appl. Phys. Lett.,* 75, 3366 (1999).

[24] MACHIKANEYAMA2002v09: H. Akai , Departement of Physics, Graduate School of Science ,Osaka University, Machikaneyama 1-1,Toyonaka 560-0043, Japan,
[25] O. Mounkachi, A. Benyoussef, A. El Kenz, E.H. Saidi, J. Magn. Magn. Mater. *J. Magn. Magn. Mater.,* 321, 2402–2406 (2009).
[26] O. Mounkachi, A. Benyoussef, E. H. Saidi, A. El Kenz, E.K. Hlil. *Journal of Physics European Physical Journal,* B 74 (4), pp. 463-466 (2010).
[27] E. Salmani, O. Mounkachi, H. Ez-Zahraouy, A. Benyoussef, M. Hamedoun, E.K. Hlil: Chemical Physics Letters, Volume 588, 19 November 2013, Pages 242-246.
[28] T. Dietl, et al, *Science,* 287, 1019 (2000).
[29] E. Salmani, A. Benyoussef, H. Ez-Zahraouy, E. H. Saidi and O. Mounkachi 2012 Chinese Phys. B 21 106601 doi:10.1088/1674-1056/21/10/106601
[30] Chattopadhyay, S., Neogi, S.K., Sarkar, A., Mukadam, M.D., Yusuf, S.M., Banerjeea, A., Bandyopadhyay, S.: 81.05.Dz, 61.72.J-, 81.40.Rs
[31] Sharma, P., Sreenivas, K.: *Appl. Phys. Lett.,* 83, 3617 (2003).
[32] Wang, Q., Sun, Q., Chen, G., Kawazoe, Y., Jena, P.: *Phys. Rev. Lett.,* 95, 225502 (2005).
[33] Xu, Q., Schmidt, H., Zhou, S., Potzger, K., Helm, M., Hochmuth,H., Lorenz, M., Setzer, A., Esquinazi, P., Meinecke, C., Grundmann, M.: *Appl. Phys. Lett.,* 92, 082508 (2008).
[34] Iu¸san, D., Sanyal, B., Eriksson, O.: *Phys. Rev.,* B 74, 235208 (2006).
[35] Sato, K., Katayama-Yoshida, H.: Jpn. *J. Appl. Phys.,* 40, L485 (2001).
[36] O. Mounkachi, A. Benyoussef, A. El Kenz, E.H. Saidi, E.K. Hlil, *J. Appl. Phys.,* 106, 093905 (2009).

Dr. Omar Mounkachi, MAScIR (Moroccan Foundation for Advanced Science, Innovation and Research).

In: Gallium Nitride
Editor: Kiera Olivia Peak
ISBN: 978-1-63321-387-6

Chapter 11

INDIUM GALLIUM NITRIDE (InGaN) FOR PHOTOVOLTAIC APPLICATION

O. Mounkachi[1,*], E. Salmani[2], H. Ez-Zahraouy[2], M. Hamedoun[1] and A. Benyoussef[1,2]

[1]Institute for Nanomaterials and Nanotechnologies, MAScIR, Rabat, Morocco
[2]LMPHE, Faculté des Sciences, Université Mohammed V-Agdal, Rabat, Morocco

Abstract

Due to their technological application, a precise prediction for the optical properties of III-nitride alloys is significantly attractive. Numerical simulation based on first-principles calculations is applied to study the structural characteristics and band-energy properties of the zincblende $Ga_{1-x}In_xN$. The purpose of this study is to explore theoretically the electronic and optical characteristics of the parent InN and GaN compounds and their ternary alloys by utilizing KKR-CPA approach. We show that density functional theory is a very powerful tool for 1-studying the local doping effects, defects effect, size and disorder effect on optical and electronic properties of these materials, and predicting properties of new materials.

Keywords: Gallium Nitride; photovoltaic application; density functional theory

Introduction

The global energy situation today is worrying. We consumed 8,353 Mtep (1 Mtep = 11 630 GWh.) in the world (17% electrical (1441 or 16 760 Mtoe TWh)). While this share is low compared to fossil fuels (coal, oil, gas), but electricity is nevertheless essential in our everyday lives, such as for the operation of computers, machinery, lighting, heating, etc. as shown in Figure 1 the word will require 56 percent more energy in 2040 than 2010.

---

* E-mail addresses: o.mounkachi@mascir.com, kachiomar@gmail.com (Corresponding author)

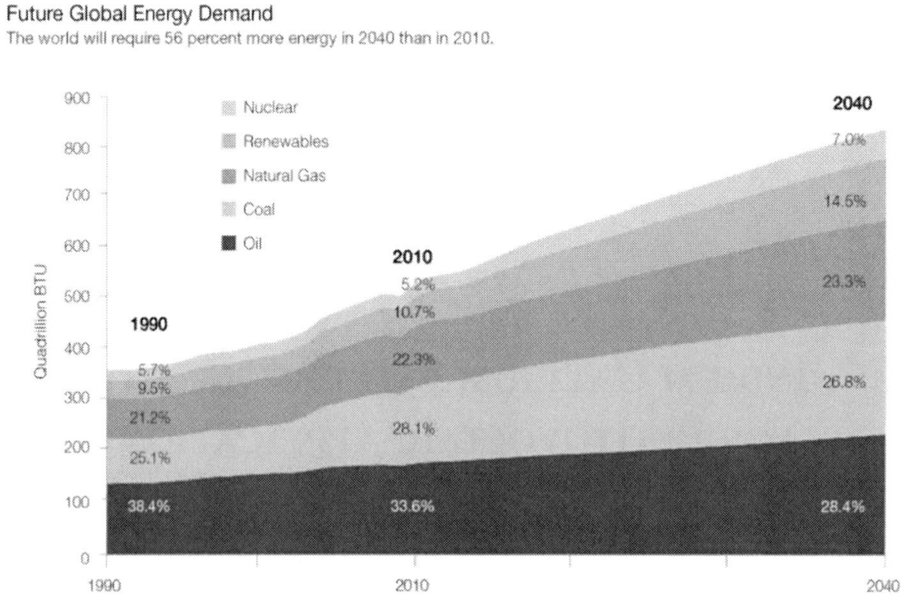

Source: EIA, International Energy Outlook 2013.

Figure 1. Future global energy demand.

The depletion of natural resources plays a major role in political energy of the states since the current state of knowledge, the oil reserves are estimated at 47 years, 60 years for gas. Besides the problems of supply, scarcity of resources causes a sharp rise in prices. For example, Brent crude raised 20 USD / barrel in 2000 to 110 USD / barrel in 2011. This is why, in this context, countries are seeking for new energy sources to diversify their energy mix. They want to spend stocks energy (oil, gas, coal, nuclear) to energy flow, renewable (hydro, wind, solar, biomass, geothermal, etc.). Among these, solar photovoltaic is emerging as a promising solution for the production of electricity. Its main advantages are the direct production of electricity without passing through a turbine and its wide potential installation on the roofs of buildings. Nevertheless, it is a complex technology, relatively expensive and suffers from low efficiency. In addition, we do not know today effectively store electricity produced in excess in order to have access in periods of low output (bad weather, night, winter, etc.). Today, photovoltaic electricity production is dominated by silicon with 95% of photovoltaic panels (monocrystalline, polycrystalline and thin film) [1]. However, these commercial panels suffer from a low average yield, about 13% (between 12 and 18% for monocrystalline, between 12 and 15% for polycrystalline and between 6 and 9% for thin layers [2]). This is why many are turning research on increasing the efficiency of silicon panels (nanostructures, thin plasmonic layer, etc.) and alternative materials with a higher yield (CdTe, CIGS, GaAs, InGaN etc.). Generally, these new materials are deposited in thin layers (a few hundred nanometers), but still have attractive yields, thereby reducing production costs. Among the new thin film materials considered for photovoltaics, InGaN alloy is an outsider. It was initially studied for light emitting diodes applications (LEDs) and UV sensors [3]. Since the early 2000's, InGaN was studied for photovoltaics application [4-7] due to its broad and flexible forbidden bad gap energy (energy from which the material responds more effectively to the incident light).

Indeed, the infrared absorbing InN with a gap of 0.7 eV (1771 nm) and the GaN absorbing ultraviolet with a gap of 3.39 eV (366 nm), the InGaN can cover a large part of the solar spectrum. It can therefore be, for example, easily design multi-junction solar cells based on InGaN simply by changing the concentration of indium in the GaN. Such structures and simplify the manufacturing processes of cells and thereby reduce costs while producing high yields of photovoltaic conversion. Today, single solar cells or multi-junctions exist in the laboratory but suffer from too high price for their industrialization. One factor increasing the cost of these components is the substrates. Indeed, InGaN cannot be manufactured on a massive scale: thin film is deposited on a substrate, traditionally sapphire. However, the latter, in addition to a high cost, has enough structural parameters different from those of the InGaN deteriorating performance optoelectronic components.

Thus, we are interested in replacing sapphire with less expensive materials. Wide direct band gap of GaN and its alloys offer a specific interest in the realization of optoelectronic devices emitting in the blue and ultraviolet light emitting diodes as (LEDs) and laser diodes (LDs) or devices detecting ultraviolet radiation.

For over a decade, the III-V nitride materials are intensively studied for optoelectronics applications in the UV and blue. The energy gap of InN is lowered to 0.7 eV paving the way for alloys can cover almost the entire solar spectrum. In particular cases, the InGaN alloy were extensively studied for photovoltaic applications due to its wide spectral coverage, its good electrical properties and resistance to high power. However, this alloy has some disadvantages such as lack of suitable substrate, hard growth percentages higher than 30% and p-doped Indium nitrides that are still inefficient. It has been reported theoretically that the fulfillment of active materials can acquire solar cells with solar energy conversion efficiency higher than 50% for InGaN alloys, if the In content is close to 40% [7].

Furthermore, for the greatest solar energy conversion efficiency, III-nitride multi-junction solar cells possessing a nearly perfect band gaps, should involve InGaN layers with higher In concentration or narrower energy band gaps.

Structural Characteristics of Nitrides

Under ambient conditions, the structure of III nitrides is wurtzite, hexagonal. This structure is defined by three parameters, the width of a hexagonal side a, the height c of the elementary mesh, and the inner u parameter describing the separation of anions (N^{3-}) and cations (Ga^{3+}) according to the c axis. The latter is defined as the length of the cation-anion binding divided by c. It is equal to 0.375 for an ideal wurtzite crystal. According to the group theory, the wurtzite structure belongs to the symmetry group P63mc (Figure 2).

The reciprocal space is also often used to describe a structure. It corresponds to the space of wave describing the periodicity of the lattice vectors. In reciprocal space, the unit cell of direct space is identified with the first Brillouin zone.

The latter can be defined as the volume bounded by surfaces from all points equidistant of the atom and its nearest neighbors. Figure 3 shows the Brillouin zone for a hexagonal structure.

Some high symmetry points of particular interest were the critical points. Among them, the center of the Brillouin zone is called the Γ points.

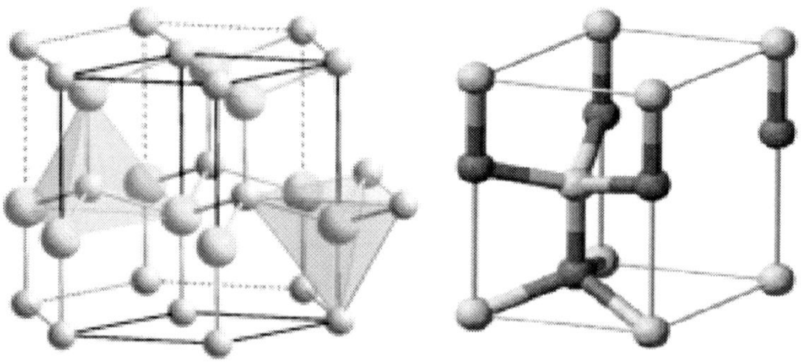

Figure 2. The wurtzite structure of III nitrides and the elementary mesh.

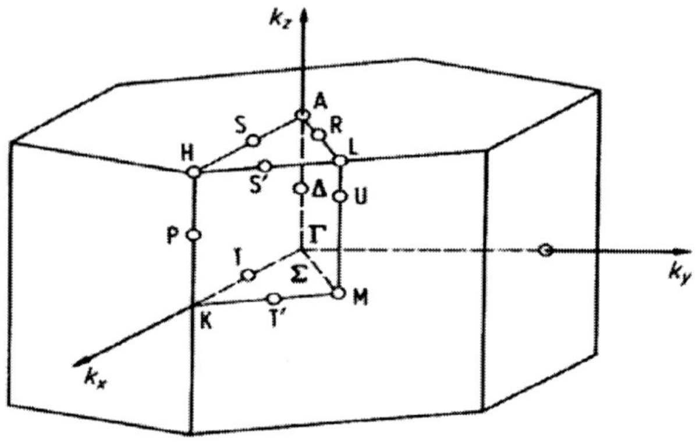

Figure 3. The Brillouin zone for a hexagonal structure.

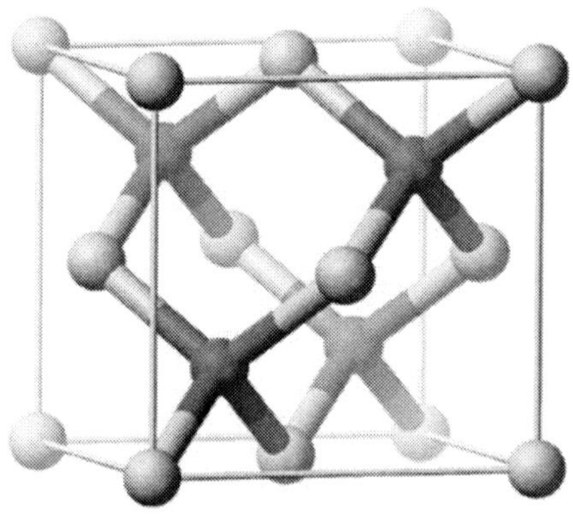

Figure 4. The zinc-blende structure of III nitrides.

Under certain conditions of pressure and temperature on a cubic substrate, it is possible to grow the III-nitrides in cubic structure in the form of zinc-blende. This structure belongs to the symmetry group F4 3m.

The stacking sequence for the (111) plane is ABCABC. The zinc-blende structure, taking the example of GaN is shown in Figure 4 in different directions.

Compared to the hexagonal structure, it has theoretically better electrical properties such as increased mobility, isotopic properties (due to the cubic symmetry) and a high optical gain.

However, these benefits have not been observed experimentally due to the difficulty in producing a uniphasic material with low defect density.

Figure 5 shows the Brillouin zone for a zinc-blende structure. Some high symmetry points of particular interest were the critical points.

Among them, the center of the Brillouin zone is called the Γ points.

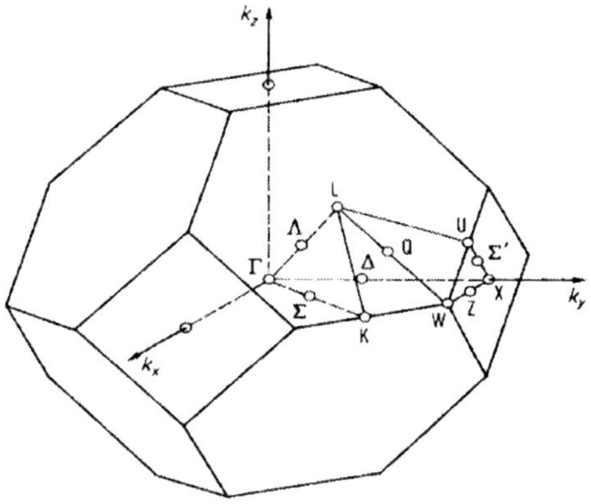

Figure 5. The Brillouin zone for a zinc-blende structure.

Photovoltaic Cell

The photovoltaic effect was discovered by Alexandre Edmond Becquerel in 1839. Photovoltaic effect is achieved by absorption of photons in a semiconductor material that generates an electric voltage. Photovoltaic solar energy derives from the direct conversion of sunlight into electricity. Photovoltaic solar cell modules the performance of this conversion of energy by producing DC power, which can be used to power a device or charge a battery.

Photovoltaic conversion was performed for the first time in the Bell Laboratory in 1954 on a single PN junction of Silicium.

As a result, many semiconductors have been used for this purpose.

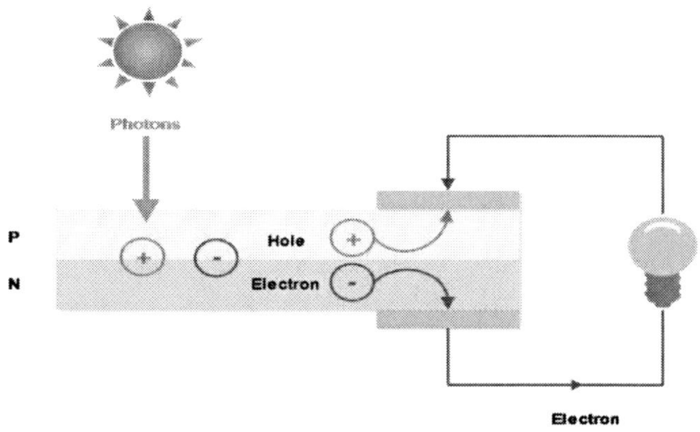

Figure 6. Photovoltaic effect.

Three basic processes involved in the photovoltaic effect (Figure 6):

- The light absorption in the material.
- Transfer of photon energy to electrical charges.
- The charge collection.

It is therefore clear that a material must have specific optical and electrical properties to enable solar conversion.

The photovoltaic effect used in solar cells can directly convert light energy from sunlight into electricity through the generation and transmission in a semiconductor material of the positive and negative electric charges due to the light effect. This material consists of two parts, one with an excess of electrons and the other an electron deficit, respectively called ***N*** type doped and ***P*** type doped. When the first is contacted with the second, the excess of electrons in the ***N*** material diffuse into ***P*** material. The initially area ***N*** doped becomes positively charged, and the initially area ***P*** doped becomes negatively charged. It thus creates an electric field between them which tends to push the electrons in the ***N*** area and holes to ***P*** area. So a ***P-N*** junction is formed. By adding the metal contacts in ***N*** and ***P*** area, a diode is obtained. When the junction is illuminated, the photon energy equal to or greater than the width of the band gap energy transfer their energy to the atoms, each passed an electron from the valence band into the conduction band and also leaves a hole capable to move, resulting in a electron-hole pair. If a load is placed across the cell, electrons from the ***N*** area join ***P*** area the holes via the outer connection, giving rise to a potential difference and an electric current flows. The principle of operation of a solar cell is illustrated in Figure 7. The incident photons create carriers in ***N***, ***P*** area and space charge regions. Behavior of these free carriers differs according to the location their creation. In electrically neutral ***P*** and ***N*** areas, minority photo carriers diffuse. Those who reach the space charge region are powered by the electric field to the region where they become majority. These carriers contribute to the current picture so they create current photo diffusion. In the space charge area, the holes created by photons electron pairs are dissociated by the electric field, the electron is propelled toward the region of the whole ***N*** type and to the region ***P*** category. Holders give rise to a current photo generation. Both contributions are added to create a resulting current ***I p h*** helping to reverse current of the photo diode. The band gap (energy gap) ***E g*** a semiconductor determines how a solar cell to light does. Table 2.3 shows the values of band gap energy for some semi-conductor. The band gap of semiconductor material determines the wavelength of the light necessary to generate electrical energy. The relationship between the band gap and the wavelength is:

$$\lambda(nm) = \frac{hc}{E_g(eV)} = \frac{1.24}{E_g(eV)}$$

where λ is the wavelength n***m***, ***h*** is Planck's constant, ***c*** is the speed of light in vacuum and ***E g*** is the energy band gap in ***e V***. In the case of gallium arsenide, the wavelength matching **1 2 4**. ***e V*** is **0. 8 7 3** n***m***.

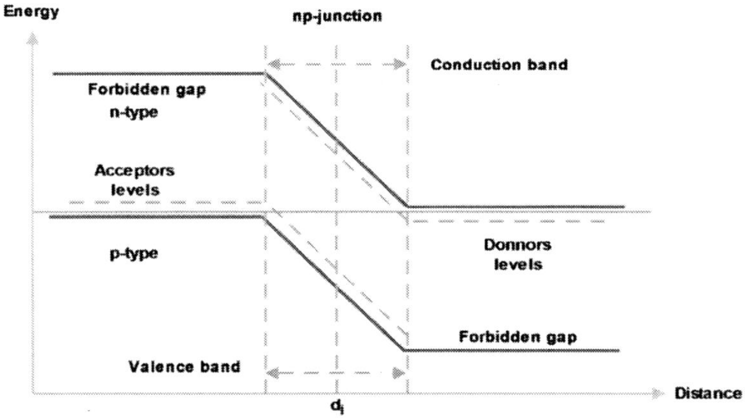

Figure 7. PN junction of photovoltaic cell.

Table 1. Band gap energy of some Semiconductor

Material	Gap(eV)@300K
Si	1.12
Ge	0.66
GaAs	1.42
InP	1.34

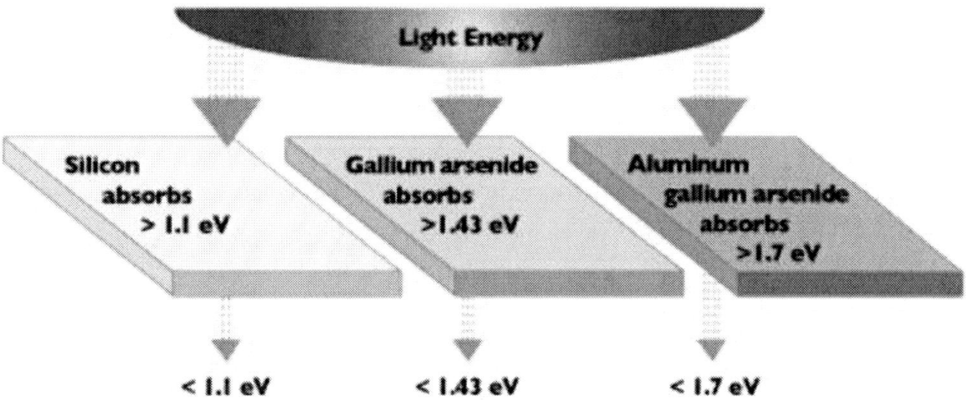

Figure 8. Effect of light energy on different band gaps Eg Band gap energy of nitrides.

Figure 8 is used to view the concept of absorption of light. When light has a higher energy $1\ 1.\ e\ V$, silicon generates electricity. Energy light below $1.\ 1\ e\ V$ is unusable. The same applies to arsenire gallium light energy than $1\ 3\ 4.\ e\ V$ excites electrons of the outer layer. And finally, arsenire aluminum gallium requires light energy greater than $1.\ 7\ e\ V$.

Due to their technological application, a precise prediction for the optical properties of III-nitride alloys is significantly attractive. To the best of our knowledge, first-principles investigations based on DFT for investigating the optical properties of InGaN alloys are barely pointed out.

The purpose of this study is to explore theoretically the electronic and optical characteristics of the parent InN and GaN compounds and their ternary alloys by utilizing KKR-CPA approach. For a fairly good description of the energy band gaps of our considered systems, the LDA-SIC approximation has been employed for handling the exchange and correlation term. In this chapter, we focus on the cubic phase of III-nitride semiconductors, namely GaN and InN and their ternary alloys because of their technological importance in the cubic nitride based devices. The crystalline growth of InGaN alloys was successfully performed particularly for the largest compositions of In in the ternary film (>30%) [8].

In this chapter, we explore the effect of In concentration on the electronic and optical properties of $In_xGa_{1-x}N$ alloys by adopting KKR-CPA method, as embodied in the MACHIKANEYAMA2002v09 package produced by Akai [9]. It is shown that for any degree of substitution for $In_xGa_{1-x}N$ has a direct band gap and no impurity state is induced in the band gap. We propose a possibility of substitution doping of In in GaN photovoltaic material to reduce the cost of materials for photovoltaic solar cells production. In the present calculations, the disordered effect was used, in $Ga_xIn_{1-x}N$, In are supposed to distribute randomly at the GaN sites. This kind of substitution disorder is treated by using the coherent-potential approximation (CPA). The CPA method can be efficiently combined with the Korringa–Kohn–Rostoker (KKR) method. The electronic structure is calculated based on the local density approximation (LDA) in the density functional theory. For a wide variety of materials, it is known that the LDA predicts reasonable electronic structure and many material designs have been proposed based on the LDA calculations. However, it also well recognized that the LDA shows systematic error in the description of the electronic structure. For example, band gap energy is often underestimated and the position of occupied *d*-states is predicted at too high energy.

Figure 9 and 10 shows the calculated band structure of GaN within LDA and LDA-SIC approximations respectively. The band gap (Eg) of GaN within LDA (Eg(LDA)= 2.80 eV) is less than that within LDA-SIC (Eg (LDA-SIC) =3.40 eV), the underestimation of the band gap by LDA is due to the lack of a discontinuity in this exchange correlation potential. Although, the band gap in LDA is always less than the experimental value, it can be corrected by LDA-SIC approximation, while the experimental band gap energy is 3.5 eV [10].

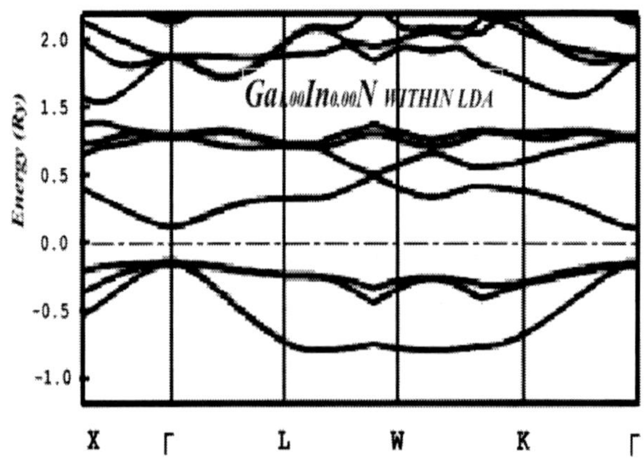

Figure 9. Band structure calculation of GaN within LDA approximations.

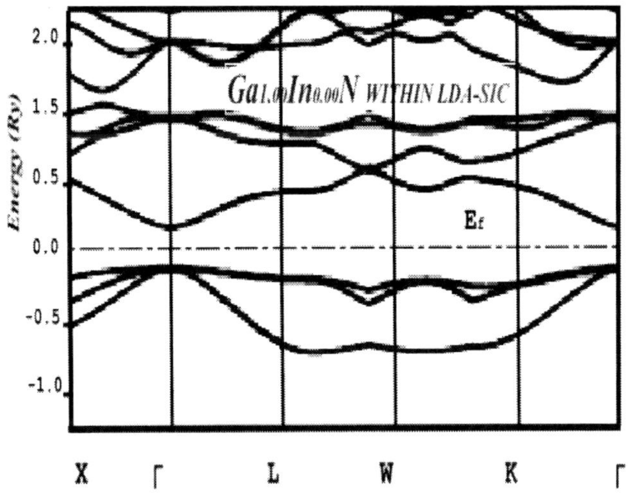

Figure 10. Band structure calculation of GaN within LDA-SIC approximations.

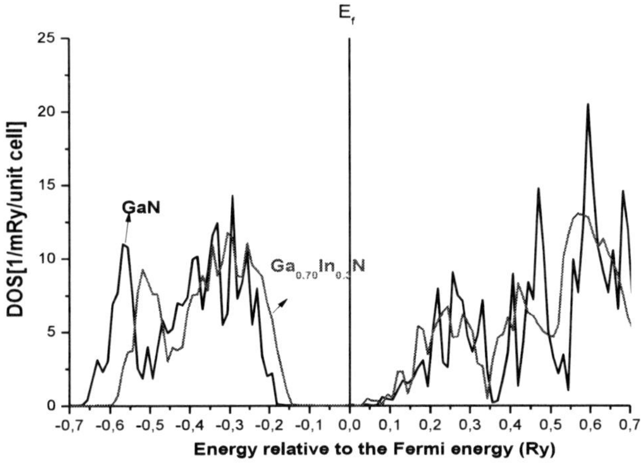

Figure 11. The total density of state of $Ga_{0.70}In_{0.30}N$ and GaN.

Doping Effect in Band Gap Energy of Nitrides

Several physical properties of materials are straightforwardly or indirectly connected to the electronic band structure. Thus, the comprehension of the band gap of a system is valuable for its systematic utilization in optoelectronic, and electromagnetic devices. The calculations of density of states (DOS) represent the major component to investigate the electronic properties of $Ga_{1-x}In_xN$ alloys. The electronic structure of these alloys was calculated with optimized structural parameters and relaxed atomic positions. To calculated band gap energy for each concentration, the DOS was computed by using the LDA-SIC approach to establish good description for the electronic structures of our considered systems. The total DOS curves for x= 0.30 and x=0 alloys are depicted in (Figure 11).

For pure GaN semiconductors, the states of the next part in the range of −8.4 and −2.02eV, arise from Ga cations. It is apparent from the DOS of $Ga_{0.70}In_{0.30}N$, that the peaks are split in the second section of the valence states. For further In content augmentation in $Ga_{1-x}In_xN$ alloys, the splitting in the peaks is enhanced and this can be attributed to the effect of lattice mismatch between GaN and InN and *d*-electrons of In.

For parent binary semiconductors, the splitting of these states will vanish because of the pure cation contribution.

The band gap energies of the zinc blende $Ga_{1-x}In_xN$ obtained with the equilibrium lattice constants are plotted in Figure 12 for further analysis. We note that the direct band gap shows a rapid variation with the concentration, which is close to a linear behaviour. The band gap varies slowly with the In composition. The above investigation has been carried out on ordered alloys, however, most of the literature has until now dealt with random alloys. The band gap bowing parameters of the zinc blende $Ga_{1-x}In_xN$ obtained in this study are close to that obtained experimentally by Goldhahn et al. [11-15], and those obtained numerically by other researchers [16]. The band gap energy of the $Ga_{1-x}In_xN$ can be depicted as a function of the indium composition x and expressed using the following formula:

$$E_g(x) = xE_{g,\text{InN}} + (1-x)E_{g,\text{GaN}} - bx(1-x)$$

where $E_g(x)$ is the band gap energy of the $Ga_{1-x}In_xN$, $E_{g;InN}$ is the band gap energy of the InN, $E_{g;GaN}$ is the band gap energy of the GaN, and b is the band gap bowing parameter of the $Ga_{1-x}In_xN$.

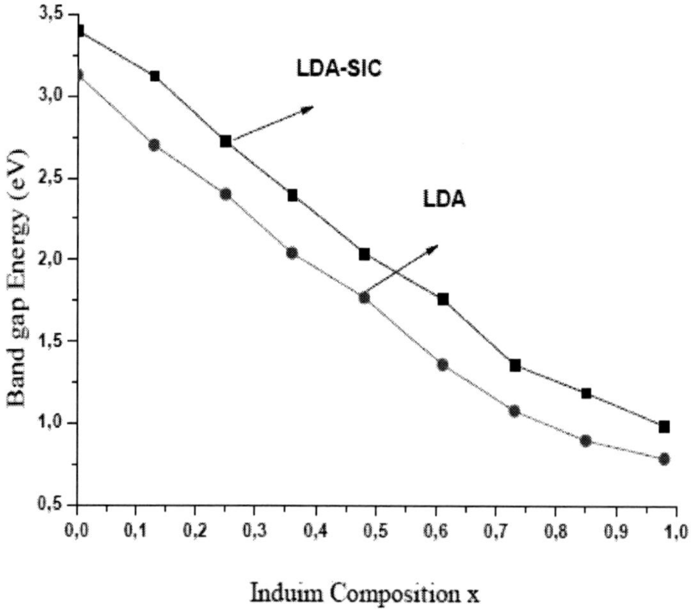

Figure 12. Variation of band gap energies of the zinc blende $Ga_{1-x}In_xN$ With LDA and LDA-SIC approximation.

Order has been reported in the wurtzite nitrides but not yet in the cubic alloys for which the growth seems to be more complicated. When the quality of samples will be improved rapidly for the cubic alloys, one can expect a considerable progress in understanding and exploiting the properties of these alloys Therefore, Figure 12 exhibits that at low In composition a prominent downward bowing can be obtained, but at immense In composition the bowing could be insignificant. Then, it is possible to characterize the alloy band gap employing a distinct bowing parameter.

It is worthwhile to indicate that at the low-In-content regime which is actually of extreme relevance for LED's and lasers; the bowing parameter could be significantly greater than the value obtained through a fit upon the thorough alloy regime. For In-rich regime (x>0.5), the system may affront some perplexities for a perfect experimental characterization.

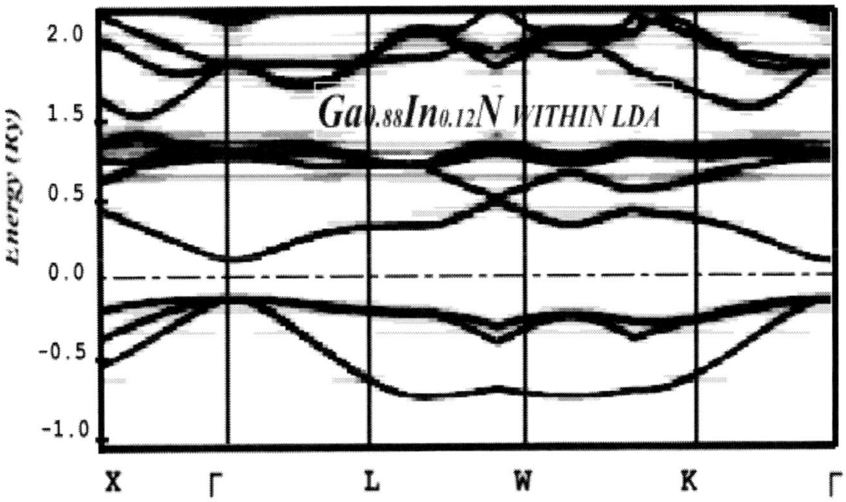

Figure 13. Band structure calculation of $Ga_{0.88}In_{0.12}N$ within LDA approximations.

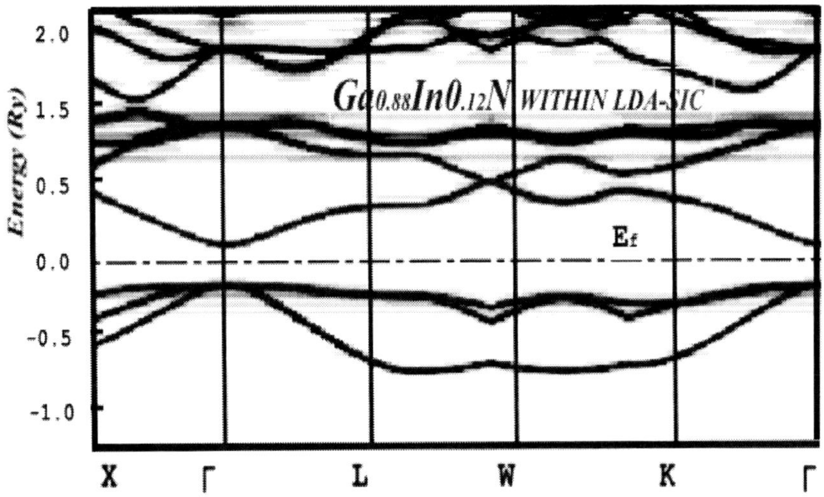

Figure 14. Band structure calculation of $Ga_{0.88}In_{0.12}N$ within LDA-SIC approximations.

The energy band structure of the zinc blende $Ga_{1-x}In_xN$ is investigated. Figures 13 and 14 shows a typical example of the calculated energy band structure of the $Ga_{0.88}In_{0.12}N$ with LDA and LDA-SIC approximations. Underestimation of the band gap energy but accurate estimation of the valence band and band gap energy is a well-known consequence of using LDA-SIC approximation. The band gap energy at the Γ point shown in Figure 13 is 2.7 eV. With LDA approximation and 3,12 eV with LDA-SIC approximation (Figure 15).

As indicated in Figure 1, the valence bands consist of two major sub-bands. The top valence band originates predominantly from the N 2p orbitals. The deep band originates predominantly from the N 2s orbitals. The highly overlapped band structures are compounds of the Ga 3*d* and In 4*d* orbitals. It is obvious that there is considerable hybridization among the Ga 3d, In 4d, and N 2s states.

Conclusion

Based on the first-principles calculations, we have designed a strategy for resource security of InGaN based photovoltaic materials. We have shown that the doping of In in GaN In works well for this purpose. It is shown that In1-xGaxN maintains a direct band gap and no impurity state appears in the band gap.

This substitution is effective up to x=1. From the calculated Bloch spectral function, it is suggested that the randomness brought by the doping does not disturb electron transport so much.

References

[1] http://www.photovoltaique.info/Les-differentes-technologies.html.

[2] http//www2.ademe.fr/servlet/KBaseShow?sort=-1andcid=96andm=3andcatid=13924.

[3] Wu, J. When group-III nitrides go infrared: New properties and perspectives. *Journal of Applied Physics.* 2009, Vol. 106, 011101.

[4] Pereira, S., et al. Compositional dependence of the strain-free optical band gap in Inx Ga1 xNlayers. *Applied Physics Letters.* 2001, Vol. 78, 2137.

[5] Bechstedt, F., et al. Energy gap and optical properties of InxGa1-xN. *Physica Status Solidi a.* 2003, Vol. 195, 628.

[6] Ruterana, P., Albrecht, M. and Neugebauer, J. *Nitride Semiconductors: Handbook on Materials and devices.* s.l.: Wiley-VCH Verlag GmbH, 2003. 978-3527403875.

[7] Wu, J., et al. Small band gap bowing in In1-xGaxN alloys. *Applied Physics Letters.* 2002, Vol. 80, 4741.

[8] R. Dahal, B. Pantha, J. Li, J. Y. Lin, H. X. Jiang, *Appl. Phys. Lett.* 94, 63505 (2009).

[9] H. Akai: http://sham.phys.sci.osaka-u.ac.jp/~kkr/

[10] H. Morkoç, S. Strite, G. B. Gao, M. E. Lin, B. Sverdlov, M. Burns, "Large-band-gap SiC, III-V nitride, and II-VI ZnSe-based semiconductor device technologies". *J. Appl. Phys.* 76 (1994) 1363.

[11] A. Tabata, L. K. Teles, L. M. R. Scolfaro, J. R. Leite, A. Kharchenko, T. Frey, D. J. As, D. Schikora, K. Lischka, J. Furthmuller, F. Bechstedt, *Appl. Phys. Lett.* 80, 769(2002).

[12] S. Pereira, M. R. Correia, E. Alves, *Adv. Mater. Forum III*, 514, 38 (2006).

[13] S. Pereira, M. R. Correia, E. Pereira, K. P. O'Donnell, C. Trager Cowan, F. Sweeney, E. Alves, *Phys. Rev.* B 64, 205311 (2001).

[14] O. Jani, I. Ferguson, C. Honsberg, S. Kurtz, *Appl. Phys. Lett.* 9, 132117 (2007).

[15] C. J. Neufeld, N. G. Toledo, S. C. Cruz, M. Iza, S. P. DenBaars, U. K. Mishra, *Appl. Phys. Lett.* 93, 143502 (2008).

[16] A. Larefl,a, A. Altujarl and S. J. Luo 2, *Eur. Phys. J. B* (2013) 86: 475 DOI: 10.1140/ epjb/e2013-40487-2.

INDEX

A

B

C

D

G

H

I

N

O

P

Q

R

S

T

U

V

W

X

Y

Z